T0201446

Cellular Biophysics and Modeling
A Primer on the Computational Biology of Excitable Cells

Cellular Biophysics and Modeling is what every neuroscientist should know about the mathematical modeling of excitable cells. Combining empirical physiology and non-linear dynamics, this text provides an introduction to the simulation and modeling of dynamic phenomena in cell biology and neuroscience. It introduces mathematical modeling techniques alongside cellular electrophysiology. Topics include membrane transport and diffusion, the biophysics of excitable membranes, the gating of voltage and ligand-gated ion channels, intracellular calcium signaling, and electrical bursting in neurons and other excitable cell types. It introduces mathematical modeling techniques such as ordinary differential equations, phase plane analysis, and bifurcation analysis of single compartment neuron models. With analytical and computational problem sets, this book is suitable for life sciences majors, in biology to neuroscience, with one year of calculus, as well as graduate students looking for a primer on membrane excitability and calcium signaling.

Greg Conradi Smith is a Professor in the Department of Applied Science and Neuroscience Program Faculty Affiliate at William & Mary, Williamsburg, VA, USA. He was co-organizer of the Cold Spring Harbor Laboratory Summer School on Computational Cell Biology (2008–2014). His research focuses on mathematical aspects of cell physiology and neuroscience.

Cellular Biophysics and Modeling

A Primer on the Computational Biology of Excitable Cells

GREG CONRADI SMITH
William & Mary, Williamsburg, VA

CAMBRIDGE
UNIVERSITY PRESS

University Printing House, Cambridge CB2 8BS, United Kingdom

One Liberty Plaza, 20th Floor, New York, NY 10006, USA

477 Williamstown Road, Port Melbourne, VIC 3207, Australia

314-321, 3rd Floor, Plot 3, Splendor Forum, Jasola District Centre, New Delhi – 110025, India

79 Anson Road, #06–04/06, Singapore 079906

Cambridge University Press is part of the University of Cambridge.

It furthers the University's mission by disseminating knowledge in the pursuit of
education, learning, and research at the highest international levels of excellence.

www.cambridge.org
Information on this title: www.cambridge.org/9781107005365
DOI: 10.1017/9780511793905

First published 2019

Printed in the United Kingdom by TJ International Ltd. Padstow, Cornwall

A catalog record for this publication is available from the British Library.

Library of Congress Cataloging-in-Publication Data
Names: Smith, Greg Conradi, 1964– author.
Title: Cellular biophysics and modeling : a primer on the computational
    biology of excitable cells / Greg Conradi Smith.
Description: Cambridge; New York, NY : Cambridge University Press, 2019. |
    Includes index.
Identifiers: LCCN 2018039283| ISBN 9781107005365 (hardback : alk. paper) |
    ISBN 9780521183055 (pbk.)
Subjects: LCSH: Excitation (Physiology)| Mathematical models. | Biophysics. |
    Cell physiology. | Computational biology.
Classification: LCC QP363 .S595 2019 | DDC 612/.014–dc23
LC record available at https://lccn.loc.gov/2018039283

ISBN 978-1-107-00536-5 Hardback
ISBN 978-0-521-18305-5 Paperback

Contents

Preface   *page* xi

1   **Introduction**   1
    1.1   Why Study Biophysics?   1
    1.2   Neurons are Brain Cells   2
    1.3   Cellular Biophysics   3
    1.4   Dynamical Systems Modeling   5
    1.5   Benefits and Limitations of Mathematical Models   6
    1.6   Minimal Models and Graphical Methods   7
    1.7   Biophysics and Dynamics Together   8
    1.8   Discussion   9
        Solutions   11
        Notes   11

Part I   Models and Ordinary Differential Equations   13

2   **Compartmental Modeling**   15
    2.1   Physical Dimensions and Material Balance   15
    2.2   A Model of Intracellular Calcium Concentration   16
    2.3   The Initial Value Problem and its Solution   17
    2.4   Checking the Solution   19
    2.5   Interpreting the Solution   19
    2.6   Calcium Dynamics and Disease   22
    2.7   Appendix: Solving $dc/dt = \jmath - kc$ with $c(0) = c_0$ 24
    2.8   Discussion   25
        Supplemental Problems   27
        Solutions   33
        Notes   39

3   **Phase Diagrams**   42
    3.1   Phase Diagram for a Single Compartment Model   42
    3.2   Stable and Unstable Steady States   44
    3.3   Phase Diagram of a Nonlinear ODE   45
    3.4   Classifying Steady States   47
    3.5   Stability Analysis Requiring Higher Derivatives   49
    3.6   Scalar ODEs with Multiple Stable Steady States   50
    3.7   Discussion   51
        Supplemental Problems   55
        Solutions   57
        Notes   58

4   **Ligands, Receptors and Rate Laws**   59
    4.1   Mass Action Kinetics   59

4.2   Reaction Order and Physical Dimensions of Rate Constants   60
4.3   Isomerization – ODEs and a Conserved Quantity   61
4.4   Isomerization – Phase Diagram and Solutions   63
4.5   Bimolecular Association of Ligand and Receptor   65
4.6   Sequential Binding   69
4.7   Sigmoidal Binding Curves   70
4.8   Binding Curves and Hill Functions   72
4.9   Discussion   74
       Supplemental Problems   75
       Solutions   77
       Notes   79

5   **Function Families and Characteristic Times**   81
5.1   Functions and Relations   81
5.2   Scaling and Shifting of Functions   82
5.3   Qualitative Analysis of Functions   84
5.4   Characteristic Times   88
5.5   Discussion   90
       Supplemental Problems   93
       Solutions   94
       Notes   96

6   **Bifurcation Diagrams of Scalar ODEs**   98
6.1   A Single-Parameter Family of ODEs   98
6.2   Fold Bifurcation   99
6.3   Transcritical Bifurcation   101
6.4   Pitchfork Bifurcations   102
6.5   Bifurcation Types and Symmetry   105
6.6   Structural Stability   106
6.7   Further Reading   108
       Supplemental Problems   109
       Solutions   110
       Notes   111

Part II   Passive Membranes   113

7   **The Nernst Equilibrium Potential**   115
7.1   Cellular Compartments and Electrical Potentials   115
7.2   Nernst Equilibrium Potential   116
7.3   Derivation of the Nernst Equation   119
7.4   Calculating Nernst Equilibrium Potentials   121
7.5   Chemical Potential   122
7.6   Discussion   124
       Supplemental Problems   129
       Solutions   130
       Notes   130

8   **The Current Balance Equation**   132
    8.1   Membrane Voltage   132
    8.2   Ionic Fluxes and Currents   132
    8.3   Ionic Currents and Voltage   133
    8.4   Applied Currents and Voltage   134
    8.5   The Current Balance Equation   135
    8.6   Constitutive Relation for Ionic Membrane Current   137
    8.7   The Phase Diagram for Voltage of Passive Membranes   139
    8.8   Exponential Time Constant for Membrane Voltage   140
    8.9   Discussion   143
           Supplemental Problems   147
           Solutions   149
           Notes   153

9   **GHK Theory of Membrane Permeation**   154
    9.1   Goldman-Hodgkin-Katz Theory – Assumptions   154
    9.2   Physical Dimensions of the GHK Current Equation   155
    9.3   The Goldman-Hodgkin-Katz Current Equation   156
    9.4   Limiting Conductances Implied by GHK Theory   157
    9.5   Derivation of the GHK Current Equation   159
    9.6   Further Reading and Discussion   161
           Supplemental Problems   164
           Solutions   165
           Notes   168

Part III   Voltage-Gated Currents   169

10 **Voltage-Gated Ionic Currents**   171
    10.1   Voltage-Dependent Gating and Permeation Block   171
    10.2   The L-Type Calcium Current I_{Ca_V} 173
    10.3   The Inward Rectifying Potassium Current I_{Kir} 176
    10.4   The Hyperpolarization-Activated Cation Current I_{sag} 177
    10.5   The Depolarization-Activated Potassium Current I_{K_V} 177
    10.6   Qualitative Features of Current-Voltage Relations   179
    10.7   Further Reading and Discussion   180
           Supplemental Problems   181
           Solutions   182
           Notes   183

11 **Regenerative Ionic Currents and Bistability**   185
    11.1   Regenerative Currents and Membrane Bistability   185
    11.2   Response of a Bistable Membrane to Applied Current Pulses   188
    11.3   Membrane Currents and Fold Bifurcations   188
    11.4   Bifurcation Diagram for the Bistable $I_{Ca_V} + I_L$ Membrane   190
    11.5   Overlaying Trajectories on the Bifurcation Diagram   191
    11.6   Bistable Membrane Voltage Mediated by I_{kir} 191
    11.7   Further Reading and Discussion   193

Supplemental Problems   197
Solutions   197
Notes   198

12 Voltage-Clamp Recording   199
    12.1 Current-Clamp and Voltage-Clamp Recording   199
    12.2 Modeling Delayed Activation of Ionic Currents   203
    12.3 Voltage Clamp and Transient Ionic Currents   206
    12.4 Modeling Transient Ionic Currents   209
    12.5 Further Reading and Discussion   211
          Supplemental Problems   213
          Solutions   213
          Notes   215

13 Hodgkin-Huxley Model of the Action Potential   216
    13.1 The Squid Giant Axon   216
    13.2 The Hodgkin-Huxley Model   219
    13.3 Excitability in the Hodgkin-Huxley Model   221
    13.4 Repetitive Spiking (Oscillations)   224
    13.5 Further Reading and Discussion   225
          Supplemental Problems   229
          Solutions   230
          Notes   230

Part IV Excitability and Phase Planes   233

14 The Morris-Lecar Model   235
    14.1 The Morris-Lecar Model   235
    14.2 The Reduced Morris-Lecar Model   237
    14.3 The Morris-Lecar Phase Plane   239
    14.4 Phase Plane Analysis of Membrane Excitability   241
    14.5 Phase Plane Analysis of Membrane Oscillations   244
    14.6 Further Reading and Discussion   248
          Supplemental Problems   249
          Solutions   251
          Notes   251

15 Phase Plane Analysis   252
    15.1 The Phase Plane for Two-Dimensional Autonomous ODEs   252
    15.2 Direction Fields of Two-Dimensional Autonomous ODEs   255
    15.3 Nullclines for Two-Dimensional Autonomous ODEs   256
    15.4 How to Sketch a Phase Plane   258
    15.5 Phase Planes and Steady States   263
    15.6 Discussion   265
          Supplemental Problems   268
          Solutions   269
          Notes   273

16   **Linear Stability Analysis**   275
     16.1   Solutions for Two-Dimensional Linear Systems   275
     16.2   Real and Distinct Eigenvalues – Saddles and Nodes   278
     16.3   Complex Conjugate Eigenvalues – Spirals   281
     16.4   Criterion for Stability   284
     16.5   Further Reading and Discussion   285
            Supplemental Problems   290
            Solutions   291
            Notes   293

Part V   Oscillations and Bursting   295

17   **Type II Excitability and Oscillations (Hopf Bifurcation)**   297
     17.1   Fitzhugh-Nagumo Model   297
     17.2   Phase Plane Analysis of Resting Steady State   300
     17.3   Loss of Stability with Increasing J (Depolarization)   303
     17.4   Analysis of Hopf Bifurcations   304
     17.5   Limit Cycle Fold Bifurcation   310
     17.6   Further Reading and Discussion   313
            Supplemental Problems   315
            Solutions   316
            Notes   317

18   **Type I Excitability and Oscillations (SNIC and SHO Bifurcations)**   319
     18.1   Saddle-Node on an Invariant Circle   319
     18.2   Saddle Homoclinic Bifurcation   323
     18.3   Square-Wave Bursting   324
     18.4   Calcium-Activated Potassium Currents as Slow Variable   328
     18.5   Further Reading and Discussion   331
            Supplemental Problems   335
            Solutions   336
            Note   337

19   **The Low-Threshold Calcium Spike**   338
     19.1   Post-Inhibitory Rebound Bursting   338
     19.2   Fast/Slow Analysis of Post-Inhibitory Rebound Bursting   342
     19.3   Rhythmic Bursting in Response to Hyperpolarization   343
     19.4   Fast/Slow Analysis of Rhythmic Bursting   344
     19.5   Minimal Model of the Low-Threshold Calcium Spike   346
     19.6   Further Reading and Discussion   349
            Solutions   351
            Notes   351

20   **Synaptic Currents**   353
     20.1   Electrical Synapses   353
     20.2   Electrical Synapses and Synchrony   355
     20.3   Chemical Synapses   356

20.4   Phase Plane Analysis of Instantaneously Coupled Cells   357
20.5   Reciprocally Coupled Excitatory Neurons   362
20.6   Further Reading and Discussion   363
        Supplemental Problems   365
        Solutions   367
        Note   367

Afterword   368
References   371
Index   380

Preface

Philosophy is written in this grand book – I mean universe – which stands continuously open to our gaze, but which cannot be understood unless one first learns to comprehend the language in which it is written. It is written in the language of mathematics, and its characters are triangles, circles and other geometric figures, without which it is humanly impossible to understand a single word of it; without these, one is wandering about in a dark labyrinth.   — **Galileo Galilei (1564–1642)**

Most students of life science accept Galileo's statement that "triangles, circles and other geometric figures" are necessary to fully understand the cosmos. But many of these students – and perhaps also their professors – have significant doubts about the relevance of mathematics to the science of life on earth.

Admittedly, biology and mathematics sometimes appear immiscible. Like oil and water, the combination does not yield a homogenous mixture of liquids, but an emulsion. When biology and mathematics are viewed as two disparate subjects in an undergraduate education, attempts to forcefully stir one into the other result in something like well-shaken Italian salad dressing, the two dispersed liquid phases having a natural tendency to separate. Because we have no surfactant to stabilize the bio-math emulsion, we shake again. In the process, the students become agitated, too!

Love of science and fear of mathematics have led many to major in biology, psychology and neuroscience. There is no shame in acknowledging this fact. Within the life sciences there are many important research questions that can be asked and answered without mathematics. Many topics covered in biology, psychology and neuroscience courses can be explained and understood without mathematical language. There are numerous scientific and health- and education-related fields that do not require mathematical aptitude, but do need intelligent and resourceful young scientists and science majors.

On the other hand, many life sciences have theoretical foundations that were developed by quantitative scientists using mathematical language (e.g., population genetics). Other life sciences, such as molecular biology and genomics, have become so complex and data rich that most practitioners would appreciate more quantitative aptitude and perspective – if not for themselves, then at least for their trainees. Contemporary life scientists who are at ease with mathematics use quantitative reasoning in the study of life on every scale: molecules, membranes, cells, networks, organisms, behavior, evolution and ecology. Both pure and applied biomedical research is replete with open scientific questions (e.g., protein folding) and technical challenges (e.g., rational drug design) whose solutions will be found by biological scientists who are comfortable with mathematics and computation.

In my opinion, mathematics is the language of all natural science, biology and neuroscience no less than astronomy, physics and chemistry. This extension of Galileo's conviction to the realm of neuroscience is, admittedly, a philosophical statement that is open to discussion. I encourage you to think it over, talk to your peers and mentors, and decide for yourself.

Certainly, this book is a combination of biology (cellular biophysics) and mathematics (dynamical systems modeling) written from a Galilean perspective. Is it a homogenous mixture or emulsion? I cannot say. But it is a sweet mix.

1 Introduction

The book to read is not the one that thinks for you, but the one which makes you think.

— **James McCosh (1811–1894)**

1.1 Why Study Biophysics?

Why should a young neuroscientist study cellular biophysics and modeling? One answer is that the electrical properties of cell plasma membrane and signal transduction are essential to cellular and systems neuroscience. Another answer is that the biophysics of individual neurons is a well-understood subject. The experimental electrophysiology and computational modeling performed over the last century has led to scientific consensus on the "right way" of thinking about electrical and chemical signaling of neurons and the collective activity of small neuronal networks. There is no such consensus on central nervous system function.

Ask 10 of the world's leading neuroscientists how the brain works – how it thinks, feels, perceives, and acts as a unified whole – and you will get 10 different answers, unless they are very narrowly framed around the biophysics and chemistry of nerve impulse conduction and synaptic transmission (Swanson, 2012).

Admittedly, cellular biophysics may seem to be a narrowly framed subject, especially if you are an undergraduate who is primarily interested in behavioral or cognitive neuroscience. On the other hand, the electrical and chemical signaling of neurons – the elementary anatomical units of the central nervous system – is absolutely essential to brain function. Cellular biophysics and modeling is foundational to cellular and systems neuroscience and a solid point of departure to more comprehensive study of brain, mind and behavior.

1.2   Neurons are Brain Cells

It goes without saying that neurons are brain cells – elementary anatomical units of the central nervous system that are polarized to mediate input/output functions: dendrites → soma → axon → synapses. Rinse, repeat. The scientific history of these foundational convictions is the subject of a book by Gordon Shepherd entitled *Foundations of the Neuron Doctrine*. The cellular structure of the central nervous system was first observed in sections of neural tissue using staining methods developed by Camillo Golgi and later perfected by Santiago Ramón y Cajal. Because a small proportion of neurons were labelled, the structure of single neurons could be resolved (see Fig. 1.1). Golgi's observations reinforced his working hypothesis that neural tissue had a reticular structure analogous to the circulatory system. Cajal concluded that each neuron was a separate entity that interacted with other neurons at synaptic junctions. The 1906 Nobel Prize in Physiology or Medicine was awarded jointly to Golgi and Cajal in recognition of their work on the structure of the nervous system.

More than a century later, neurons (and glia) are still brain cells. However, some aspects of the neuron doctrine are difficult to sustain. Today's neuroscience students

…could well conclude that the nerve cell is not the unit of function that is of primary interest to them. The units of contemporary studies are the packets of transmitter molecules, the channels and receptors by means of which nerve cells communicate with each other. Although many

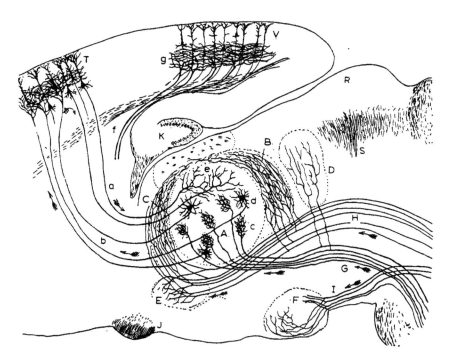

Figure 1.1   A sagittal section through the rat brain as drawn by Ramón y Cajal. The section shows thalamocortical neurons that project to cortex (d and b), and cortical pyramidal neurons that project to the thalamus (T, a, e), and much more. Reproduced from Cajal's 1906 Nobel lecture *The structure and connexions of neurons*. © The Nobel Foundation 1906. *Question*: Where is the hippocampus?

nerve cells are polarized in the classical sense, others are not: parts of the classical neuronal output system, the axons, can serve as receptors, and the classical receptor portions of the dendrites can serve as effectors. ... There are many examples of nerve cells linked to each other by specialized gap junctions that provide electrical coupling and allow the passage of small molecules from one nerve cell to another. ... There are nerve fibres that are produced by the fusion of processes from several cells. ... These facts are all contrary to the neuron doctrine, as originally expressed (Guillery, 2005).

> *Discuss*   Neurons are discrete anatomical components of the central nervous system, but does this imply that neurons are the elementary physiological units subserving the computations performed by the brain?[1]

1.3   Cellular Biophysics

Biophysics is a thriving scientific discipline, but it can be difficult to define. In *Biophysical Journal*, Olaf S. Andersen recounts that the centerpiece of biophysical research

> ... in the early part of the twentieth century was neuro- and muscle physiology, disciplines that lend themselves to quantitative analysis and in which most of the investigators had trained in biology or medicine. In the latter half of the century, an increasing number of biophysicists were trained in chemistry, physics, or mathematics, which led to the development of the modern generation of optical and electron microscopes, fluorescent probes ... as well as the computational methods that, by now, have become indispensable tools in biophysical research (Andersen, 2016).

What is biophysics? Archibald Vivian Hill[2] emphasized that physical instrumentation in a biological laboratory does not a biophysicist make. Rather, it is the biophysical *mindset* that is important. Biophysicists attempt to understand biological structure, organization and function using the ideas and methods of physics and physical chemistry (Hill, 1956).

I use *cellular biophysics* as a flexible term for quantitative, physical and physico-chemical approaches to the complex phenomena of cell biology and neuroscience. These include, but are not limited to, the electrical properties of the plasma membrane of neurons (e.g., voltage- and ligand-gated ion channels), cell signal transduction (e.g., ionotropic and metabotropic receptors, intracellular calcium responses), and aspects of biochemistry and cell biology (e.g., metabolic oscillations, microtubule dynamics and cell motility).

Cellular biophysics is somewhat, but not entirely, distinct from *molecular* biophysics, e.g., the use of nuclear magnetic resonance (NMR) to determine the structure of macromolecules. Both subjects are inherently interdisciplinary and highly dependent upon physical techniques. The experimental methods specifically relevant to cellular biophysics include, but are not limited to, voltage-clamp electrical recordings, confocal microfluorimetry, and fluorescence resonance energy transfer (FRET).

A wide variety of small molecules contribute to the electrical and chemical signaling of excitable cells such as neurons and myocytes. Examples include: neurotransmitters (glutamate, glycine, gamma-aminobutyric acid = GABA), hormones (epinephrine = adrenaline, vasopressin, cortisol, estrogen), lipids (PIP_2 = phosphatidylinositol 4,5-bisphosphate, ceramide, sphingosine) and second-messengers (cyclic adenosine monophosphate = cAMP, inositol trisphosphate = IP_3, diacylglyc-erol = DAG, calcium = Ca^{2+}).

The macromolecules that dominate cellular biophysics are the integral and peripheral membrane proteins involved in cell signaling and membrane transport. Integral membrane proteins are permanently attached to the cell plasma membrane (e.g., cell surface receptors) or intracellular membranes. Integral membrane proteins of the cell plasma membrane often span the lipid bilayer and have significant extracellular and intracellular (as well as transmembrane) domains. Peripheral membrane proteins attach (sometimes fleetingly) to integral membrane proteins or the inner leaflet of the lipid bilayer (e.g., phospholipase C, which hydrolyzes PIP_2 into the second messengers IP_3 and DAG).

Fig. 1.2 summarizes the classes of transmembrane proteins that are most rele-vant to our study of cellular biophysics. These include neurotransmitter receptors, ion channels, and transporters (e.g., pumps and exchangers). An ionic channel is a membrane protein with an aqueous pore; when the channel is permissive, certain ions may pass through the pore, crossing from one side of the cell plasma membrane to the other. Examples include non-gated potassium "leak" channels, potassium channels that are gated by the binding of cytosolic calcium, voltage-gated sodium channels, and

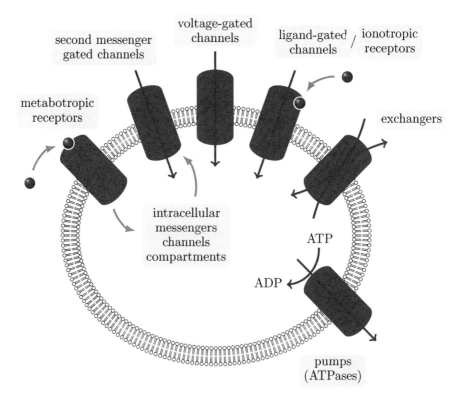

Figure 1.2   Integral membrane proteins involved in membrane transport include neurotransmitter receptors, ion channels and transporters (exchangers and pumps).

glutamate receptors (ligand-gated ion channels permeable to both sodium and potassium). Pumps are transporters that couple the hydrolysis of ATP to the movement of one or more ions against concentration or electrochemical gradients. Examples include the sodium-potassium ATPase and the sarco-endoplasmic reticulum calcium ATPase (SERCA).[3] Exchangers are transporters that use the concentration gradient of one type of ion to catalyze the movement of another (e.g., the sodium-calcium exchanger).

This primer does not cover the molecular and structural biology of ion channels, transporters and pumps. Rather, our emphasis is the *functional role* of ionic channels in the electrophysiology of neurons, myocytes and other excitable cells. Because the physiological states of living cells measured using biophysical techniques are complicated functions of time, our study necessarily involves the mathematics of dynamical systems.

> *Discuss*   What are some examples of physiological signals (electrical or chemical) that are routinely measured by physicians and experimental biologists? What is the role of these physiological measurements in medical diagnosis?

1.4   Dynamical Systems Modeling

In mathematics, a dynamical system is a rule that describes how the location of a point in geometrical space changes as a function of time. When this concept is used to model dynamical phenomena of cellular biophysics, the "point" is some aspect of the state of a living cell (e.g., the membrane potential or the concentration of a chemical species) in the "space" of all realizable states (e.g., a concentration may not be negative). While the rule for the time-evolution of a dynamical system can take many different forms, a natural choice is to specify the rule using one or more differential equations.

Differential equations are used by engineers and physicists to describe and predict change in the physical world, that is, the time-dependent dynamics of inanimate objects. In physics (classical mechanics) Newton's second law of motion describes the relationship between the net force acting on a body[4] and its motion using the familiar equation, $F = ma$, where $F(t)$ is the force applied, m is the mass of the body (constant), and $a(t)$ is the body's acceleration. Acceleration is rate of change of velocity, that is, $a = dv/dt$; and momentum (p) is mass times velocity $p = mv$. Consequently, Newton's second law can be rewritten as a differential equation[5] that describes how the body's momentum changes in response to a force, $dp/dt = F(t)$. From the dynamical systems perspective, this differential equation is a rule that determines how the momentum $p(t)$ (the state of the system) changes in response to the time-dependent force $F(t)$.

In a similar manner, many contemporary biologists apply dynamical systems to the study of life (e.g., ecologists model population dynamics of food webs, and public health experts model epidemics of communicable disease). Just as chemical engineers use differential equations to describe and predict the kinetics of chemical reactions, biochemists use differential equations to analyze and simulate metabolic pathways. Modeling with differential equations is paramount in systems physiology and pharmacokinetics. The approach has long been employed to study the flow, between different organs in the body, of dissolved gases, nutrients, drugs, hormones and radio-isotopes.

When it comes to the physiology of excitable cells – those cell types that can generate action potentials in response to depolarization, such as neurons and cardiac myocytes – the dynamical systems perspective has been with us for over half a century. Alan Hodgkin and Andrew Huxley's Nobel Prize winning studies[6] of the action potential in the squid giant axon included dynamical systems modeling with differential equations. Today, Hodgkin-Huxley-style modeling is the dominant framework for studying and analyzing ion channel kinetics and the physiological consequences of differentially expressed ionic currents in neurons and other excitable cells (e.g., myocytes, pancreatic beta cells, saccular hair cells). The dynamical systems perspective is a key ingredient of a historically accurate and scientifically rigorous understanding of the physiology of excitable cells.

Discuss Putting dynamics and differential equations to one side, what other types of mathematics are important to the life sciences?

1.5   Benefits and Limitations of Mathematical Models

Life scientists study experimental model systems such as nematodes (roundworms) and drosophila (fruit flies) because this practice facilitates scientific discovery. Experimental model systems are developed and utilized because they lend themselves to investigation (e.g., due to a technical or ethical advantage). Similarly, mathematical models in the life sciences are idealizations of biological phenomena that have certain advantages. Like experimental model systems, mathematical models also have limitations.

This book emphasizes how a conceptual model – e.g., a verbal description or a cartoon summary of a hypothesis – can be converted into a dynamical systems model composed of one or more differential equations (Fig. 1.3). When this process of bringing cartoons to life[7] is mastered, one learns that mathematical models are natural manifestations of conceptual models, but with significant advantages. While the implications of a conceptual model are often unclear, mathematical models may be analyzed by hand calculations, graphical techniques and computer simulation to clarify the implications of the parent conceptual model. As we will see, conclusions often depend on model parameters (e.g., rate constants) or aspects of the conceptual model that were indeterminate or obscure. In this way, mathematical modeling often sharpens scientific hypotheses by highlighting the subtleties hidden within conceptual models.

The limitations of mathematical models are obvious to experimental scientists and are often emphasized. We are often reminded that a mathematical model is an abstraction and not a biological reality. Of course, this point should always be kept in mind, just as we should never forget that conceptual models are also abstractions!

The idealization that occurs in biological modeling is an advantage as well as a limitation. Unlike biological reality, a mathematical object can in principle be completely understood. Theorists and experimentalists agree that differential equations are usually easier to interrogate than the corresponding tangible systems that they represent. The analysis of mathematical models often results in qualitative "take home messages" that are heeded by experimental scientists.

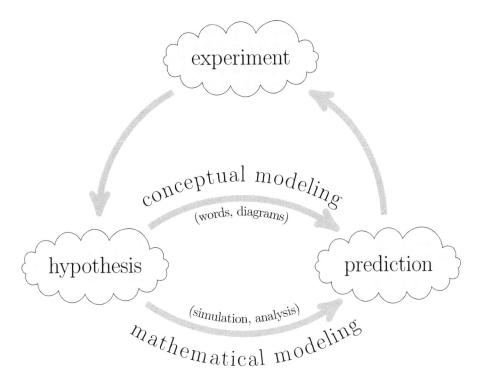

Figure 1.3 The cycle of experiment, hypothesis and prediction in the biological sciences. Moving from hypothesis to prediction requires either conceptual or mathematical modeling.

Discuss   Mendel's laws are fundamental to our understanding of the genetics of inheritance. What aspects of Mendel's laws are idealizations of biological reality? Is Mendelian genetics an example of conceptual modeling or mathematical modeling? How about the Hardy-Weinberg model?

1.6   Minimal Models and Graphical Methods

Mathematical models come in many varieties and are constructed for different purposes. Highly realistic mathematical models of living cells can be so extremely complex that detailed analysis is time consuming and impractical even with the assistance of sophisticated software packages and high-performance computers. Although physiological realism is an important goal, computational models that are nearly as difficult to understand as the corresponding experimental system are not particularly useful.

To see this, imagine a pedestrian tourist who requires a map to help explore a city. A good map has a clear correspondence to the city, but this agreement is reduced, abstract and approximate, because the map must be easily carried and read. A perfectly realistic map with exact correspondence to the city would be the *size* of a city and completely useless, because reading the map would be no less difficult than exploring the city on foot.

In the same way, a helpful mathematical model is physiologically realistic to some degree, but not so complex that it cannot be analyzed and well understood. A mathematical model should be simple enough to be comprehended by the theoretician and realistic enough that this understanding is relevant to the experimentalist. Mathematical models that display this balance between realism and complexity are often referred to as **minimal models**. A mathematical model that is understood to be oversimple but is nevertheless of interest is a **toy model**.

One of the benefits of minimal models is that they can be analyzed, sometimes through pencil and paper calculations, using the qualitative theory of dynamical systems. This geometrical approach to analyzing differential equations is highly visual, easy to learn, and produces intuition about dynamical systems that is of benefit to the life scientist. Learning this graphical way of thinking about mathematical models is well worth the effort, even if one does not intend to pursue mathematics further, because these tools allow one to quickly glean the ambiguities and take home messages of conceptual models. With the help of special purpose software packages, these graphical techniques can also be used to analyze complex mathematical models.

Discuss   The phrase "Everything should be made as simple as possible, but not simpler" is often attributed to the theoretical physicist Albert Einstein. What exactly did he mean by this? How is this maxim relevant to mathematical modeling in biology?

1.7   Biophysics and Dynamics Together

Cellular biophysics and differential equations are both challenging subjects. Teaching both simultaneously might seem like an ill-fated pedagogical choice. Attempting to learn both subjects at the same time might feel like a daunting task. However, there are benefits to communicating mathematical and biological ideas in an integrated manner (Fig. 1.4).

First, teaching dynamics and biophysics at the same time shows students how mathematical modeling is actually used in cell physiology and neuroscience. Golgi and Cajal did not use dynamics in their investigations of the microscopic structure

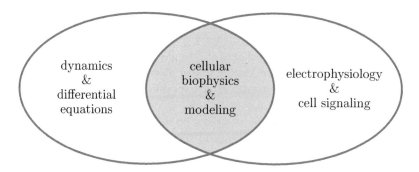

Figure 1.4   *Cellular Biophysics and Modeling* is an exploration of the physiology of excitable cells and the mathematical language of dynamics (differential equations).

of the brain. But differential equations were used by Alan Hodgkin and Andrew Huxley to explain the initiation and propagation of action potentials in the squid giant axon. Contemporary cellular and systems neuroscience is highly quantitative, and the cognitive and behavioral neurosciences are becoming more quantitative with each passing year.

Second, mathematical modeling is best learned through the experience of many specific, concrete examples that are relevant to one's scientific interests. For biology and neuroscience students learning about membrane excitability and cell signaling, this means that physiological examples and exercises are preferred. Although we mentioned Newtonian mechanics above, you will not find any oscillating pendula in this book! This leads us to the third, and perhaps most important, benefit of learning dynamics and electrophysiology at the same time: deeper understanding.

We choose to teach cellular biophysics and differential equations in an integrated manner because, frankly, the mathematical perspective is required to achieve a deep understanding of the dynamical phenomena that comprise electrophysiology and cell signaling. The electrical properties of individual neurons and myocytes and, by extension, the function of the central nervous system cannot be understood without the language of mathematics. The possibility that one may be teaching undergraduate neuroscience majors does not change this. In the words of Bob Marley, we must "tell the children the truth."

1.8   Discussion

Students new to neuroscience will benefit from reading the classic *Scientific American* article "The organization of the brain" (Nauta and Feirtag, 1979) and the introductory chapters of Swanson (2012) and Schneider (2014).

For further reading on the neuron doctrine see the monograph Shepherd (2015) as well as articles by Glickstein (2006), Bullock et al. (2005) and Guillery (2005). The Nobel lectures of Cajal (1906) and Golgi (1906) are available at www.nobelprize.org.

The history of theory in several areas of biology is discussed by Shou et al. (2015). The short *Nature* essay "Bringing cartoons to life" by Tyson (2007) argues that because cells are dynamical systems, mathematical tools are required to understand the relationship between molecular interactions and physiological consequences.

My answer to "What is biophysics" draws heavily from Andersen (2016). The Biophysical Society web page is www.biophysics.org.

Cells and Computers

In his lectures on mathematical modeling of the cell cycle, computational cell biologist John Tyson asks students to consider the similarities and differences between an individual cell and a computer. Computers and cells are different in that computers are manufactured while cells are self-reproducing, but similar in that both obey the laws of thermodynamics. Computers process an input and produce an output. Do cells function in this manner? If so, how are the processing steps occurring within cells and computers similar or different?

How Many Neurons?

The number of neurons in the human central nervous system is estimated to be somewhere between ten billion (10^{10}) and one trillion (10^{12}) (Nauta and Feirtag, 1979). Truly, these are astronomical numbers, as there are thought to be 200–400 billion stars in the Milky Way galaxy. Imagine a neuron represented by a coarse grain of sand with linear dimension of half a millimeter. The volume of a cubical grain of sand would be

$$(0.5\,\text{mm})^3 = (0.5 \times 10^{-3}\,\text{m})^3 = 1.25 \times 10^{-10}\,\text{m}^3\,.$$

A sand castle representing a human brain composed of 10^{10} grains of sand (lower estimate) would have a volume of

$$(10^{10})(1.25 \times 10^{-10}\,\text{m}^3) = 1.25\,\text{m}^3\,,$$

that is, about a cubic meter. Although it is not difficult to visualize this amount of sand, placing the grains side by side in linear array would yield a row fifty thousand kilometers long, which is greater than the circumference of the earth.[8]

> **Problem 1.1**   The upper estimate of 10^{12} neurons per brain is 100-fold greater than the lower estimate. Repeat the above calculation using this value. What is the fold increase in the height of the cubical sandcastle?

> **Problem 1.2**   Estimate the number of neurons in a cubical human brain under the assumption that a neuron may be represented by a cube that is 10 micrometers on a side.

Know Your Neurons

Cajal's drawing of a sagittal section of a rat brain includes thalamocortical and cortical pyramidal neurons (Fig. 1.1). Are these neuron types also present in the human brain? How many morphological types of neurons can you recall and/or visually identify after a Google image search?

Is Neuroscience Hard or Soft?

In the so-called hard sciences (e.g., physics), there are a small number of causative variables, the language of mathematics formalizes hypotheses, and predictions of hypotheses can be rigorously derived. In the soft sciences (e.g., psychology), numerous variables are in play, most of which are difficult to quantify. In the soft sciences, hypotheses are often not formalized using mathematics, the predictions of hypotheses are debatable, and empirical observations rarely lead to strong conclusions. On the other hand, one could argue that the soft sciences are in some ways harder (more difficult) than the hard sciences, because the phenomena addressed are more complex and less easily understood. Is neuroscience a soft science or hard science? Why might some scientists and historians of science find this soft/hard terminology problematic?

Of Math and Majors

To what degree does mathematics play an important role in the science and social science majors at your college or university? Does your answer correlate with the reputation of each subject for rigor and difficulty? To what extent did mathematics play a role in your choice of major?

The Effectiveness of Mathematics

When scientists drill down into physical science, the black gold of mathematics is discovered again and again. In fact, the relationship between the empirical and (applied) formal sciences is so tight, and experienced so routinely by mathematical scientists, that we tend to lose sight of the wonder of this connection. Why does mathematics play an essential role in physical theory?

Solutions

Figure 1.1   The hippocampus is labelled with a K.

Problem 1.1   The fold increase in height is $10^{2/3} \approx 4.64$.

Problem 1.2   An estimate of the number of neurons per brain would be given by the ratio of the brain volume and the neuron volume,

$$\frac{\text{brain volume}}{\text{cell volume}} = \frac{(10\,\text{cm})^3}{(10\,\mu\text{m})^3} = \frac{(0.1\,\text{m})^3}{(10^{-5}\,\text{m})^3} = \frac{10^{-3}\,\text{m}^3}{10^{-15}\,\text{m}^3} = 10^{12},$$

that is, one trillion, consistent with the upper estimate in Nauta and Feirtag (1979).

Notes

1. Complex synaptic structures known as glomeruli suggest that subcellular regions of a neuron can operate as distinct functional units (Sherman and Guillery, 2004).
2. The English physiologist who shared the 1922 Nobel Prize in Physiology or Medicine for his discovery "relating to the production of heat in the muscle."
3. The organelle known as the endoplasmic reticulum (ER) in neurons and many other cell types is, among other things, an intracellular calcium store. Sarcoplasmic reticulum (SR) is a specialized version of ER that occurs in skeletal and cardiac muscle cells.
4. For simplicity, F, a and p are assumed to be scalars as opposed to vectors. We are ignoring rotation and assuming a rigid body constrained to move in one dimension. $F(t)$ is a force applied to the body's center of mass.
5. The equation $dp/dt = F(t)$ is an example of an *ordinary* differential equation (ODE) as opposed to a *partial* differential equation (PDE).

6. Together with John Carew Eccles (1903–1997), Andrew Fielding Huxley (1917–2012) and Alan Lloyd Hodgkin (1914–1998) won the 1963 Nobel Prize in Physiology or Medicine "for their discoveries concerning the ionic mechanisms involved in excitation and inhibition in the peripheral and central portions of the nerve cell membrane." For historical perspective see Schwiening (2012).

7. "Bringing cartoons to life" is the title of a highly recommended *Nature* essay by John Tyson (2007).

8. The circumference of the earth is about 40,000 kilometers (4×10^7 m). The row of sand grains would be at least $10^{10} \cdot 0.5\,\text{mm} = 5 \times 10^7\,\text{m}$ in length.

PART I
Models and Ordinary Differential Equations

2 Compartmental Modeling

In this chapter we construct a simple model of calcium concentration changes in a living cell. Specifying the calcium fluxes of interest – and how these fluxes depend on intracellular calcium concentration – leads to an ordinary differential equation (ODE) initial value problem. The model exhibits the phenomenon of exponential relaxation.

2.1 Physical Dimensions and Material Balance

Experimental measurements based on live-cell imaging typically yield signals that are proportional to concentration rather than the number of molecules. Let $c(t)$ represent the time-dependent concentration of a molecule of interest in a cell of constant volume (v). Concentration is number density, that is, number of molecules (n) divided by the cell volume (v),

$$\text{concentration} \ominus c = \frac{n}{v} \ominus \frac{\#}{\text{volume}} = \frac{\#}{\text{length}^3}, \tag{2.1}$$

where # indicates something is being counted (molecules) and the symbol $\ominus$ is read "has physical dimensions of."[1] In the second equality we use $\text{volume} = \text{length}^3$.

Assuming constant cell volume (v is not a function of time), differentiating both sides of Eq. 2.1 with respect to time yields

$$\frac{\#/\text{length}^3}{\text{time}} \ominus \frac{dc}{dt} = \frac{1}{v} \cdot \frac{dn}{dt} \ominus \frac{1}{\text{length}^3} \cdot \frac{\#}{\text{time}}. \tag{2.2}$$

The rate of change of the concentration of molecules is proportional to the rate of change of the number of molecules within the cell.

Let us assume (for the moment) that it is not possible for the molecule of interest to cross the cell's plasma membrane. In that case, changes in the number of molecules can only occur through chemical reaction (production or degradation of the species, see Fig. 2.1). This assumption may be formalized by writing,

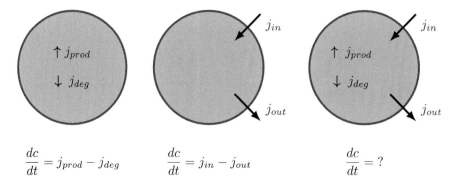

$$\frac{dc}{dt} = j_{prod} - j_{deg} \qquad \frac{dc}{dt} = j_{in} - j_{out} \qquad \frac{dc}{dt} = ?$$

Figure 2.1 Left: A closed cell model with production and degradation but no plasma membrane fluxes. Middle and right: Two open cell models.

$$\underbrace{\frac{\text{concentration}}{\text{time}}}_{} \ominus \quad \frac{dc}{dt} = j_{prod} - j_{deg} , \tag{2.3}$$

where j_{prod} and j_{deg} are positive quantities and concentration $= \#/\text{volume} = \#/\text{length}^3$. Because the physical dimensions of both sides of this equation must be the same, j_{prod} (production) and j_{deg} (degradation) must have physical dimensions of concentration/time.

   Discuss   The production and degradation terms have the same physical dimensions; otherwise, it would be nonsensical to subtract them. To see this, ask yourself "What is 5 meters minus 3 seconds?"[2]

   Eq. 2.3 is a **material balance equation**[3] that restates, in mathematical language, both the physical law of conservation of matter and our assumption of a closed compartment. Whatever the production (j_{prod}) and degradation (j_{deg}) terms happen to be, this differential equation will be satisfied. A material balance equation that includes plasma membrane fluxes could take the form

$$\frac{dc}{dt} = j_{prod} - j_{deg} + j_{in} - j_{out} . \tag{2.4}$$

In this case, the rate of change of the number of molecules in the cell is influenced by four processes: production (j_{prod}), degradation (j_{deg}), influx (j_{in}) and efflux (j_{out}).

2.2   A Model of Intracellular Calcium Concentration

Let $c(t)$ denote the time-dependent concentration of intracellular calcium ions in an open cell of constant volume (see Fig. 2.2). To account for calcium fluxes across the plasma membrane we write,

$$\underbrace{\frac{\text{conc}}{\text{time}}}_{} \ominus \quad \frac{dc}{dt} = j_{in} - j_{out} \ominus \underbrace{\frac{\text{conc}}{\text{time}}}_{}, \tag{2.5}$$

where we have assumed no production or degradation of calcium in the cell interior and we have written conc as an abbreviation for concentration $= \#/\text{length}^3$. Let us assume a constant calcium influx rate,

$$j_{in} = J , \tag{2.6}$$

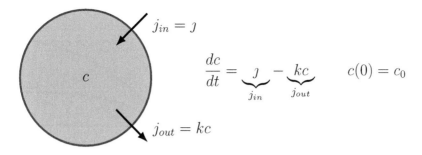

$$\frac{dc}{dt} = \underbrace{J}_{j_{in}} - \underbrace{kc}_{j_{out}} \qquad c(0) = c_0$$

Figure 2.2 Left: Diagram of an open cell model with constant calcium influx rate and an efflux rate proportional to intracellular calcium concentration. Right: The corresponding ODE initial value problem (see Eq. 2.8).

where the parameter J ⊕ conc/time is nonnegative ($J \geq 0$).[4] Let us further assume that the efflux rate is proportional to the intracellular calcium concentration ($j_{out} = kc$) where the proportionality constant k is positive ($k > 0$) and has physical dimensions of 1/time,

$$\frac{\text{conc}}{\text{time}} \oplus j_{out} = k \cdot c \oplus \frac{1}{\text{time}} \cdot \text{conc.} \tag{2.7}$$

Substituting these **constitutive relations**[5] (Eqs. 2.6 and 2.7) into the material balance equation (Eq. 2.5), we obtain

$$\frac{dc}{dt} = J - kc, \tag{2.8}$$

an **ordinary differential equation**[6] (ODE) with two parameters, J and k. This differential equation is our first model – a mathematical restatement of our assumptions (constant influx rate and efflux rate proportional to concentration). Using this model we will answer questions such as (1) "What intracellular calcium concentration will ultimately be attained?" and (2) "How quickly?" The answers will depend on the parameters J and k.

2.3   The Initial Value Problem and its Solution

Associating an initial condition to Eq. 2.8 gives the following ODE **initial value problem**,

$$\frac{dc}{dt} = J - kc \qquad c(0) = c_0, \tag{2.9}$$

where time (t) is the **independent variable**, c is the **dependent variable**, and c_0 is an **initial value** (the calcium concentration at time $t = 0$). Solving Eq. 2.9 means finding a function of time $c(t)$ for $t \geq 0$ with derivative dc/dt equal to $J - kc(t)$. The function $c(t)$ must also evaluate to c_0 when $t = 0$. Although there is no reason why this should be obvious to you right now, the function that satisfies these criteria is

$$c(t) = (c_0 - J/k)e^{-kt} + J/k. \tag{2.10}$$

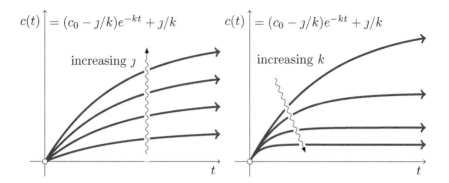

Figure 2.3   Solutions of the open cell model (Fig. 2.2) exponentially relax with rate constant k from an initial value $c_0 = 0$ to a limiting value of $\jmath/k$.

This is the solution of the ordinary differential equation initial value problem that is our model of calcium concentration dynamics (Eq. 2.9).

In order to understand and visualize this solution, let us assume (for the moment) that the intracellular calcium concentration is initially zero ($c_0 = 0$). Using this initial value (open circles), Fig. 2.3 plots the solution $c(t)$ for several different values of $\jmath$ and k. Take a moment to convince yourself that the arrows in Fig. 2.3 point in the correct direction. For example, the left plot suggests that when $c_0 = 0$ and k is fixed, increasing $\jmath$ increases $c(t)$ for any t. Does this seem correct? That is, if you imagine plotting Eq. 2.10 using several different increasing values for $\jmath$, would you expect these plots to stack on top of one another as they do in Fig. 2.3 (left)?

The dynamics of intracellular calcium concentration plotted in Fig. 2.3 are examples of **exponential relaxation**. Take note, this is *not* exponential *growth*, which would look like

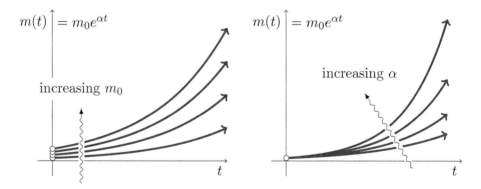

for positive m_0 and α. The dynamics of exponential growth (above) are what you might expect from, say, the unconstrained multiplication of bacteria in a flask with sufficient nutrients. Plots of exponential growth are convex (positive second derivative), because the rate of change of bacterial biomass – given by $m'(t) = \alpha m_0 e^{\alpha t}$ – is an increasing function of time. In the exponential *relaxation* predicted by our compartmental model (Fig. 2.3), the plots are concave (negative second derivative). Our compartmental model predicts that the rate of change of intracellular calcium concentration – given by $c'(t)$ – is a decreasing function of time.

2.4   Checking the Solution

To this point a compartmental model of intracellular calcium dynamics (Eq. 2.9) was constructed, and its solution was presented, plotted and discussed (Eq. 2.10 and Fig. 2.3). However, I have purposefully delayed the difficult part, which is showing you how to *solve* the ODE initial value problem that is our model. This is done in Section 2.7, which you may read now, provided you have mastered single variable calculus. If you do not feel like integrating right now, you may skip Section 2.7 on your first reading of this chapter.[7]

What is most important, at this point, is understanding what it means for Eq. 2.10 to solve Eq. 2.9. It means two things. First, the function $c(t)$ satisfies the ordinary differential equation. Secondly, the function $c(t)$ satisfies the initial condition. The first point may be confirmed by substituting the solution (Eq. 2.10) into the differential equation, as follows:

$$\frac{d}{dt}[c(t)] \overset{?}{=} J - k[c(t)]$$

$$\frac{d}{dt}[(c_0 - J/k) e^{-kt} + J/k] \overset{?}{=} J - k[(c_0 - J/k) e^{-kt} + J/k].$$

Next, we differentiate and simplify to check the equality,

$$(c_0 - J/k) \frac{d}{dt}[e^{-kt}] \overset{?}{=} J - k(c_0 - J/k) e^{-kt} - k \cdot J/k$$

$$(c_0 - J/k) e^{-kt} (-k) \overset{\text{yes}}{=} -k(c_0 - J/k) e^{-kt}. \checkmark$$

The check mark indicates that the equality has been shown to be true (multiplication is associative). The second point may be confirmed by substituting the function $c(t)$ into the initial condition and checking the equality, as follows:

$$c(0) \overset{?}{=} c_0$$

$$(c_0 - J/k) \underbrace{e^{-k \cdot 0}}_{1} + J/k \overset{?}{=} c_0$$

$$c_0 - J/k + J/k \overset{\text{yes}}{=} c_0. \checkmark$$

These two calculations show that Eq. 2.10 solves Eq. 2.9.

2.5   Interpreting the Solution

Our compartmental model is an ODE that formalizes the assumptions of constant influx rate J and efflux rate k that is proportional to concentration (Eqs. 2.6 and 2.7). The solution of the ODE is an exponential relaxation whose various terms and factors involve the model parameters, J and k. By considering how these parameters influence a plot of the solution, we learn what dynamics of intracellular calcium concentration are expected.

Question 1: What is the Concentration Ultimately Attained?

What intracellular calcium concentration is ultimately attained? The answer is found by taking the limit as $t \to \infty$ of the model's solution (Eq. 2.10),

$$\lim_{t \to \infty} c(t) = \left(c_0 - \frac{J}{k}\right) \underbrace{\lim_{t \to \infty} e^{-kt}}_{0} + \frac{J}{k} = \frac{J}{k}, \tag{2.11}$$

where we have used $k > 0$ to evaluate $\lim_{t \to \infty} e^{-kt} = 0$. This analysis shows that the intracellular calcium concentration $c \to J/k$ as $t \to \infty$. The calcium concentration ultimately attained is given by the *ratio* of the calcium influx rate (J) and the calcium efflux proportionality constant (k).

The above analysis is 100% correct; however, there is an easier way. Assuming that intracellular calcium concentration ultimately attained is constant, this limiting value must result in no further change ($dc/dt = 0$), that is, a balance of calcium influx (J) and efflux (kc). Let us denote this unknown constant intracellular calcium concentration that leads to no further change as $c(t) = c_{ss}$ (ss for steady state). In that case, an equation for the constant c_{ss} may be derived by setting the left side of Eq. 2.8 to zero ($dc/dt = 0$) and replacing c by c_{ss} on the right side,

$$0 = J - kc_{ss}. \tag{2.12}$$

Solving this equation for c_{ss} gives

$$\text{conc} \ominus c_{ss} = \frac{J}{k} \ominus \frac{\text{conc/time}}{1/\text{time}} = \text{conc}, \tag{2.13}$$

in agreement with Eq. 2.11. This ratio has physical dimensions of concentration, as required.

This second approach to determining the concentration ultimately attained is *easier*, because it did not involve taking a limit (no calculus required).[8] Instead, we reasoned our way to a simple algebra problem (Eq. 2.12) for the steady state intracellular calcium concentration.

Take a moment to interpret the relationship of the steady state concentration (c_{ss}) to the model parameters (J, k) given by Eq. 2.13. I would say that the model predicts the following.

- Increasing the calcium influx rate (J) increases the calcium concentration ultimately attained (c_{ss}).
- Increasing the calcium efflux proportionality constant (k) decreases the limiting concentration (c_{ss}).

Do you find these results intuitive?

Before moving on, we rewrite the time-dependent model solution (Eq. 2.10) to highlight our understanding that the steady state calcium concentration is $c_{ss} = J/k$,

$$c(t) = (c_0 - c_{ss})\,e^{-kt} + c_{ss}\,. \tag{2.14}$$

The left panel of Fig. 2.4 plots this solution for fixed c_{ss} and various initial values c_0.

Question 2: How Quickly?

Given our modeling assumptions, how quickly does the intracellular calcium concentration change? When the calcium concentration is zero, the rate of change of calcium concentration is J (to see this, substitute $c = 0$ into Eq. 2.9). Increasing the calcium influx rate J also increases the steady state calcium concentration c_{ss} (see Fig. 2.3).

Assuming c_{ss} is fixed using $J = kc_{ss}$, Fig. 2.4 (right) shows the dynamics of calcium concentration for different influx rate constants (k). Decreasing k slows the exponential relaxation to c_{ss}. Because the exponential function e^{-kt} in Eq. 2.14 involves the product kt and $k \Leftrightarrow 1/\text{time}$, the parameter k is referred to as an **exponential rate constant**. The reciprocal of k, which has physical dimensions of time,

$$\text{time} \Leftrightarrow \tau = \frac{1}{k} \Leftrightarrow \frac{1}{1/\text{time}}\,,$$

is an **exponential time constant**. Substituting $k = 1/\tau$ into Eq. 2.14 makes explicit both the steady state and exponential time constant,

$$c(t) = (c_0 - c_{ss})\,e^{-t/\tau} + c_{ss}\,. \tag{2.15}$$

Exponentially relaxing functions of time may be either decreasing or increasing. To illustrate, Fig. 2.5 (left) plots Eq. 2.15 with $c_0 > c_{ss}$, while the right panel uses $c_0 < c_{ss}$. In both cases, the quantity that is decaying away, that is, the quantity suppressed by the decreasing exponential factor $e^{-t/\tau}$, is $c(t) - c_{ss}$, the deviation of the calcium concentration from the limiting value. This deviation may be either positive (left) or negative (right), but does not change sign as the exponential relaxation proceeds. To be

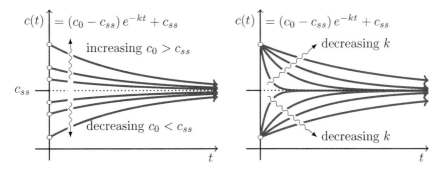

Figure 2.4 Left: $c(t)$ increases or decreases depending on whether c_0 is less than or greater than c_{ss}. Right: Decreasing the calcium efflux rate constant (k) slows the relaxation process.

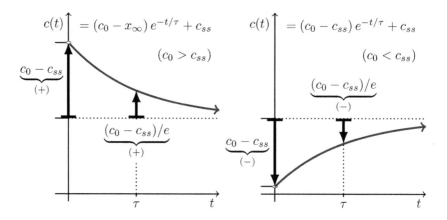

Figure 2.5   Exponential relaxation when the initial value c_0 is greater or less than the asymptotic value (c_{ss}). The exponential time constant τ is the time required for e-fold decay of the deviation from steady state.

precise, the deviation $c(t) - c_{ss}$ undergoes an e-fold decrease during the time interval $t = 0$ to $t = \tau$, as can be confirmed using Eq. 2.15:

$$\frac{c(\tau) - c_{ss}}{c(0) - c_{ss}} = \frac{(c_0 - c_{ss})\,\overbrace{e^{-\tau/\tau}}^{e^{-1}}}{\underbrace{(c_0 - c_{ss})\,e^{-0/\tau}}_{1}} = \frac{1}{e}\,.$$

Multiplying both sides of Eq. 2.5 by $c(0) - c_{ss}$ we obtain

$$c(\tau) - c_{ss} = \frac{c_0 - c_{ss}}{e} \approx \frac{c_0 - c_{ss}}{3}\,.$$

The exponential time constant τ is the time required for the deviation from steady state $(c - c_{ss})$ to become about $1/3$ of its original value (because $e = 2.71 \approx 3$).

2.6   Calcium Dynamics and Disease

Because intracellular calcium plays an important role in a variety of physiological processes, it is not surprising that rates of calcium influx and efflux may be modified in disease. For example, changes in intracellular calcium signaling are observed in transgenic mouse models of the neurodegenerative disease amyotrophic lateral sclerosis (ALS) (Julien and Kriz, 2006).

Fig. 2.6 shows results published in *Experimental Neurology* that report disregulated calcium influx in cultured cortical neurons obtained from transgenic G93A mice that are well characterized and highly utilized in ALS research. G93A refers to a single amino acid replacement (glycine to alanine at position 93) in Cu^{2+}, Zn^{2+} superoxide dismutase (SOD1), a homodimeric 32 kDa metalloprotein, thought to be relevant to

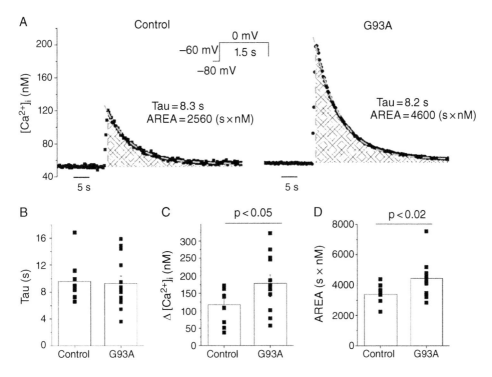

Figure 2.6   A: Depolarization-induced calcium influx in Control and G93A (transgenic mouse model of ALS) cortical neurons. B: Exponential time constants τ obtained by fitting the relaxation of calcium concentration to $c(t) = c_{ss} + (c_0 - c_{ss})e^{-t/\tau}$ were not significantly different. C and D: Statistically significant differences in the calcium concentration increase and area under the curve. Reprinted from *Experimental Neurology*, **247**, Pieri, M, Caioli, S, Canu, N, Mercuri, NB, Guatteo, E, and Zona, C, Over-expression of N-type calcium channels in cortical neurons from a mouse model of amyotrophic lateral sclerosis, pp. 349–358, Copyright 2013, with permission from Elsevier.

understanding an autosomal dominant familial form of ALS. Pieri et al. (2013) used microfluorimetry to study the dynamics of intracellular calcium concentration (vertical axis labelled $[Ca^{2+}]_i$) exhibited by Control and G93A cultured cortical neurons. In these experiments, the exponential time constant for relaxation of intracellular calcium concentration is almost identical in Control and G93A neurons ($\tau \approx 8$ s). Pieri et al. (2013) interpreted this to mean that the G93A did not influence the efflux rate constant k, which is given by $1/\tau \approx 0.125\,s^{-1}$. On the other hand, the rapid increase in intracellular calcium concentration at the onset of stimulation is larger in the G93A neurons than Control, consistent with higher calcium influx rate $_J$. Pieri et al. (2013) used pharmacological and biochemical methods to show that this excess of calcium influx in G93A cortical neurons is mediated by over-expression of voltage-gated plasma membrane calcium channels.[9]

If you do not feel like doing integrals, read Section 2.8 next.

2.7   Appendix: Solving $dc/dt = J - kc$ with $c(0) = c_0$

Analytical solution methods for the ODE initial value problem Eq. 2.9 include "guess and check" and "separation of variables."

Method 1: Guess and Check

If one makes an educated guess that the solution $c(t)$ is an appropriately shifted and scaled exponential function, then a candidate solution is

$$c(t) = \alpha e^{\beta t} + \gamma, \tag{2.16}$$

where α, β and γ are constants to be determined. Substituting Eq. 2.16 into the differential equation, $c' = J - kc$, yields

$$\left[\alpha e^{\beta t} + \gamma\right]' = J - k\left[\alpha e^{\beta t} + \gamma\right]. \tag{2.17}$$

The prime (') indicates differentiation so c' is just another way of writing dc/dt. Taking the derivative of the left side and distributing the k on the right side, we obtain

$$\alpha \beta e^{\beta t} = J - k\alpha e^{\beta t} - k\gamma. \tag{2.18}$$

Equating the coefficients[10] of the exponential terms gives $\alpha\beta = -k\alpha$ so $\beta = -k$. Equating the constant terms gives $0 = J - k\gamma$ so $\gamma = J/k$. Thus,

$$c(t) = \alpha e^{-kt} + J/k \tag{2.19}$$

solves the differential equation.

Next, we choose α to satisfy the initial condition,

$$c(0) = c_0 \implies \alpha \underbrace{e^{-k \cdot 0}}_{1} + J/k = c_0 \implies \alpha + J/k = c_0 \implies \alpha = c_0 - J/k.$$

Substituting $\alpha = c_0 - J/k$ into Eq. 2.19 gives Eq. 2.10.

Method 2: Separation of Variables

The method of **separation of variables** begins by multiplying both sides of the differential equation (Eq. 2.9) by dt and dividing both sides by $J - kc$,

$$\frac{dc}{J - kc} = dt.$$

Next, integrate both sides of this equation from the initial values ($\hat{t} = 0$ and $\hat{c} = c_0$) to arbitrary final values $\hat{t} = t$ and $\hat{c} = c$,[11]

$$\int_{c_0}^{c} \frac{d\hat{c}}{J - k\hat{c}} = \int_{0}^{t} d\hat{t}. \tag{2.20}$$

The integral on the right side evaluates to

$$\hat{t}\Big|_{0}^{t} = t - 0 = t. \tag{2.21}$$

For the integral on the left, one may use an integral table or symbolic integrator to obtain

$$\int \frac{dc}{J - kc} = -\frac{\ln(J - kc)}{k} + \text{constant}.$$

Using this antiderivative, the left side of Eq. 2.20 evaluates to

$$\int_{c_0}^{c} \frac{d\hat{c}}{J - k\hat{c}} = -\frac{\ln(J - k\hat{c})}{k}\bigg|_{c_0}^{c} = -\frac{1}{k}[\ln(J - kc) - \ln(J - kc_0)]$$

$$= -\frac{1}{k}\ln\left(\frac{J - kc}{J - kc_0}\right). \tag{2.22}$$

Substituting Eqs. 2.21 and 2.22 into Eq. 2.20, we obtain

$$-\frac{1}{k}\ln\left(\frac{J - kc}{J - kc_0}\right) = t \;\Rightarrow\; \ln\left(\frac{J - kc}{J - kc_0}\right) = -kt.$$

Exponentiating both sides of this expression gives

$$\frac{J - kc}{J - kc_0} = e^{-kt} \;\Rightarrow\; J - kc = (J - kc_0)e^{-kt} \;\Rightarrow\; kc = (kc_0 - J)e^{-kt} + J.$$

Dividing by k gives Eq. 2.10.

2.8   Discussion

Devaney et al. (2006, pp. 1–20) and Shiflet and Shiflet (2006, pp. 71–110) are suggested readings on dynamics problems leading to first-order differential equations. For relevant mathematical history read *e: The Story of a Number* (Maor, 2009).

Extensive and Intensive Quantities

The number of molecules in a cell (n) and the volume of a cell (v) are **extensive** physical quantities, because their values are proportional to system size. The number of molecules of an aggregate system composed of two identical subsystems doubles the original number. The volume also doubles. Conversely, concentration (c) is an **intensive** physical quantity that does not scale with system size.

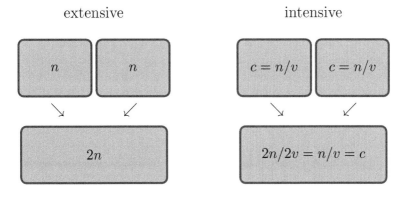

For a cell of constant volume, the number of calcium ions is proportional to concentration,

$$\# \ominus n = vc \ominus \frac{\#}{\text{length}^3} \cdot \text{length}^3,$$

and the rates of change are related as follows:

$$\frac{\#}{\text{time}} \ominus \frac{dn}{dt} = v\frac{dc}{dt} \ominus \text{length}^3 \cdot \frac{\#/\text{length}^3}{\text{time}}.$$

By rescaling the dependent variable, an ODE model for concentration can be put in terms of number of molecules, and vice versa, as illustrated:

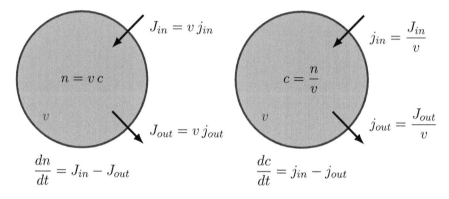

$$\frac{dn}{dt} = J_{in} - J_{out} \qquad\qquad \frac{dc}{dt} = j_{in} - j_{out}$$

The ratio of any two extensive quantities is intensive. For another example, Gibbs free energy is an extensive quantity with physical dimensions of energy = mass · length^2/time^2 and SI unit of joules. Chemical potential (also called molar Gibbs free energy) is an intensive quantity with SI unit of joules/mole and physical dimensions of energy/#.

Fundamental and Derived Physical Dimensions

All physical quantities have a physical dimension that is independent of the units of measurement, e.g., velocity = length/time. Physical dimensions clarify the equations of biophysics, the modeling process, and unfamiliar physical quantities.

This chapter introduced the notation # for number of molecules or amount in moles, 1/time (rate), #/time (rate of change in amount), length^3 (volume), conc = #/length^3 (concentration), and conc/time = #/(length^3 · time). These *derived* physical dimensions are products of integer powers of *fundamental* physical dimensions (see Table 2.1).

To see how physical dimensions clarify mathematical modeling, consider the following pedantic but important observation. We have referred to the quantities j_{in} and j_{out} as influx and efflux rates, and j_{prod} and j_{deg} as production and degradation rates. In spite of the suffix of in*flux* and ef*flux*, these terms do not have physical dimensions of

Table 2.1 Fundamental physical dimensions

Physical dimension	Abbreviation[12]	SI unit[13]	Abbreviation
number of molecules[14]	#	mole	mol
time	time	second	s
mass	mass	kilogram	kg
length	length	meter	m
electric charge	charge	coulomb	C
temperature	temp	kelvin	K
luminous intensity	lumin	candela	cd

$$\text{flux} = \frac{\#}{\text{time} \cdot \text{length}^2}.$$

Rather, these quantities have physical dimensions of $j \ominus \text{conc/time}$, that is, they are *flow rates expressed as the rate of change of concentration in a compartment with constant volume.* A bona fide flow rate has physical dimensions of #/time. This may be obtained from $j \ominus \text{conc/time}$ by multiplying by the cell volume (v), as follows

$$\text{flow rate} = \frac{\#}{\text{time}} \ominus J = vj \ominus \frac{1}{\text{length}^3} \cdot \frac{\#/\text{length}^3}{\text{time}}.$$

In our compartmental model, this flow rate is the product of membrane surface area ($= \text{length}^2$) and the calcium flux across the membrane that is mediated by calcium channels,

$$\frac{\#}{\text{time}} \ominus J = A \cdot \bar{J} \ominus \text{length}^2 \cdot \underbrace{\frac{\#}{\text{time} \cdot \text{length}^2}}_{\text{flux}}.$$

Combining these two expressions shows that the net *flow rate expressed in terms of the rate of change of concentration* ($j = j_{in} + j_{out}$) is proportional to the net flux of calcium across the plasma membrane ($\bar{J}$), as follows

$$\frac{\text{conc}}{\text{time}} \ominus j = \frac{J}{v} = \frac{A}{v} \cdot \bar{J} \ominus \underbrace{\frac{\text{length}^2}{\text{length}^3}}_{1/\text{length}} \cdot \underbrace{\frac{\#}{\text{time} \cdot \text{length}^2}}_{\text{flux}},$$

where $A/v \ominus 1/\text{length}$ is a surface-to-volume ratio.

Supplemental Problems

Problem 2.1 A cell of radius $10\,\mu\text{m}$ has a cAMP concentration of $100\,\text{nM}$. How many cAMP molecules is this?

Problem 2.2 If time has units of seconds (s) and concentration has units of micromolar (μM), what are the units of j_{prod} and j_{deg} in Eq. 2.3?

Problem 2.3   What is the common feature of biological, chemical and physical processes that leads to the dynamical phenomenon of exponential relaxation?

Problem 2.4   Let $f(x) = e^x$ and $g(x) = e^{-x}$. Calculate the first and second derivatives of $f(x)$ and $g(x)$ and plot all six functions on one graph.

Problem 2.5   Rewrite the ODE $dc/dt = \jmath - kc$ in terms of $c_{ss} = \jmath/k$ and $\tau = 1/k$.

Problem 2.6   Verify that $m(t) = m_0\, e^{\alpha t}$ is the solution to $dm/dt = \alpha m$ with initial condition $m(0) = m_0$ by (a) substitution and differentiation, and (b) separation of variables and integration. Plot $m(t)$ for $\alpha \leq 0$ and compare with the example of exponential growth on p. 18. Describe $m(t)$ when $\alpha = 0$.

Problem 2.7   Verify that $y = y_0\, e^{-t/\tau}$ is the solution to $dy/dt = -y/\tau$ with initial condition $y(0) = y_0$.

Problem 2.8   For each ODE below, find the steady state solution that satisfies $dy/dt = 0$.

(a) $\dfrac{dy}{dt} = 14 - 2y$ (b) $\dfrac{dy}{dt} = 125 - y^3$ (c) $\dfrac{dy}{dt} = -\dfrac{y - 0.1}{0.05}$

Problem 2.9   The numerical value of an exponential time constant can be identified by comparing an exponential function with $e^{-t/\tau}$ and solving for τ. In the case of $ae^{-2kt} + b$ with $k = 5\,\mathrm{s}^{-1}$, what is τ in units of seconds?

Problem 2.10   In our discussion of exponential time constants, we suggested that $e \approx 2.718$ is nearly 3 and $1/e \approx 0.3679$ is nearly $1/3$. Determine the accuracy of these approximations. Hint: If x_* is an approximation for x, the absolute error is $s = |x - x_*|$ and the relative error is $r = s/|x|$ $(x \neq 0)$.

Problem 2.11   Are the following physical quantities extensive or intensive? Mass, electric charge, temperature, viscosity, heat capacity, specific heat capacity, length, chemical potential, energy, mass density, momentum.

Problem 2.12   If x has physical dimensions of length and $f(x)$ has physical dimensions of mass, what are the physical dimensions of $f'(x)$?

Problem 2.13   Explain why the physical dimensions of the integral $\int_a^b f(x)\, dx$ is the product of the physical dimensions of $f(x)$ and x.

Problem 2.14   In a physical equation of the form $y = e^x$, x is usually a ratio of two quantities with identical physical dimensions. Explain why.

Problem 2.15   Consider the ODE initial value problem $\frac{dx}{dt} = -kx$ where $x(0) = x_0$ and $x_0, k > 0$. The solution $x(t) = x_0 e^{-kt}$. What is the time constant (τ) for this exponential decay in terms of k? Is τ a function of x_0? What is the half life $(\tau_{1/2})$ of this exponential decay? How are these concepts similar? How are they different?

Problem 2.16 The rate of IP$_3$ metabolism depends on the concentration of IP$_3$ and calcium (Sims and Allbritton, 1998). At low IP$_3$ concentration (100 nM) and high calcium concentration (≥ 1 µM), IP$_3$ is metabolized predominantly by inositol 1,4,5-trisphosphate 3-kinase with a half life of $\tau_{1/2} = 60$ s. What is the exponential time constant (τ) that corresponds to this reported half life ($\tau_{1/2}$)?

Problem 2.17 At low concentrations of IP$_3$ and calcium (both ≤ 400 nM), the activity of the inositol 1,4,5-trisphosphate 5-phosphatase was comparable to the 3-kinase. Assuming a constant production rate of IP$_3$, one might model this situation with the following ODE initial value problem,

$$\frac{d[IP_3]}{dt} = J - k_1[IP_3] - k_2[IP_3] \qquad [IP_3](0) = [IP_3]_0$$

where $[IP_3]$ denotes the concentration of IP$_3$. Assuming the respective IP$_3$ degradation rates under these conditions are $k_1 = 0.01$ s^{-1} and $k_2 = 0.03$ s^{-1}, calculate the exponential time constant. Assuming $J = 0.2$ µM s^{-1}, calculate the limiting IP$_3$ concentration.

Problem 2.18 Consider the exponential decay $x(t) = x_0 e^{-kt}$. Show that the shifted function $x(t - t_0)$ is proportional to $x(t)$, that is, $x(t - t_0) = \eta x(t)$, where the proportionality constant η depends on t_0. Does any other continuous function have this property?

Problem 2.19 When solving $dc/dt = J - kc$ we assumed that the calcium influx rate is the constant J. However, cells that are activated (e.g., through the binding of neurotransmitter to cell surface receptors) often exhibit an increase in calcium influx rate. If the calcium influx rate is constant both before and after a sudden increase, the situation may be modeled using the Heaviside step function,

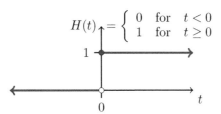

$$H(t) = \begin{cases} 0 & \text{for} \quad t < 0 \\ 1 & \text{for} \quad t \geq 0 \end{cases}$$

An instantaneous jump in calcium influx rate from J_0 to J_1 at time $t = t_0$ is written,

$$J(t) = J_0 + \underbrace{(J_1 - J_0)}_{\Delta J} H(t - t_0) = \begin{cases} J_0 & \text{for} \quad t < t_0 \\ J_1 & \text{for} \quad t \geq t_0 \end{cases} \tag{2.23}$$

where $\Delta J \ominus$ conc/time is the change. Eq. 2.23 is plotted below for $0 < J_0 < J_1$:

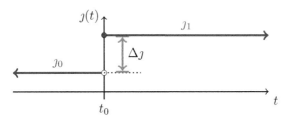

Assuming that the calcium concentration leading up to $t = t_0$ is given by the steady state consistent with J_0, the model response to the step increase in calcium influx rate is the following piecewise continuous function,

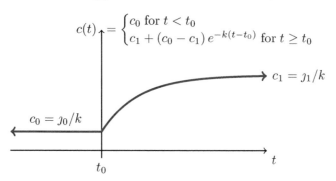

$$c(t) = \begin{cases} c_0 \text{ for } t < t_0 \\ c_1 + (c_0 - c_1)\, e^{-k(t-t_0)} \text{ for } t \ge t_0 \end{cases}$$

where $c_0 = J_0/k$ and $c_1 = J_1/k > c_0$. (a) What is the derivative of the solution at $t = t_0$? (b) Sketch a piecewise continuous solution corresponding to a step *decrease* in the influx rate ($J_1 < J_0$, $\Delta J < 0$).

Problem 2.20   The solution to $dc/dt = J(t) - c/\tau$ is plotted below using a square pulse increase in calcium influx rate, as may occur when an agonist is applied to media bathing cells and subsequently washed out. The blue and gray curves use the same values for J_0 and J_1, but two different values of τ.

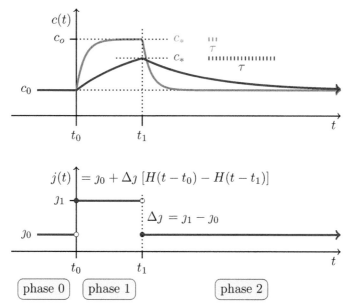

The model response has three phases defined by the times that the influx rate changes (t_0 and t_1). (a) Sketch model responses that would occur for pulses of a fixed duration $t_1 - t_0$, fixed τ, but several different ΔJ (both positive and negative). (b) Sketch model responses to square pulses of different duration $t_1 - t_0$, fixed τ, and fixed $\Delta J > 0$. (c) How do the responses sketched in (a) and (b) change for smaller and larger τ? (d) Find the piecewise differentiable solution $c(t)$.

Problem 2.21 In Problem 2.20, the relationship between exponential time constant (τ) and the pulse duration ($t_1 - t_0$) determines the degree to which the steady state (c_1) associated with the influx rate (j_1) during the pulse is achieved at pulse end (when $t = t_1$ and $c = c_*$). Writing $\Delta t = t_1 - t_0 > 0$, observe that a very short duration pulse leads to a small response. In fact, $c_* \to c_0$ as $t_1 \to t_0$ (see illustration above), as can be shown by taking the limit

$$\lim_{\Delta t \to 0} c_* = \lim_{\Delta t \to 0} \left[c_1 + (c_0 - c_1) e^{-\Delta t / \tau} \right] = c_1 + (c_0 - c_1) = c_0, \tag{2.24}$$

for fixed $c_0 = \tau j_0$ and $c_o = \tau j_1$. When these quantities do not depend on Δt, a pulse of zero duration leads, unsurprisingly, to zero response. A more interesting limit to consider is illustrated below (left).

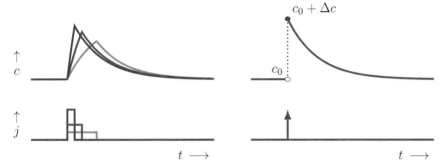

Here the influx rate increase is a square pulse with short duration *but fixed area*. That is, the pulse has height Δj that is inversely proportional to its duration (Δt),

$$\Delta j = \frac{\Delta c}{\Delta t}. \tag{2.25}$$

In this expression, the proportionality constant is denoted Δc because it has physical dimensions of concentration,

$$\text{conc} \ominus \Delta c = \Delta j \, \Delta t \ominus \frac{\text{conc}}{\text{time}} \cdot \text{time}.$$

The illustration above suggests that the concentration at the end of such short/strong pulses approaches the limiting value $c_* \to c_0 + \Delta c$. Show that this is the case by recalculating Eq. 2.24 using Eq. 2.25.

Problem 2.22 When the calcium influx rate $j(t)$ is a *sequence* of impulses (short/strong pulses as in Problem 2.21), the response is

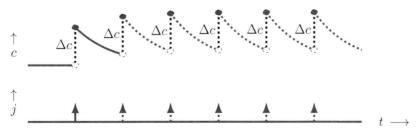

Sketch your expectations for the model response to impulses of the same magnitude (step size Δc) arriving 2 and 4 times faster.

Problem 2.23   The single compartment model presented above may be extended in many ways, e.g., by including equations for multiple chemical species and constitutive relations for the kinetics of chemical reactions (see Chapter 4). It is also possible to include a second compartment to represent intracellular calcium stores, such as the endoplasmic reticulum (ER). The illustration below shows a closed cell model for cytosolic (c_{cyt}) and ER (c_{er}) calcium concentration,

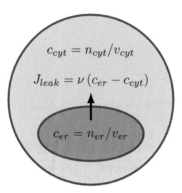

where v_{cyt} and v_{er} are the volumes of these compartments. The arrow represents the net flux across the ER membrane. The number of calcium ions in each compartment is given by $n_{cyt} = v_{cyt}c_{cyt}$ and $n_{er} = v_{er}c_{er}$. The material balance equations are

$$\frac{\#}{\text{time}} \ominus \quad \frac{dn_{cyt}}{dt} = J_{leak} \qquad \frac{dn_{er}}{dt} = -J_{leak} \, , \qquad (2.26)$$

where $J_{leak} \ominus$ #/time denotes the (net) number of calcium ions moving from the ER into the cytosol per unit time. The direction of the arrow indicates that positive flux corresponds to calcium leaving the ER and entering the cytosol. (a) What are the corresponding ODEs in terms of concentration? (b) Let n_T denote the total number of calcium ions in the cell, that is, $n_T = n_{cyt} + n_{er}$, differentiate this equation, and use Eq. 2.26 to show that $dn_T/dt = 0$. Constant n_T means that $n_{cyt} + n_{er}$ is a combination of independent variables that does not change as a function of time (a conserved quantity).[15] Using $n_{er} = n_T - n_{cyt}$, the ODE for n_{er} may be eliminated,

$$\frac{dn_{cyt}}{dt} = J_{leak}(n_{cyt}, n_{er}) \quad \text{where} \quad n_{er}(t) = n_T - n_{cyt}(t). \qquad (2.27)$$

In this way the two compartment closed cell model is reduced to an ODE with one dependent variable.

Assume the passive leak out of the ER into the cytosol is proportional to the difference between the ER and cytosolic calcium concentration, that is,

$$J_{leak} = v \left(c_{er} - c_{cyt} \right) , \qquad (2.28)$$

where $v > 0$. (c) What are the physical dimensions for v? (d) When $c_{er} > c_{cyt}$, is J_{leak} inward our outward? (e) Use Eq. 2.28 to rewrite Eq. 2.27 as an ODE for c_{cyt}. (f) Associate the initial condition $c_{cyt}(0) = c_{cyt}^0$. Assuming $c_{er}(0) = (n_T - v_{cyt}c_{cyt}^0)/v_{er} > c_{cyt}^0$, sketch a representative solution. What is the steady state value and exponential time constant for c_{cyt}? Hint: Use algebra to put the ODE into the familiar form, $dc_{cyt}/dt = J - kc_{cyt}$, where J and k are functions of v, n_T, v_{cyt} and v_{er}. (g) Assuming constant total cell volume $v = v_{cyt} + v_{er}$, what value of v_{cyt}/v_{er} leads to the largest τ?

Solutions

Problem 2.1   A spherical cell of radius $1\,\mu m$ has volume

$$v = \frac{4}{3}\pi(1\,\mu m)^3 = \frac{4}{3}\pi \cdot 1\,\mu m^3 \approx 4.188\,\mu m^3\,.$$

Using the fact that $1\,\mu m^3 = 10^{-15}$ liter, we continue

$$v = 4.188\,\mu m^3 \cdot \frac{10^{-15}\,\text{liter}}{1\,\mu m^3} = 4.188 \times 10^{-15}\,\text{liter}.$$

A concentration of $c = 100\,\text{nM} = 10^{-7}\,\text{mol/liter} = 10^{-7} \cdot 6.02 \times 10^{23}\,\text{liter}^{-1}$ corresponds to $n = cv$ molecules, that is,

$$n = cv = 6.02 \times 10^{16}\,\text{liter}^{-1} \cdot 4.188 \times 10^{-15}\,\text{liter} \approx 252\,.$$

Problem 2.2   In Eq. 2.3, j_{prod} has units of μMs^{-1} and j_{deg} has units of $1/s$.

Problem 2.4

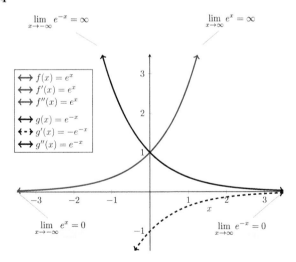

Problem 2.5

$$\frac{dc}{dt} = J - kc = k\left(\frac{J}{k} - c\right) = \frac{J/k - c}{1/k} = \frac{c_{ss} - c}{\tau} = -\frac{c - c_{ss}}{\tau}\,.$$

Problem 2.7

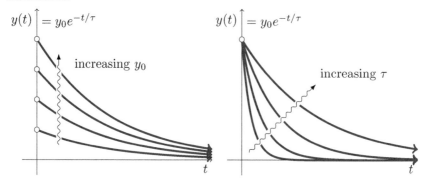

Problem 2.8   The steady states are $y(t) = y_{ss}$ where $y_{ss} = 7$ (a), 5 (b), and 2 (c).

Problem 2.9   Equate e^{-2kt} with $e^{-t/\tau}$ and solve for τ, as follows:

$$-2kt = -t/\tau \;\Rightarrow\; 1/\tau = 2k \;\Rightarrow\; \tau = 1/(2k) = 1/(2 \cdot 5\,\mathrm{s}^{-1}) = 100\,\mathrm{ms}\,.$$

Problem 2.10   The relative errors of substituting 3 and $1/3$ for e and $1/e$ are about 10%: $(3 - e)/e = 0.1036$ and $(1/e - 1/3)/(1/e) = 0.0939$. In fact, 3 is a better approximation to π than e, as $(\pi - 3)/\pi = 0.0451$. The rational approximation $22/7 \approx 3.1429$ to π is well known and has a relative error of 4.03×10^{-4}. The rational approximation $19/7 \approx 2.7143$ to e has a relative error of 1.47×10^{-3}.

Problem 2.11   *Extensive*: mass, length, heat capacity, energy, momentum, electric charge. *Intensive*: temperature, viscosity, chemical potential, electrical resistivity, melting point, pressure, specific heat capacity, mass density.

Problem 2.12 $f(x) \ominus$ mass and $x \ominus$ length so $f'(x) = df/dx \ominus$ mass/length.

Problem 2.13   The physical dimensions must be those of "area under the curve."

Problem 2.14   The derivative of $y = e^x$ is $dy/dx = e^x$, $dy/dx = y$ and thus $y \ominus dy/dx$. This can only occur for dimensionless x.

Problem 2.15   Left: The half life of the exponential decay $x(t) = x_0 e^{-kt}$ is $\tau_{1/2} = \log 2/k$, the time for a decay to $x_0/2$. Right: The exponential time constant $\tau = 1/k$ is the time for a decay to x_0/e.

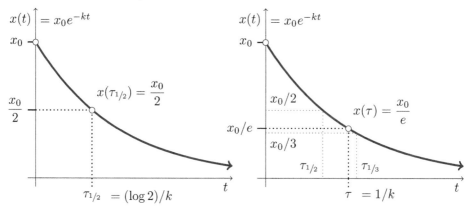

To find the half life, equate $x(t = \tau_{1/2})$ with the initial value x_0 divided by two,

$$x_0 e^{-k\tau_{1/2}} = \frac{x_0}{2} \Rightarrow \cdots \Rightarrow \tau_{1/2} = \frac{\ln 2}{k}.$$

For the exponential time constant, equate $x(t = \tau)$ with x_0 divided by e,

$$x_0 e^{-k\tau} = \frac{x_0}{e} \Rightarrow \cdots \Rightarrow \tau = \frac{1}{k}.$$

The exponential time constant is larger than the half life: $\tau/\tau_{1/2} = 1/\ln 2 \approx 1.443$.

Problem 2.19   (a) $c'(t) = 0$ for $t < t_0$; however, for $t > t_0$, $c'(t) = -k\,(c_0 - c_1)\,e^{-k(t-t_0)}$. Taking the limit from the right gives $\lim_{t\downarrow t_0} c'(t) = -k\,(c_0 - c_1) = J_1 - J_0 > 0$. (b) Below is a solution using $J_1 > J_0$:

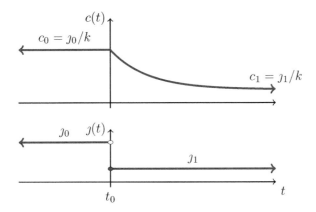

Problem 2.20   (a) Below are responses for different values of ΔJ:

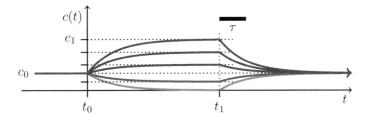

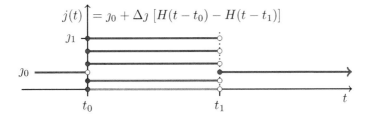

(b) Model responses to a square pulse of different duration $t_1 - t_0$:

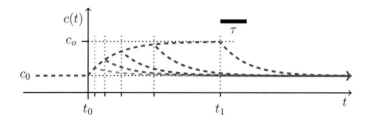

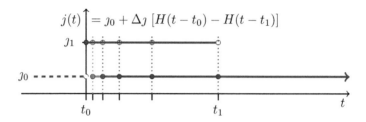

When the duration of phase 1 is long compared to the exponential time constant τ, the concentration c_* at the beginning of phase 2 will be approximately c_1. This is not the case for pulses of short duration. (c) The multipart figure below shows different step amplitudes Δj (bottom row) that are either positive (blue) or negative (gray). The responses of the compartmental model are shown above the dotted line. Within the 4×4 matrix of panels, each column uses a different pulse duration $t_1 - t_0$ and each row uses a different time constant τ.

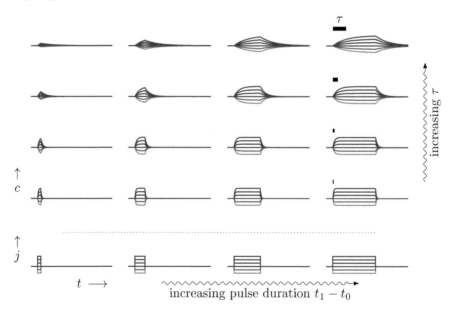

(d) The model has a piecewise differentiable solution,

$$c(t) = \begin{cases} c_0 & t < t_0 \\ c_1 + (c_0 - c_1)\, e^{-(t-t_0)/\tau} & t_0 \le t < t_1 \\ c_0 + (c_* - c_0)\, e^{-(t-t_1)/\tau} & t_1 \le t, \end{cases}$$

where $c_0 = \tau J_0$, $c_1 = \tau J_1$, and

$$c_* = c_1 + (c_0 - c_1)e^{-(t_1-t_0)/\tau} . \qquad (2.29)$$

Problem 2.21 Rewriting Eq. 2.29 as suggested gives

$$c_* = c_0 + \Delta_J \tau \left[1 - e^{-\Delta t/\tau}\right] = c_0 + \Delta c \frac{\tau}{\Delta t} \left[1 - e^{-\Delta t/\tau}\right] .$$

In the limit of a short-duration fixed-area pulse,

$$\lim_{\Delta t \to 0} c_* = \lim_{\Delta t \to 0} \left\{c_0 + \Delta c \frac{\tau}{\Delta t} \left[1 - e^{-\Delta t/\tau}\right]\right\} = c_0 + \Delta c , \qquad (2.30)$$

where we have used $(1 - e^{-x})/x \to 1$ as $x \to 0$, a limit that can be confirmed using L'Hopital's rule, $(1 - e^{-x})' = e^{-x}$ and $x' = 1$.

Problem 2.22 For impulses arriving 2 and 4 times faster, the responses are:

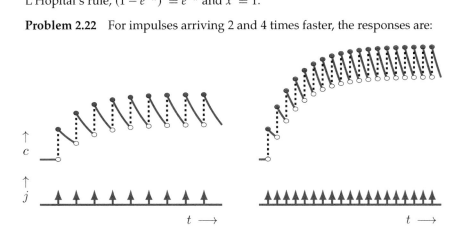

Assume evenly spaced pulses arrive at $t_1, t_2, \ldots$ where $t_{i+1} = t_i + T$ where T is the time between pulses. That is, $T \ \ominus \ $ time is the stimulation period and $f = 1/T \ \ominus \ $ time^{-1} is the stimulation frequency.[16] It can be shown that

$$c(t) = c_0 + \Delta c \sum_{i=1}^{\infty} H(t - t_i)e^{-(t-t_i)/\tau} \quad \text{where} \quad 0 \le t < \infty .$$

One may estimate the limiting periodic response by observing that the average influx rate during the repetitive stimulation is

$$\frac{\text{conc}}{\text{time}} \ \ominus \ j_{in}^{avg} = \int_0^T j(t)\, dt = \frac{\Delta c}{T} = f \Delta c \ \ominus \ \text{time}^{-1} \cdot \text{conc} ,$$

while the average calcium efflux rate is

$$\frac{\text{conc}}{\text{time}} \ \ominus \ j_{out}^{avg} = \int_0^T \frac{c(t)}{\tau}\, dt = \frac{1}{\tau} \int_0^T c(t)\, dt = \frac{c_{avg}}{\tau} \ \ominus \ \frac{\text{conc}}{\text{time}} .$$

For balance of average influx and efflux, $j_{in}^{avg} = j_{out}^{avg} \Rightarrow f \Delta c = c_{avg}/\tau$. Solving for the average calcium concentration, we find $c_{avg} = f \tau \Delta c \ \ominus \ \text{time}^{-1} \cdot \text{time} \cdot \text{conc}$.

Problem 2.23 (a) $v_{cyt}\, dc_{cyt}/dt = J_{leak}$ and $v_{er}\, dc_{er}/dt = -J_{leak}$. (b) $dn_T/dt = dn_{cyt}/dt + dn_{er}/dt = J_{leak} - J_{leak} = 0$. (c) The parameter v has physical dimensions of length^3/time,

$$\frac{\#}{\text{time}} \ominus J_{leak} = v\left(c_{er} - c_{cyt}\right) \ominus \frac{\text{length}^3}{\text{time}} \cdot \frac{\#}{\text{length}^3}. \tag{2.31}$$

Eq. 2.31 is related to Fick's law that flux due to diffusion is proportional to the concentration gradient dc/dx,

$$\text{flux} = \frac{\#}{\text{time} \cdot \text{length}^2} \ominus J_{diff} = -D\frac{dc}{dx} \ominus \frac{\text{length}^2}{\text{time}} \cdot \frac{\text{conc}}{\text{length}}, \tag{2.32}$$

where x is the spatial variable and D is the diffusion coefficient. The negative sign is important: a concentration gradient decreasing with increasing x (from left to right, $dc/dx < 0$) leads to positive flux $J_{diff} > 0$, that is, net diffusive transport in the direction of increasing x. If the ER membrane is an interface of thickness L over which the concentration drop $c_{er} - c_{cyt}$ occurs, then according to Fick's law,

$$\text{flux} = \frac{\#}{\text{time} \cdot \text{length}^2} \ominus J_{diff} = -D\frac{c_{er} - c_{cyt}}{L} \ominus \frac{\text{length}^2}{\text{time}} \cdot \frac{\#/\text{length}^3}{\text{length}}.$$

Defining the membrane **permeability** (P) as

$$\frac{\text{length}}{\text{time}} \ominus P = \frac{D}{L} \ominus \frac{\text{length}^2/\text{time}}{\text{length}},$$

rewrite the diffusion flux with the permeability as the proportionality constant,

$$\text{flux} \ominus J_{diff} = -P\left(c_{er} - c_{cyt}\right) \ominus \frac{\text{length}}{\text{time}} \cdot \frac{\#}{\text{length}^3}.$$

Multiplying both sides by surface area (A) yields Eq. 2.28,

$$\underbrace{\text{length}^2 \cdot \text{flux}}_{\frac{\#}{\text{time}}} \ominus \underbrace{A J_{diff}}_{J_{leak}} = -\underbrace{A P}_{v}\left(c_{er} - c_{cyt}\right) \ominus \underbrace{\text{length}^2 \cdot \frac{\text{length}}{\text{time}}}_{\frac{\text{length}^3}{\text{time}}} \cdot \frac{\#}{\text{length}^3}.$$

(d) Outward. $c_{er} > c_{cyt}$ implies $J_{leak} > 0$, which is a net flux in the direction of the outward pointing arrow in the model diagram. (e) The resulting scalar ODE is

$$v_{cyt}\frac{dc_{cyt}}{dt} = v(c_{er} - c_{cyt}) \quad \text{where} \quad c_{er} = \frac{n_T - v_{cyt}c_{cyt}}{v_{er}}. \tag{2.33}$$

Substituting the algebraic expression for c_{er} (right) into the ODE (left) and associating an initial condition, we obtain the ODE initial value problem,

$$v_{cyt}\frac{dc_{cyt}}{dt} = v\left(\frac{n_T - v_{cyt}c_{cyt}}{v_{er}} - c_{cyt}\right) \qquad c_{cyt}(0) = c_{cyt}^0, \tag{2.34}$$

where $n_T = v_{cyt}c_{cyt}^0 + v_{er}c_{er}^0$ is constant, i.e., n_T is a conserved quantity determined by initial concentrations c_{cyt}^0 and c_{er}^0. (f) Algebraic manipulation of Eq. 2.34 yields

$$\frac{dc_{cyt}}{dt} = v\underbrace{\frac{n_T}{v_{cyt}v_{er}}}_{J} - v\underbrace{\left(\frac{v_{cyt} + v_{er}}{v_{cyt}v_{er}}\right)c_{cyt}}_{k} \qquad c_{cyt}(0) = c_{cyt}^0. \tag{2.35}$$

Using an initial value $c_{cyt}^0 < c_{er}(0)$,

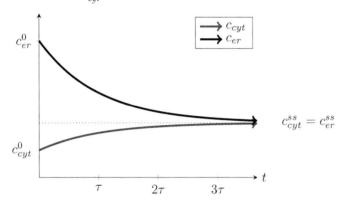

The limiting value for c_{cyt} solves $dc_{cyt}/dt = 0$, which implies $J_{leak} = v(c_{er}^{ss} - c_{cyt}^{ss}) = 0 \Rightarrow c_{cyt}^{ss} = c_{er}^{ss}$ and

$$c_{cyt}^{ss} = \frac{J}{k} = \frac{v\, n_T/\left(v_{cyt}v_{er}\right)}{v\left(v_{cyt} + v_{er}\right)/\left(v_{cyt}v_{er}\right)} = \frac{n_T}{v_{cyt} + v_{er}}.$$

The time constant for equilibration of c_{cyt} and c_{er} is

$$\tau = \frac{1}{k} = \frac{1}{v}\left(\frac{v_{cyt}v_{er}}{v_{cyt} + v_{er}}\right).$$

(g) Observe that $\tau = \bar{v}/(2v)$ where $\bar{v} = 2v_{cyt}v_{er}/(v_{cyt} + v_{er}) = 2/(1/v_{cyt} + 1/v_{er})$ is the harmonic mean of the cytosolic and ER volumes. For constant $v = v_{cyt} + v_{er}$, $\bar{v}$ and τ are maximized when $v_{cyt} = v_{er}$.

Notes

1. This symbol was to be read "has units of" in Louis DeFelice's classic text *Introduction to Membrane Noise* (1981).
2. These physical quantities cannot be subtracted (or summed), because one has physical dimensions of length and the other has physical dimensions of time.
3. Also called a *mass balance equation* or, generically, a *conservation law*.
4. We write j without a dot to remind ourselves that j is a constant parameter, not a flux that remains to be specified.
5. The adjective *constitutive* indicates that each specified relation forms an essential component (constituent) of the mathematical model. This is not

to be confused with the use of constitutive to mean continuous production or expression, as when a biochemical process is described as constitutively active. Constitutive relations are also called *rate laws*.

6. Eq. 2.8 is a *first-order* differential equation. Classification of ODEs is discussed at the end of Chapter 3.

7. Find time soon to review differentiation and integration of polynomial functions.

8. That is, Eq. 2.13 may be derived *without solving the differential equation*. The ability to analyze steady states of ODEs without actually solving for their time-dependent solutions will serve us well as models become more complex.

9. Specifically, N-type voltage-gated Ca^{2+} channels, Pieri et al. (2013). Abstract: "Voltage-gated Ca^{2+} channels (VGCCs) mediate calcium entry into neuronal cells... [We show that] N-type Ca^{2+} channels are over expressed in G93A cultured cortical neurons and in motor cortex of G93A mice compared to Controls. In fact, by western blotting, immunocytochemical and electrophysiological experiments, we observe higher membrane expression of N-type Ca^{2+} channels in G93A neurons compared to Controls. G93A cortical neurons filled with calcium-sensitive dye Fura-2 show a net calcium entry during membrane depolarization that is significantly higher compared to Control. Analysis of neuronal vitality following the exposure of neurons to a high K concentration (25 mM, 5 h) [resulting in chronic depolarization], shows a significant reduction of G93A cellular survival compared to Controls. N-type channels are involved in the G93A higher mortality because ω-conotoxin GVIA (1 μM), which selectively blocks these channels, is able to abolish the higher G93A mortality when added to the external medium..."

10. Equality is only possible for all t if the coefficients of $e^{\beta t}$ agree.

11. The circumflexes in Eq. 2.20 indicate that $\hat{c}$ and $\hat{t}$ are "dummy variables" that appear in the integrals only as placeholders. I use a circumflex ($\hat{c}$ and $\hat{t}$) rather than a prime (c' and t') to reserve the prime for differentiation as in $f'(x) = df/dx$. Dummy variables disappear upon evaluating the antiderivative at the lower and upper limits of integration, so any symbol with no previous meaning will do. For example, $\int_{c_0}^{c} \frac{d\lambda}{J-k\lambda} = \int_{0}^{t} d\xi$ is equivalent to Eq. 2.20. Using $\hat{c}$ and $\hat{t}$ emphasizes the physical interpretation of the dummy variables.

12. In this primer, the fundamental physical dimensions are denoted by #, time, mass, length, charge, temp, and lumin. We hope that readers who are new to the concept of physical dimensions appreciate our use of readable abbreviations such as temp, mass, length, etc. as opposed to the standard single character symbols N = number of molecules, T = time, M = mass, L = length, Q = electric charge, Θ = temperature, and J = luminous intensity, given the common usage of T for temperature, L for liter, M for molar, J for joule.

13. The International System of Units is a widely used system of measurement. SI is an abbreviation of Système International d'Unités (French).
14. Sometimes referred to as "amount."
15. n_T is determined by the initial values: $n_T = n_{cyt}(0) + n_{er}(0)$.
16. A natural frequency is related to a period via $f = 1/T$ whereas an angular frequency is related through $\omega = 2\pi f = 2\pi/T$.

3 Phase Diagrams

Phase diagrams are an intuitive graphical technique for the qualitative analysis of differential equations. Although phase diagram analysis is most useful for nonlinear ODEs, we introduce the technique using a linear ODE, namely, the single compartment model developed in Chapter 2.

3.1 Phase Diagram for a Single Compartment Model

Beginning with the conservation law for material balance in an open cell model (Fig. 2.1), the assumption of a constant plasma membrane influx rate and an efflux rate that is proportional to the intracellular $[Ca^{2+}]$ resulted in the ODE initial value problem,

$$\frac{dc}{dt} = J - kc \qquad c(0) = c_0, \tag{3.1}$$

where c denotes the cytosolic $[Ca^{2+}]$. Fig. 3.1 shows the **phase diagram** for this ODE: a graph with the independent variable c on the horizontal axis and the rate of change dc/dt on the vertical axis. That is, the phase diagram is a plot of the right side of Eq. 3.1, which in this case is the straight line $f(c) = J - kc$ that intersects the vertical axis at J and intersects the horizontal axis at J/k.

The horizontal axis of the phase diagram (Fig. 3.1) graphically represents the possible values – in a mathematical sense – of the independent variable of the ODE. However, some values of c are nonphysical (gray region), because concentrations are nonnegative ($0 \leq c < \infty$). For any state of the system – that is, for any particular value of c – the phase diagram graphically represents the corresponding rate of change dc/dt.

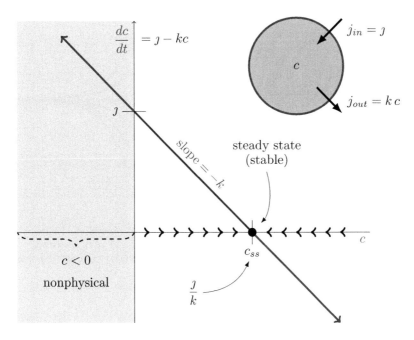

Figure 3.1 The phase diagram for a single compartment model with constant influx rate $j_{in} = j$ and efflux rate proportional to calcium concentration $j_{out} = kc$. Arrows on the phase line indicate the flow of solutions toward the stable steady state ($c_{ss} = j/k$).

The sign of dc/dt (negative, zero, or positive) indicates whether c, the intracellular $[Ca^{2+}]$, will decrease, remain unchanged, or increase.

The vertical intercept of Fig. 3.1 shows that $dc/dt = j$ when $c = 0$, that is, when the intracellular $[Ca^{2+}]$ is zero, the rate of change of the intracellular $[Ca^{2+}]$ is equal to the influx rate j ⊝ conc/time. For all $c < j/k$, the blue line is above the horizontal axis ($dc/dt > 0$). The rightward facing arrows on the phase line indicate that the rate of change of intracellular $[Ca^{2+}]$ is positive on the interval $0 \le c < j/k$. When $c > j/k$, the line representing dc/dt is below the horizontal axis ($dc/dt < 0$). The leftward facing arrows on the phase line indicate that the rate of change of the intracellular $[Ca^{2+}]$ is negative for $j/k < c < \infty$.

The blue line intersects the horizontal axis of Fig. 3.1 at $c = j/k$. Because the rate of change of the intracellular $[Ca^{2+}]$ is zero ($dc/dt = 0$) for this value of c, choosing the initial condition $c_0 = j/k$ for the ODE leads to the constant solution $c(t) = j/k$. To see this, substitute $c = j/k$ into Eq. 3.1 to obtain

$$\frac{dc}{dt} = j - k \cdot \frac{j}{k} = j - j = 0.$$

This solution is referred to as a **fixed point** or **steady state** of the compartmental model. When drawing the phase diagram, we label this important point on the horizontal axis with c_{ss} (ss is an abbreviation for steady state).

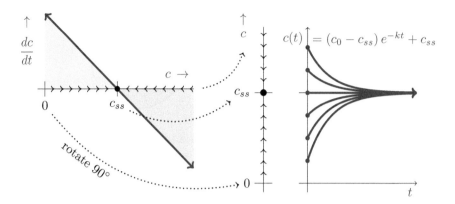

Figure 3.2   Left: A simplified phase diagram for Eq. 3.1 (cf. Fig. 3.1). Right: Different initial values (c_0) lead to solutions consistent with the flow on the phase line (black arrows). *Question*: (i) What is the sign of dc/dt in the gray region of the phase diagram? (ii) For initial value $c(0) = c_0 = 0$, is the initial rate of change positive or negative?

3.2   Stable and Unstable Steady States

Fig. 3.2 (left) shows a simplified version of the phase diagram of the compartmental model (Fig. 3.1). Rotating the phase line 90° in the counterclockwise direction, one sees the correspondence between model solutions using various initial conditions and the flow on the phase line. The plotted solutions are exponentially relaxing functions

$$c = (c_0 - c_{ss}) e^{-kt} + c_{ss} \tag{3.2}$$

with c_0 varied while j and k (and thus $c_{ss} = j/k$) are fixed. Beginning from the initial state c_0, the intracellular calcium concentration asymptotically approaches c_{ss}, either from above or from below, with exponential time constant $\tau = 1/k$ (Fig. 3.2, right). For initial conditions greater than the steady state value ($c_0 > c_{ss}$), solutions decrease over time and asymptotically approach the steady state from above ($c(t) \downarrow c_{ss}$ as $t \to \infty$). For initial conditions $c_0 < c_{ss}$, solutions increase over time and asymptotically approach the steady state from below ($c(t) \uparrow c_{ss}$ as $t \to \infty$). When the initial condition is equal to the steady state value ($c_0 = c_{ss}$), the solution is constant.

For comparison, Fig. 3.3 (left) shows a phase diagram for the ODE $dy/dt = ay - b$ where the parameters a and b are positive and, consequently, dy/dt is a positively sloped line with negative y-intercept. As in Fig. 3.2, the phase line (horizontal axis) represents every possible value of the independent variable y and the sign of dy/dt determines the direction of flow on the phase line. In particular, the value of y that leads to zero rate of change satisfies $0 = ay_{ss} - b$ and thus the steady state is $y_{ss} = a/b$. Because $dy/dt > 0$ for $y > y_{ss}$ and, conversely, $dy/dt < 0$ for $y < y_{ss}$, the arrows on the phase line point away from the steady state. For any initial condition greater than the steady state value ($y_0 > y_{ss}$), the solution will increase without bound ($y(t) \to \infty$ as $t \to \infty$). For initial conditions less than the steady state value ($y_0 < y_{ss}$), solutions will decrease without bound ($y(t) \to -\infty$ as $t \to \infty$). In fact, given that a and b are

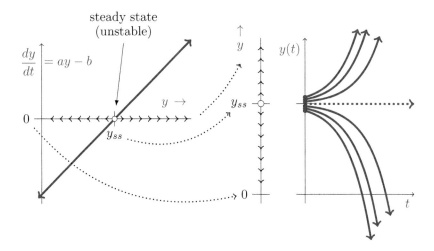

Figure 3.3   Left: The phase diagram for $dy/dt = ay - b$ where $a, b > 0$ is a positively sloped line with negative y-intercept. The direction of flow on the phase line is away from the steady state, that is, $dy/dt > 0$ for $y > y_{ss}$ and $dy/dt < 0$ for $y < y_{ss}$; consequently, the steady state $y_{ss} = b/a$ is unstable (in contrast to Fig. 3.2).

positive, the only solution of $dy/dt = ay - b$ that remains bounded as $t \to \infty$ is the constant solution $y(t) = y_{ss}$.

The steady states in Figs. 3.2 and 3.3 are referred to as **stable** and **unstable**, respectively. A steady state is *stable* when the flow on the phase line indicates that solutions in proximity of a steady state forever remain in proximity of that steady state. If this is not the case, the steady state is classified as *unstable*.[1] Filled circles (●) will be used to indicate stable steady states, while open circles (○) will represent unstable steady states.

Problem 3.1   In Fig. 3.4 (right), note the sigmoidal shape of the solution with initial condition $y(0)$ just slightly greater than the unstable steady state $y_{ss} = -a$ (see ⋆). Why is the initial rate of change of this solution so slow?

3.3   Phase Diagram of a Nonlinear ODE

All scalar ODEs take the form $dy/dt = f(y, t)$, but only scalar **autonomous** ODEs can be written as

$$\frac{dy}{dt} = f(y), \tag{3.3}$$

where the function $f(y)$ does not explicitly depend on time. For example, $dy/dt = 1 - y^3$ is an autonomous ODE, while $dy/dt = t - y^3$ and $dy/dt = \cos(t)$ are not.

The phase diagram technique may be applied to any autonomous scalar ODE. For example, consider the initial value problem,

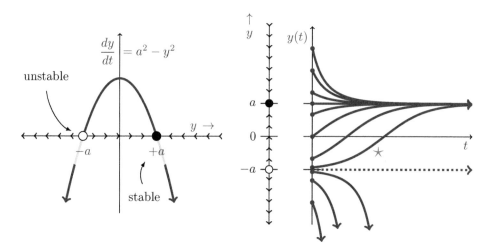

Figure 3.4   Left: Phase diagram and representative solutions of Eq. 3.4. *Question:* (i) Explain why solutions that cross the horizontal axis of the right plot do so with identical slopes. (ii) At what value of y is the rate of change of y greatest?

$$\frac{dy}{dt} = a^2 - y^2 \qquad y(0) = y_0 , \tag{3.4}$$

where a is a positive constant ($a > 0$). While an analytical solution that expresses $y(t)$ as a function of a, y_0 and t can be derived by separation of variables (see Problem 3.6), it is more enlightening (and far easier) to sketch the phase diagram, identify steady state solutions, and analyze the qualitative dynamics as a flow on the phase line.

To find the steady states of Eq. 3.4, set dy/dt to zero and factor the right side of the ODE,

$$0 = a^2 - y_{ss}^2 = (a + y_{ss})(a - y_{ss}) \;\Rightarrow\; y_{ss} = -a \text{ or } a . \tag{3.5}$$

There are two steady state solutions, $y_{ss} = a$ and $y_{ss} = -a$. If the dependent variable y represented a concentration, then $y_{ss} = -a$ would be nonphysical.

Fig. 3.4 (left) shows the phase diagram for Eq. 3.4. The horizontal axis is y and the vertical axis is dy/dt, that is $f(y) = a^2 - y^2$, a concave[2] parabola with maximum at $y = 0$. The steady states $y_{ss} = -a$ and $y_{ss} = a$ are intersections of this parabola with the horizontal axis. The arrows sketched on the phase line indicate that the direction of flow is increasing when $-a < y < a$ and decreasing when $y < -a$ or $y > a$. The steady state $y_{ss} = a$ is stable, because the flow on the phase line is directed toward it (on both sides), that is, any solution near $y_{ss} = a$ will forever remain near. Conversely, the steady state at $y_{ss} = -a$ is unstable because the flow is directed away from the fixed point; solutions that begin near $y_{ss} = -a$ move away from this point.

The flow on the phase line indicates that solutions of Eq. 3.4 with initial condition in the range $y_0 < -a$ decrease without bound. Conversely, solutions with initial condition $-a < y_0 < \infty$ will asymptotically approach the stable steady state $y_{ss} = a$. For this reason, the steady state $y_{ss} = a$ is referred to as an **attractor** with a **basin of attraction** given by $(-a, \infty)$, which is the interval of the real number line shown below.

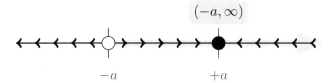

The unstable steady state $y_{ss} = -a$ is a **repellor**.

> **Problem 3.2** When *interval notation* is applied to the real numbers, the sets $[a, \infty)$ and (a, ∞) have different meanings. Using the symbol $\mathbb{R}$ to denote the real numbers, the set $(-a, \infty) = \{y \in \mathbb{R} : -a < y < \infty\}$ while $[-a, \infty) = \{y \in \mathbb{R} : -a \leq y < \infty\}$. Explain why the basin of attraction of $y_{ss} = a$ is *not* $[-a, \infty)$.

3.4 Classifying Steady States

When the flow on the phase line is directed (on both sides) toward a steady state, that steady state is stable. After considering the phase diagrams we have encountered so far, you might conclude that the flow on the phase line has this property when the function $f(y)$ crosses the horizontal axis with negative slope. And you would be correct! Provided the right side of $dy/dt = f(y)$ is continuous and differentiable, $f'(y_{ss}) < 0$ is a necessary and sufficient condition for stability of y_{ss}. Conversely, a point on the phase line for which $f(y_{ss}) = 0$ and $f'(y_{ss}) > 0$ is an unstable steady state. For this reason, it is often straightforward to ascertain the stability of steady states without sketching a phase diagram.

For example, after determining that the steady states of Eq. 3.4 are $y_{ss} \in \{-a, a\}$ (Eq. 3.5), one may classify the steady states as stable or unstable by calculating the derivative of the right side of the ODE $dy/dt = f(y)$ with respect to the independent variable y,

$$f'(y) = [a^2 - y^2]' = -2y. \tag{3.6}$$

Evaluating this derivative at each steady state, we find

$$y_{ss} = -a : \quad f'(-a) = -2(-a) = 2a > 0 \Rightarrow \text{unstable}$$
$$y_{ss} = a : \quad f'(a) = -2a < 0 \Rightarrow \text{stable},$$

where the inequalities (and stability analysis) use the fact that the parameter a is positive ($a > 0$). This analytical stability analysis is consistent with the geometric analysis that emphasized the direction of flow on the phase line (Fig. 3.4). With practice one may become proficient at both the graphical and analytical methods of finding steady states to nonlinear scalar ODEs and classifying them according to their stability (see Fig. 3.5).

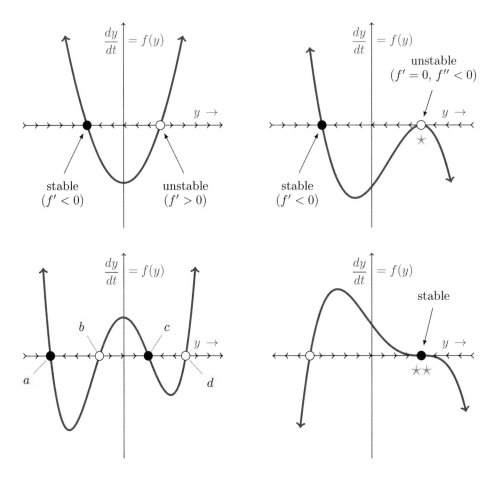

Figure 3.5 Phase diagrams for nonlinear scalar ODEs. The sign of $dy/dt = f(y)$ determines the direction of flow on the phase line (increasing or decreasing y). When $f(y)$ is continuous and differentiable, the sign of its derivative $f'(y)$ evaluated at y_{ss} determines whether a steady state is stable ($f'(y_{ss}) < 0$, negative slope) or unstable ($f'(y_{ss}) > 0$, positive slope). When $f'(y_{ss}) = 0$ (gray stars, ⋆) the stability can be read off the direction of flow on the phase line or by the sign of higher derivatives. *Question*: Consider the phase diagram of the bottom left panel. What are the basins of attraction of the steady states $y_{ss} = a$ and $y_{ss} = c$?

Problem 3.3   Analytically determine the steady states of the following differential equation,

$$\frac{dy}{dt} = -y^2 - 3y + 10,$$ (3.7)

and draw a phase diagram summarizing the qualitative behavior of solutions. Use filled and open circles for stable and unstable equilibria, respectively.

3.5   Stability Analysis Requiring Higher Derivatives

Fig. 3.5 (top right) shows a phase diagram for

$$\frac{dy}{dt} = -y^3 + 2y^2 + 4y - 8, \tag{3.8}$$

an ODE with two steady state solutions. These steady states may also be found by setting dy/dt to zero,

$$0 = -y_{ss}^3 + 2y_{ss}^2 + 4y_{ss} - 8, \tag{3.9}$$

and solving this cubic polynomial for y_{ss}. Because the phase diagram indicates real-valued roots, we are encouraged to factor Eq. 3.9,

$$0 = y_{ss}^3 - 2y_{ss}^2 - 4y_{ss} + 8 = y_{ss}^2(y_{ss} - 2) - 4(y_{ss} - 2)$$
$$= (y_{ss} - 2)(y_{ss}^2 - 4) = (y_{ss} + 2)(y_{ss} - 2)^2, \tag{3.10}$$

where the factor $(y_{ss} - 2)^2$ corresponds to the repeated root indicated by the $\star$ in Fig. 3.5 (top right). The two values of y_{ss} that solve Eq. 3.10 are the two steady states of Eq. 3.8,

$$y_{ss} = -2 \text{ or } 2. \tag{3.11}$$

Algebraic stability analysis begins by noting that the derivative of $f(y) = -y^3 + 2y^2 + 4y - 8$ (the right side of Eq. 3.8) is

$$f'(y) = -3y^2 + 4y + 4.$$

Evaluating this derivative at $y_{ss} = -2$ and $y_{ss} = 2$, we find

$$y_{ss} = -2: \quad f'(-2) = -3(-2)^2 + 4(-2) + 4 = -16 < 0 \Rightarrow \text{stable}$$
$$y_{ss} = 2: \quad f'(2) = -3(2)^2 + 4(2) + 4 = 0 \Rightarrow \text{inconclusive.}$$

Determining the stability of $y_{ss} = 2$ requires the second derivative, $f''(y) = -6y + 4$, that evaluates to $f''(2) = -8 < 0$. The negative second derivative indicates that $f(y)$ is a concave parabola with a maximum that is tangent to the horizontal axis. This implies that the steady state at $y_{ss} = 2$ is unstable, as noted in Fig. 3.5 (top right, $\star$). The flow on the phase line near the steady state $y_{ss} = 2$ is as follows:

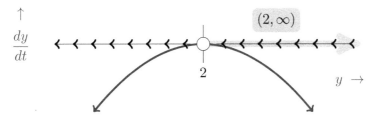

The steady state may be further classified as *semistable*, in recognition of the basin of attraction $(2, \infty)$ located on the right, but not the left. Semistable steady states are unstable.

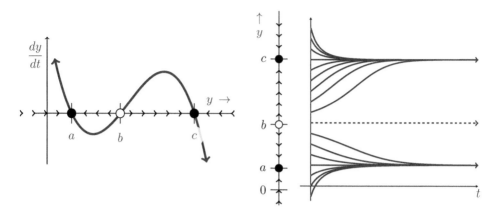

Figure 3.6   Left: Phase diagram for Eq. 3.12 that illustrates bistability. Right: Different initial conditions lead to solutions whose qualitative dynamics are consistent with the flow on the phase line (black arrows). Solutions with initial condition $y_0 > b$ asymptotically approach the steady state $y_{ss} = c$, while solutions with initial condition $y_0 < b$ asymptotically approach $y_{ss} = a$. The unstable steady state $y_{ss} = b$ separates two basins of attraction, $y \in (-\infty, b)$ and $y \in (b, \infty)$.

3.6   Scalar ODEs with Multiple Stable Steady States

Fig. 3.6 (left) shows a phase diagram for another scalar autonomous ODE,

$$\frac{dy}{dt} = f(y) = -(y - a)(y - b)(y - c) \qquad y(0) = y_0, \qquad (3.12)$$

where $0 < a < b < c$. Because the polynomial is already factored, it is clear that $y_{ss} = a$ or b or c. In set theory notation, we write

$$y_{ss} \in \{a, b, c\}$$

which is read "y_{ss} *is an element of* the set $\{a, b, c\}$." The flow on the phase line on the intervals $(-\infty, a)$, (a, b), (b, c), and (c, ∞) is easily established by considering the sign of each term of $f(y)$, for example, for $y \in (b, c)$,

$$\frac{dy}{dt} = - \underbrace{(y - a)}_{+} \underbrace{(y - b)}_{+} \underbrace{(y - c)}_{-} > 0,$$

because $y > a$, $y > b$ and $y < c$. Reasoning in this manner, we see that $y_{ss} = a$ and $y_{ss} = c$ are stable steady states, while $y_{ss} = b$ is an unstable steady state. To perform the stability analysis analytically, differentiate $f(y)$ to obtain

$$f'(y) = -(y - b)(y - c) - (y - a)(y - c) - (y - a)(y - b).$$

Next, evaluate $f'(y)$ at each steady state. This is straightforward because, in each case, only one of the three terms is nonzero,

$$f'(a) = -\underbrace{(a-b)}_{-}\underbrace{(a-c)}_{-} < 0 \Rightarrow \text{stable}$$

$$f'(b) = -\underbrace{(b-a)}_{+}\underbrace{(b-c)}_{-} > 0 \Rightarrow \text{unstable}$$

$$f'(c) = -\underbrace{(c-a)}_{+}\underbrace{(c-b)}_{+} < 0 \Rightarrow \text{stable.}$$

We conclude that Eq. 3.12 has two stable steady states ($y_{ss} = a$ and $y_{ss} = c$) and one unstable steady state ($y_{ss} = b$). This ODE is **multistable** because it has *more than one* stable steady state. More precisely, the ODE is **bistable**, because there are *exactly two* stable steady states.

Fig. 3.6 (right) shows representative solutions of Eq. 3.12 that are typical of an ODE that exhibits **bistability**. The stable steady states $y_{ss} = a$ and $y_{ss} = c$ are attractors with basins of attraction given by $(-\infty, b)$ and (b, ∞).

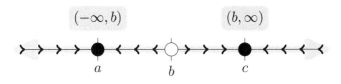

The unstable steady state $y_{ss} = b$ that separates these basins of attraction is a repellor. For all initial values $y_0 \in (-\infty, b)$, the solution $y(t)$ asymptotically approaches $y_{ss} = a$ as $t \to \infty$. For initial values $y_0 \in (b, \infty)$, $y(t) \to c$ as $t \to \infty$.

Bistable dynamics are like a switch that remains in the "off" or "on" state indefinitely, until a finger pushes the system over the unstable steady state between off and on. Another example of bistable dynamics is an object that in principle can be balanced perfectly, but in practice always falls to one or the other side of the tipping point. Bistability and related threshold phenomena are important and ubiquitous in cellular and systems neuroscience.

3.7   Discussion

The Distinction Between Lyaponov and Asymptotic Stability

In this chapter we have introduced the concept of the stability of a steady state of a dynamical system. When $f(y)$ is a polynomial, the number of steady states of $dy/dt = f(y)$ is a nonnegative integer less than or equal to the polynomial's degree, because every steady state is a root, but some roots may occur with multiplicity. Polynomial functions are continuous and differentiable; consequently, steady states for which $f'(y_{ss}) < 0$ are stable, because this condition implies that $f(y)$ traverses the horizontal axis with negative slope at the point y_{ss}.

When $f(y)$ is not a polynomial the situation may be more complex. For example, consider

$$\frac{dy}{dt} = \begin{cases} y^2 - 1 & |y| > 1 \\ 0 & \text{otherwise.} \end{cases} \tag{3.13}$$

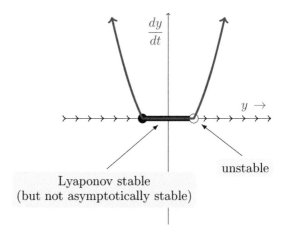

Figure 3.7 Phase diagram for Eq. 3.13 with an uncountably infinite number of steady states that are Lyaponov stable but not asymptotically stable, i.e., the interval $y_{ss} \in [-1, 1)$. The steady state $y_{ss} = 1$ is not Lyaponov stable and is thus unstable.

The steady states of this ODE are a closed interval of the real number line, namely, $y_{ss} \in [-1, 1]$. This interval includes an uncountably infinite number of points, each of which is a steady state.

Our process for classifying these steady states as stable or unstable has been to ask the question, "Do solutions in proximity of the steady state remain in proximity of the steady state?" This intuitive notion of stability is referred to as **Lyaponov stability**. A steady state is *Lyaponov stable* precisely when the answer to the above question is "Yes." Using this definition of Lyaponov stability, we may classify all the points in the half-open interval $y_{ss} \in [-1, 1)$ as stable steady states; however, $y_{ss} \in [1]$ is unstable, because $dy/dt > 0$ for $y > 1$ (see Fig. 3.7).

There is a related but distinct conception of stability known as **asymptotic stability** that is stricter than "nearby solutions remain nearby." A steady state y_{ss} is *asymptotically stable* if it is Lyaponov stable *and* if all solutions that start near y_{ss} converge to y_{ss} as $t \to \infty$.[3]

All of the stable steady states shown as filled circles in Fig. 3.5 are asymptotically stable as well as Lyaponov stable, because $f(y_{ss}) = 0$ and $f'(y_{ss}) < 0$ implies convergence from the left and right. On the other hand, none of the Lyaponov stable steady states of Fig. 3.7 satisfy the stronger conditions of asymptotic stability.[4]

Steady States, Fixed Points and Equilibria

The terms *steady state* and *fixed point* have precisely the same meaning in the context of dynamical systems. Mathematical scientists favor the term steady state, while mathematicians use fixed point.

Sometimes the word **equilibrium** is used as a synonym for fixed point. This may lead to confusion because in chemical reaction theory an important distinction is made between the concepts of steady state and equilibrium. We will avoid the terms *equilibrium* and *equilibria* unless referring to chemical equilibrium or an analogous phenomena.

One way to clarify the distinction between the terms *steady state* and *equilibrium* is to write the material balance equation for an open cell model that includes influx, efflux, production, and degradation,

$$\frac{dc}{dt} = j_{prod}(c) - j_{deg}(c) + j_{in}(c) - j_{out}(c),$$ (3.14)

so that each of the four terms is potentially a function of c. The concentration c_{ss} is a steady state of Eq. 3.14 provided this value simultaneously satisfies **global balance**, that is,

global balance: $0 = j_{prod}(c_{ss}) - j_{deg}(c_{ss}) + j_{in}(c_{ss}) - j_{out}(c_{ss}).$

The term global balance refers to all four fluxes balancing when considered together (i.e., globally).

The term **detailed balance** indicates an equilibrium condition where the forward and reverse rates of each process are equal. With respect to Eq. 3.14,

detailed balance: $\begin{cases} j_{prod}(c_{ss}) = j_{deg}(c_{ss}) \\ j_{in}(c_{ss}) = j_{out}(c_{ss}). \end{cases}$

Detailed balance is a stronger condition than global balance. Detailed balance implies global balance,

detailed balance/equilibrium $\Rightarrow$ global balance/steady state,

but global balance does not imply detailed balance. For example, consider a cell that imports glucose at constant rate $j_{in} = J$, exports glucose at the rate $j_{out} = k_{out}c$, metabolizes glucose at the rate $j_{deg} = k_{deg}c$, but has no intracellular production of glucose ($j_{prod} = 0$). We will assume k_{deg} and k_{out} are nonnegative and write the material balance equation (Eq. 3.14) as

$$\frac{dc}{dt} = 0 - k_{deg}c + J - k_{out}c = J - \left(k_{deg} + k_{out}\right)c.$$

The second equality makes it easy to see the steady state is $c_{ss} = J/(k_{deg} + k_{out})$. However, the import and export processes are not in equilibrium at steady state, because equilibrium ($j_{in} = j_{out}$) implies

$$J = k_{out}c_{ss} \Rightarrow J = k_{out} \cdot \frac{J}{k_{deg} + k_{out}} \Rightarrow k_{deg} + k_{out} = k_{out} \Rightarrow k_{deg} = 0,$$

which contradicts our assumption that $k_{deg} > 0$. A steady state that does not satisfy detailed balance is a **nonequilibrium steady state**.

Classification of ODEs

The most general form of a first-order scalar ODE is the implicit expression,

$$\Psi\left(y', y, t\right) = 0 \quad \text{(first-order)} \tag{3.15}$$

where Ψ is a function of the independent variable t, the dependent variable y, and the derivative $y' = dy/dt$. The most general form of a second-order scalar ODE is

$$\Psi\left(y'', y', y, t\right) = 0 \quad \text{(second-order)}$$

where $y'' = d^2y/dt^2$. The ODEs in this text will usually be written in explicit form, that is,

$$y' = f(y, t) \quad \text{(first-order)} \tag{3.16}$$

$$y'' = f(y', y, t) \quad \text{(second-order)}. \tag{3.17}$$

When an ODE takes this form and the function f does not involve the independent variable t,[5] the ODE is **autonomous**,

$$y' = f(y) \quad \text{(first-order, autonomous)} \tag{3.18}$$

$$y'' = f(y', y) \quad \text{(second-order, autonomous)}. \tag{3.19}$$

The phase diagram method is applicable to any explicit, first-order, autonomous ODE, because all that is required is a plot of $f(y)$ in the (y, y')-plane.

Scalar ordinary differential equations are also classified as **linear** or **non-linear**. Linear first- and second-order ODEs can be put in the form,

$$y' = q(t)y + r(t) \quad \text{(first-order, linear)} \tag{3.20}$$

$$y'' = p(t)y' + q(t)y + r(t) \quad \text{(second-order, linear)}, \tag{3.21}$$

where $p(t)$, $q(t)$ and $r(t)$ are continuous functions. In these expressions, $r(t)$ is called a **source term** and when $r(t) = 0$ these ODEs are **homogenous**. Note that a first-order linear autonomous ODE can be written as $y' = \alpha y + \beta$ where α and β are constants; solutions take the form $y(t) = \gamma e^{\alpha t} + \delta$ where γ and δ are constants. The phase diagram of a first-order linear autonomous ODE is a line in the (y, y')-plane, but $y' = \alpha y + \beta$ is not homogenous when $\beta \neq 0$.

The phase diagram technique works for any first-order autonomous ODE (whether linear or nonlinear). On the other hand, there are many ODEs that do not lend themselves to the phase diagram method, e.g., the linear first-order nonautonomous ODE $y' = t^2 y$.

> **Problem 3.4** Classify the following scalar ODEs as linear or nonlinear, autonomous or nonautonomous, first-order or second-order, and homogenous or nonhomogenous.
>
> 1. $y' = t^2 y$
> 2. $y' = -y + 1$
> 3. $y' = t^3 y^2$
> 4. $y' = a(t)y + b(t)$
> 5. $(y')^3 = \sqrt{y}$
> 6. $y'' + p(t)y' + q(t) = 0$
> 7. $y'' = 0$

Superposition Property of Linear Homogenous ODEs

Linear homogenous ODEs such as $y' = q(t)y$ and $y'' = p(t)y' + q(t)y$ have a **superposition property**, meaning that solutions may be scaled by a constant and summed to yield new solutions,

$$y_1(t) \text{ and } y_2(t) \text{ are solutions} \Rightarrow y(t) = \alpha y_1 + \beta y_2 \text{ is a solution,}$$

where α and β are arbitrary constants.[6] For example, the first-order linear homogenous ODE $dy/dt = q(t)y$ is not autonomous, but solutions superpose. To see this, note that $y_1(t)$ and $y_2(t)$ are solutions precisely when $dy_1/dt = q(t)y_1$ and $dy_2/dt = q(t)y_2$. Let us write $y_*(t) = \alpha y_1 + \beta y_2$ and check that $y_*(t)$ satisfies the ODE:

$$\frac{dy_*}{dt} \overset{?}{=} q(t)y_*$$

$$\frac{d}{dt}[\alpha y_1 + \beta y_2] \overset{?}{=} q(t)[\alpha y_1 + \beta y_2]$$

$$\alpha\frac{dy_1}{dt} + \beta\frac{dy_2}{dt} \overset{\text{yes}}{=} q(t)\alpha y_1 + q(t)\beta y_2 . \checkmark$$

The final step uses $dy_1/dt = q(t)y_1 \Rightarrow \alpha dy_1/dt = \alpha q(t)y_1$ and similarly for y_2.

An ODE that satisfies the superposition property is linear, but linear ODEs may have solutions that do not superpose. For example, solutions of the first-order linear autonomous ODE $dy/dt = -y + 1$ do not superpose. To see this, assume $dy_1/dt = -y_1 + 1$ and $dy_2/dt = -y_2 + 1$. Then check if $y_*(t) = \alpha y_1(t) + \beta y_2(t)$ satisfies the ODE:

$$\frac{dy_*}{dt} \overset{?}{=} -y_* + 1$$

$$\frac{d}{dt}[\alpha y_1 + \beta y_2] \overset{?}{=} -[\alpha y_1 + \beta y_2] + 1$$

$$\alpha\frac{dy_1}{dt} + \beta\frac{dy_2}{dt} \overset{?}{=} -\alpha y_1 - \beta y_2 + 1.$$

Because y_1 and y_2 solve the ODE, $\alpha dy_1/dt = \alpha(-y_1 + 1) = -\alpha y_1 + \alpha$ and $\beta dy_2/dt = -\beta y_2 + \beta$. We substitute and continue,

$$-\alpha y_1 + \alpha - \beta y_2 + \beta \overset{?}{=} -\alpha y_1 - \beta y_2 + 1$$

$$\alpha + \beta \neq 1.$$

Solutions of $dy/dt = -y+1$ do *not* superpose, because $\alpha + \beta \neq 1$ for arbitrary constants α and β.

Supplemental Problems

Problem 3.5 For each of the following scalar ODEs,

1. $\dfrac{dy}{dt} = y^2 - 6$

2. $\dfrac{dy}{dt} = 3 - (y-4)^2$

3. $\dfrac{dy}{dt} = y(y-1)(5-y)$

4. $\dfrac{dy}{dt} = -y^3 + 7y^2 - 10y$

(a)   Find all steady state solutions, y_{ss}, that satisfy $dy/dt = 0$.
(b)   Analytically calculate the derivatives $f'(y_{ss})$. Use the sign of this quantity to classify each equilibrium as stable, unstable, or semistable.
(c)   Sketch a qualitatively correct phase diagram.

Problem 3.6   The analytical solution of Eq. 3.4 can be derived by separation of variables

$$\frac{dy}{dt} = a^2 - y^2$$

$$\frac{dy}{a^2 - y^2} = dt$$

$$\int_{y_0}^{y} \frac{d\hat{y}}{a^2 - \hat{y}^2} = \int_0^t d\hat{t}$$

$$\frac{\ln(\hat{y} + a)}{2a} - \frac{\ln(\hat{y} - a)}{2a}\Big|_{y_0}^{y} = \hat{t}\Big|_0^t$$

$$\vdots$$

$$\frac{(y - a)}{(y + a)} = \frac{(y_0 - a)}{(y_0 + a)}e^{-2at}.$$

To simplify the algebraic steps that remain, define $g(t)$ as follows,

$$g(t) = \frac{(y_0 - a)}{(y_0 + a)}e^{-2at},$$

and continue the calculation,

$$y - a = (y + a)g(t)$$
$$y = yg(t) + ag(t) + a$$

$$\vdots$$

$$y = a\frac{1 + \dfrac{(y_0 - a)}{(y_0 + a)}e^{-2at}}{1 - \dfrac{(y_0 - a)}{(y_0 + a)}e^{-2at}}.$$

Thus, the solution of the ODE initial value problem is found to be

$$y(t) = a\frac{y_0 + a + (y_0 - a)e^{-2at}}{y_0 + a - (y_0 - a)e^{-2at}}.$$

Note that setting $t = 0$ in this solution gives $y(0) = y_0$ as required, and that $y(t) \to a$ as $t \to \infty$. Although it is important to note that this analytical work is not necessary to understand qualitative aspects of solutions of this ODE (see Fig. 3.4), mathematically inclined students ought to complete the missing steps in this derivation.

Problem 3.7 A two compartment model of intracellular Ca^{2+} signaling that includes a passive leak out of the ER and active re-uptake by a "linear" SERCA-like Ca^{2+}-ATPase results in the following ODE,

$$\frac{dc}{dt} = \underbrace{v_l\,(c_{er} - c)}_{J_l} - \underbrace{v_p c}_{J_p}$$

where $c_{er} = (c_T - c)/\alpha$ and $\alpha = \Omega_{er}/\Omega$. In these equations, c is the cytosolic Ca^{2+} concentration, c_{er} is the ER Ca^{2+} concentration, and Ω and Ω_{er} are the cytosolic and ER volumes, respectively.

(a) Find the steady state cytosolic Ca^{2+} concentration (c_{ss}).
(b) By evaluating an appropriate derivative, show that this steady state is stable.

Problem 3.8 Consider the following ordinary differential equation,

$$\frac{dy}{dt} = \alpha y - \beta y^2 \tag{3.22}$$

where $\alpha > 0$ and $\beta > 0$.

(a) Eq. 3.22 is a scalar ODE. Is it linear? autonomous? homogenous?
(b) Assuming y has physical dimensions of concentration, what are the physical dimensions of β?
(c) Draw a phase diagram for Eq. 3.22. Label both axes and indicate the direction of flow on the phase line (i.e., the horizontal axis) and whether each steady state is stable or unstable.
(d) Analytically determine the steady state values of y and classify each as stable or unstable. Check that your answer is consistent with your phase diagram.

Problem 3.9 The following scalar ODE is known as the logistic equation for population growth,

$$\frac{dw}{dt} = r\left(1 - \frac{w}{K}\right)w \quad \text{where} \quad r > 0,\ K > 0.$$

(a) Analytically show that the constant solution $w_{ss} = K$ is stable.
(b) Is the logistic equation a scalar ODE? Is it linear? autonomous? homogenous?
(c) If t is time measured in years and w is measured in $\mu g/ml$, what are the units of r and K?

Solutions

Figure 3.2 (i) $dc/dt < 0$ in the gray region. (ii) Positive.

Figure 3.4   (i) The horizontal axis corresponds to $y = 0$ and when $y = 0$ the slope is $dy/dt = a^2$. (ii) The left plot shows that $y = 0$ leads to the maximum rate of change.

Figure 3.5   The basins of attraction are $(-\infty, b)$ for $y_{ss} = a$ and (b, d) for $y_{ss} = c$.

Problem 3.1   The initial rate of change is slow because $dy/dt = a^2 - y^2 \approx 0$ when $y(0) \approx y_{ss} = -a$ (see Fig. 3.4, left).

Problem 3.3   Setting the left side of Eq. 3.7 to zero and factoring the right side, we obtain $0 = (y + 5)(y - 2)$; thus, steady state solutions are given by $y_{ss} = -5$ (unstable) and $y_{ss} = 2$ (stable).

Problem 3.4   1–5 are first order; 6,7 are second order. 1,2,4,6,7 are linear; 3,5 are nonlinear. 2,5,7 are autonomous. 2 is not homogenous.

Notes

1.   Asymptotic stability versus Lyaponov stability is discussed on p. 51.
2.   Synonymously, concave down.
3.   See Glendinning (1994, Chapter 2) for a rigorous treatment of stability concepts.
4.   This includes the steady state $y_{ss} = -1$, which is Lyaponov stable but not asymptotically stable (convergence from the left but not from the right).
5.   That is, the partial derivative $\partial f / \partial t = 0$.
6.   This can be written as two conditions that must both hold:

    1. $y_1(t)$ is a solution $\Rightarrow y(t) = \alpha y_1$ is a solution
    2. $y_1(t)$ & $y_2(t)$ are solutions $\Rightarrow y(t) = y_1 + y_2$ is a solution

    where α is an arbitrary constant.

4 Ligands, Receptors and Rate Laws

The mass action rate law provides the production and degradation rates needed to model chemical reactions of cellular signaling. We derive an ODE model for the association of a ligand (e.g., a neurotransmitter such as glutamate) and a receptor (e.g., a postsynaptic glutamate receptor). Analysis of these differential equations leads to equilibrium relations and binding curves that are relevant to synaptic transmission and quantitative pharmacology.

4.1 Mass Action Kinetics

This chapter shows how compartmental models may account for multiple chemical species and chemical reactions. The constitutive relation most commonly used for this purpose is **mass action kinetics**. This rate law states that *reaction rates are proportional to the product of reactant concentrations.* Consider the following irreversible elementary reactions,[1]

$$\text{reaction 1:} \quad 2A + B \rightarrow A + C \tag{4.1}$$

$$\text{reaction 2:} \quad 2C \rightarrow A + 3B. \tag{4.2}$$

Mass action kinetics suggests,

$$\text{rate of reaction 1:} \quad j_1 = k_1 [A]^2 [B] \tag{4.3}$$

$$\text{rate of reaction 2:} \quad j_2 = k_2 [C]^2, \tag{4.4}$$

where [A], [B] and [C] represent the concentrations of chemical species A, B and C. For a closed, well-stirred compartment of constant volume, a material balance equation may be written for each chemical species, as follows:

$$\frac{d[A]}{dt} = s_1^a j_1 + s_2^a j_2 \tag{4.5a}$$

Table 4.1 Stoichiometric coefficients for Eqs. 4.1 and 4.2

Reaction	Species	Reactant	Product	Product–Reactant
1	A	2	1	$s_1^a = -1$
1	B	1	0	$s_1^b = -1$
1	C	0	1	$s_1^c = +1$
2	A	0	1	$s_2^a = +1$
2	B	0	3	$s_2^b = +3$
2	C	2	0	$s_2^c = -2$

$$\frac{d[B]}{dt} = s_1^b j_1 + s_2^b j_2 \tag{4.5b}$$

$$\frac{d[C]}{dt} = s_1^c j_1 + s_2^c j_2 \, . \tag{4.5c}$$

In each equation, the net production rate is the sum of two terms (one for each reaction). Each term is the product of a chemical flux (j_1 or j_2) and a stoichiometric coefficient (s_1^a, s_2^a, s_1^b, etc.). In the equation whose left side is $d[A]/dt$, the constant $s_1^a = 1 - 2 = -1$ is the change in the stoichiometric coefficient for species A in reaction 1. Similarly, $s_2^a = 1 - 0 = 1$ is the change in the stoichiometric coefficient for species A in reaction 2 (see Table 4.1).

Substitution of the stoichiometric coefficients and reaction rates into Eq. 4.5 yields the following system of ODEs:

$$\frac{d[A]}{dt} = -k_1[A]^2[B] + k_2[C]^2 \tag{4.6a}$$

$$\frac{d[B]}{dt} = -k_1[A]^2[B] + 3k_2[C]^2 \tag{4.6b}$$

$$\frac{d[C]}{dt} = k_1[A]^2[B] - 2k_2[C]^2 \, . \tag{4.6c}$$

After specifying three initial values, one for each chemical species ($[A](t = 0) = [A]_0$, and so on), these ODEs may be solved to predict the dynamics of the concentrations $[A](t)$, $[B](t)$ and $[C](t)$.

4.2   Reaction Order and Physical Dimensions of Rate Constants

Elementary chemical reactions are described as first-order, second-order, and so on, depending on the number of *reactants*,

$$A \rightarrow \qquad \text{first-order}$$
$$A + B \rightarrow \qquad \text{second-order}$$
$$2A \rightarrow \qquad \text{second-order}$$

$$A + B + C \rightarrow \qquad \text{third-order}$$
$$A + 2B \rightarrow \qquad \text{third-order}$$
$$3A \rightarrow \qquad \text{third-order.}$$

A reaction with constant rate (i.e., not a function of chemical species concentration) is *zeroth order*.

In a compartmental model of chemical kinetics such as Eq. 4.6, the reaction rates $j_1 = k_1[A]^2[B]$ and $j_2 = k_2[C]^2$ have physical dimensions of conc/time. As a consequence, the physical dimensions of reaction rate constants k_1 and k_2 must be different. Eq. 4.6a implies the following balance between the physical dimensions of $d[A]/dt$ and the rate of reaction 2,

$$\frac{\text{conc}}{\text{time}} \ominus \frac{d[A]}{dt} \ominus j_2 = k_2 \cdot [C]^2 \ominus \frac{1}{\text{conc} \cdot \text{time}} \cdot \text{conc}^2,$$

which implies that $k_2 \ominus \text{conc}^{-1}\text{time}^{-1}$. Similarly, $d[A]/dt \ominus j_1 = k_1[A]^2[B]$ implies that $k_1 \ominus \text{conc}^{-2}\text{time}^{-1}$ (diligent readers will check this).[2] Indeed, physical dimensions of reaction rate constants depend, in a predictable way, on reaction order:

$$A \xrightarrow{k} \qquad \text{first-order rate constant} \qquad k \ominus \text{time}^{-1}$$
$$A + B \xrightarrow{k} \qquad \text{second-order rate constant} \qquad k \ominus \text{conc}^{-1}\text{time}^{-1}$$
$$A + B + C \xrightarrow{k} \qquad \text{third-order rate constant} \qquad k \ominus \text{conc}^{-2}\text{time}^{-1}.$$

A zeroth-order rate constant has physical dimensions of conc/time. First-order rate constants are sometimes called *unimolecular* rate constants, while second-order rate constants are *biomolecular*.

> **Problem 4.1**   In the open cell model of intracellular calcium concentration (Section 2.2), the influx rate was given by $j_{in} = J$ and the efflux rate was given by $j_{out} = kc$. Which of these processes is first order, and which is zeroth order?

4.3   Isomerization – ODEs and a Conserved Quantity

Isomerization reactions transform one molecule into another with the same number of atoms. The reaction diagram for a pair of molecules that reversibly isomerize is

$$(\text{reactant}) \quad A \overset{\alpha}{\underset{\beta}{\rightleftharpoons}} B \quad (\text{product}). \tag{4.7}$$

The harpoons notation indicates that α is the rate constant for the forward reaction (reactant-to-product), while β is the rate constant for the reverse reaction (product-to-reactant). Assuming mass action kinetics, the forward reaction rate is $\alpha[A]$. This reaction leads to a loss of one A and gain of one B. The reverse reaction proceeds at rate $\beta[B]$ with a loss of one B and gain of one A. Thus, the ODE system corresponding to Eq. 4.7 is

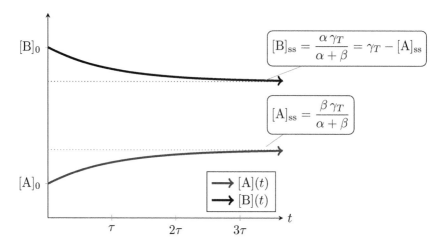

Figure 4.1   Solution of the ODE model of isomerization (Eq. 4.18).

$$\frac{\text{conc}}{\text{time}} \ominus \frac{d[A]}{dt} = -\alpha[A] + \beta[B] \ominus \frac{1}{\text{time}} \cdot \text{conc} \tag{4.8a}$$

$$\frac{d[B]}{dt} = \alpha[A] - \beta[B], \tag{4.8b}$$

where α and β are first-order (unimolecular) rate constants. Fig. 4.1 shows a numerically calculated solution of these coupled ODEs.

Because we have not yet discussed the analysis of systems of coupled ODEs, we appear to be at an impasse. However, it is possible to reduce Eq. 4.8 to a scalar ODE. Observe that summing Eqs. 4.8a and 4.8b leads to cancelation of the right sides,

$$\frac{d[A]}{dt} + \frac{d[B]}{dt} = -\alpha[A] + \beta[B] + \alpha[A] - \beta[B] = 0.$$

Using the summation rule for derivatives, this expression may be written as

$$\frac{d}{dt}([A] + [B]) = 0. \tag{4.9}$$

The zero rate of change in Eq. 4.9 indicates that $[A] + [B]$ is **conserved**, that is, the quantity $[A] + [B]$ does not change value as the reaction dynamics proceed. Let us denote this conserved quantity as

$$\gamma_T = [A] + [B] \tag{4.10}$$

where the subscript T stands for *total* concentration of A and B. Because γ_T is constant, we can substitute

$$[B] = \gamma_T - [A] \tag{4.11}$$

in Eq. 4.8a, to obtain a first-order, linear ODE,

$$\frac{d[A]}{dt} = -\alpha[A] + \beta \underbrace{(\gamma_T - [A])}_{[B]}. \tag{4.12}$$

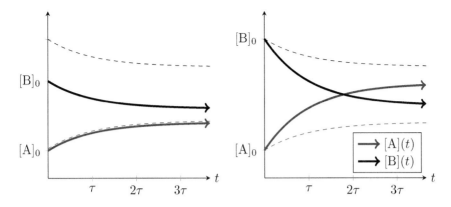

Figure 4.2 Isomerization reaction kinetics of Fig. 4.1 are replotted (dashed lines) so they may be compared with solutions that use modified parameters. Left: The total concentration $\gamma_T = [A]_0 + [B]_0$ (Eq. 4.10) is decreased without changing $[A]_{ss}$. Right: The forward rate constant α is decreased (Eq. 4.7). This increases $[A]_{ss}$ and decreases $[B]_{ss}$ (see Eqs. 4.14 and 4.15).

The ODE for [B] is now superfluous, because we may solve Eq. 4.12 for [A] and, subsequently, determine [B] using Eq. 4.11.

Observe that Eq. 4.12 has a single stable steady state found by solving

$$0 = -\alpha[A]_{ss} + \beta\,(\gamma_T - [A]_{ss}) \tag{4.13}$$

for $[A]_{ss}$. The result is

$$\text{conc} \ominus [A]_{ss} = \frac{\beta}{\alpha + \beta}\,\gamma_T \ominus \frac{1/\text{time}}{1/\text{time}} \cdot \text{conc}, \tag{4.14}$$

where the dimensionless ratio $\beta/(\alpha + \beta)$ takes values between 0 and 1. Using Eq. 4.10 we calculate the steady state for [B],

$$[B]_{ss} = \gamma_T - [A]_{ss} = \gamma_T - \frac{\beta}{\alpha + \beta}\,\gamma_T = \left(1 - \frac{\beta}{\alpha + \beta}\right)\gamma_T = \frac{\alpha}{\alpha + \beta}\,\gamma_T. \tag{4.15}$$

Fig. 4.2 shows how the kinetics of isomerization and the steady state concentrations of A and B depend on the parameters α, β and γ_T.

Problem 4.2 Show that $[A]_{ss} = \beta\gamma_T/(\alpha + \beta)$ and $[B]_{ss} = \alpha\gamma_T/(\alpha + \beta)$ sum to γ_T, consistent with Eq. 4.10.

4.4 Isomerization – Phase Diagram and Solutions

Fig. 4.3 shows the phase diagram for Eq. 4.12, which is straightforward to sketch after rewriting the ODE as follows:

$$\frac{d[A]}{dt} = \underbrace{\beta\gamma_T}_{J} - \underbrace{(\alpha + \beta)}_{k}[A] \qquad [A](t = 0) = [A]_0. \tag{4.16}$$

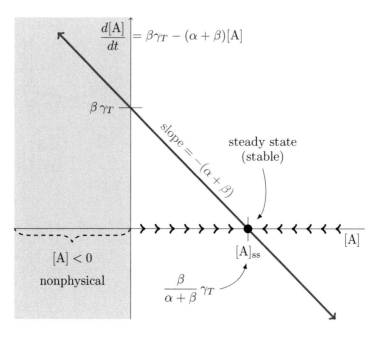

Figure 4.3   The phase diagram for the linear ODE (Eq. 4.12) obtained from the isomerization reaction (Eq. 4.7) and the law of mass action.

The solution is the exponential relaxation

$$[A](t) = ([A]_0 - [A]_{ss})\, e^{-t/\tau} + [A]_{ss} \tag{4.17}$$

where $[A]_0$ is the initial value and the steady state $[A]_{ss}$ is given by Eq. 4.14. The exponential time constant is

$$\text{time} \,\ominus\, \tau = \frac{1}{k} = \frac{1}{\alpha + \beta} \;\ominus\; \frac{1}{1/\text{time}}\,.$$

Problem 4.3   Use Eq. 4.10 to show that $[B](t)$ is $[B](t) = ([B]_0 - [B]_{ss})\, e^{-t/\tau} + [B]_{ss}$ with $[B]_0 = \gamma_T - [A]_0$ and $[B]_{ss}$ as in Eq. 4.15.

This steady state is a **chemical equilibrium**. This can be confirmed by writing the condition for equilibrium (i.e., forward and reverse rates balancing), and checking that this equation is satisfied, as follows:

$$\alpha[A]_{ss} \overset{?}{=} \beta[B]_{ss}$$
$$\alpha\, \frac{\beta}{\alpha + \beta}\, \gamma_T \overset{\text{yes}}{=} \beta\, \frac{\alpha}{\alpha + \beta}\, \gamma_T \cdot \checkmark$$

The **equilibrium constant** (K^{eq}) for the isomerization reaction is the steady state product/reactant concentration ratio, a dimensionless number equal to the ratio of the forward (α) and reverse (β) reaction rate constants,

$$K_{eq} \stackrel{\text{def}}{=} \frac{\text{products}}{\text{reactants}} = \frac{[B]_{ss}}{[A]_{ss}} = \frac{\dfrac{\alpha}{\alpha+\beta}\gamma_T}{\dfrac{\beta}{\alpha+\beta}\gamma_T} = \frac{\alpha}{\beta} \ominus \frac{\text{time}^{-1}}{\text{time}^{-1}}.$$

The change in Gibbs free energy associated with this reaction (product minus reactant) is given by $\Delta G = G_b - G_a$. These quantities are related,

$$K_{eq} = \alpha/\beta = e^{-\Delta G/k_B T}, \tag{4.18}$$

where k_B is Boltzmann's constant and T is the absolute temperature. Looking at it the other way, the free energy of the reaction is the following function of the rate constants α and β:

$$\ln K_{eq} = \ln(\alpha/\beta) = -\Delta G/k_B T \;\Rightarrow\; \Delta G = k_B T \ln(\beta/\alpha).$$

In Fig. 4.2 (left), the forward reaction is favored, $\alpha > \beta$, $\beta/\alpha < 1$, $\ln(\beta/\alpha) < 0$ and, consequently, the free energy of the reaction is negative ($\Delta G < 0$). This is consistent with the product B being at higher concentration than the reactant A. In Fig. 4.2 (right), $\alpha < \beta$ and $\Delta G > 0$.

> **Problem 4.4**   What is the equilibrium constant for an isomerization reaction for which the equilibrium concentrations of the isomers are equal?

4.5   Bimolecular Association of Ligand and Receptor

Consider a postsynaptic receptor for the neurotransmitter glutamate, an ion channel that opens in response to the binding of glutamate that is released into a synaptic cleft by a presynaptic neuron (see Fig. 4.4). This process can be modeled as a **bimolecular association reaction**,

$$L + R \underset{k_-}{\overset{k_+}{\rightleftharpoons}} LR, \tag{4.19}$$

where L is the free ligand (glutamate), R is the postsynaptic receptor without ligand bound, and LR is the receptor bound to glutamate (see Fig. 4.5). In this kinetic scheme, k_+ is a bimolecular **association rate constant**, and k_- is a unimolecular **dissociation rate constant**.

> **Problem 4.5**   What are the physical dimensions of k_+ and k_-? When concentration and time are measured in micromolar (μM) and seconds (s), what are their units?

The ODEs for Bimolecular Association

The stoichiometric coefficients in Eq. 4.19 are $-1L$, $-1R$, $+1LR$ for the forward reaction (association) and have opposite sign for dissociation. Thus, the material balance equations take the form $d[L]/dt = -j_+ + j_-$, $d[R]/dt = -j_+ + j_-$, and $d[LR]/dt = j_+ - j_-$,

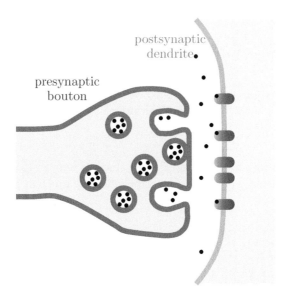

Figure 4.4   Release of neurotransmitter from vesicles in presynaptic bouton and binding of neurotransmitter to receptors on postsynaptic dendrite.

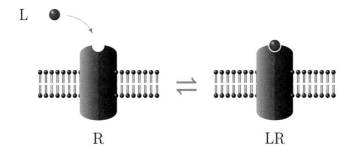

Figure 4.5   Diagram of bimolecular association between ligand (L) and a cell surface receptor (R). LR is the ligand-receptor complex.

where j_+ and j_- are chemical fluxes for the forward and reverse reactions. Assuming mass action kinetics, $j_+ = k_+[L][R]$ and $j_- = k_-[LR]$ and, consequently,

$$\frac{d[L]}{dt} = -k_+[L][R] + k_-[LR] \tag{4.20a}$$

$$\frac{d[R]}{dt} = -k_+[L][R] + k_-[LR] \tag{4.20b}$$

$$\frac{d[LR]}{dt} = k_+[L][R] - k_-[LR]. \tag{4.20c}$$

Associating the initial concentration $[L](t = 0) = [L]_0$ and similarly for $[R]_0$ and $[LR]_0$ yields an initial value problem for ligand receptor binding.

Steady State Concentrations and Equilibrium Constants

Our analysis of ligand-receptor binding will focus on the steady state concentrations of ligand ($[L]_{ss}$), free receptor ($[R]_{ss}$), and bound receptor ($[LR]_{ss}$) predicted by Eq. 4.20. Setting derivatives to zero gives $0 = -j_+ + j_-$ and shows that the steady state concentrations balance the forward and reverse reaction rates,

$$j_+ = k_+[L]_{ss}[R]_{ss} = k_-[LR]_{ss} = j_- \, . \tag{4.21}$$

Evidently, the steady state corresponds to chemical equilibrium.[3] The equilibrium constant, defined as the ratio of product-to-reactant concentrations,

$$\frac{1}{\text{conc}} \ominus K_{eq} \overset{\text{def}}{:=} \frac{\text{products}}{\text{reactants}} = \frac{[LR]_{ss}}{[L]_{ss}[R]_{ss}} \ominus \frac{\text{conc}}{\text{conc} \cdot \text{conc}} \, ,$$

has physical dimensions of conc^{-1}. Rearranging Eq. 4.21 we observe that the equilibrium constant is a ratio of rate constants,

$$\frac{1}{\text{conc}} \ominus K_{eq} = \frac{[LR]_{ss}}{[L]_{ss}[R]_{ss}} = \frac{k_+}{k_-} \ominus \frac{\text{conc}^{-1}\,\text{time}^{-1}}{\text{time}^{-1}} \, .$$

Because K_{eq} becomes larger when the association rate constant is increased (products favored), it is an equilibrium *association* constant. The **dissociation constant** (denoted by K) is the inverse of the association constant,

$$\text{conc} \ominus K = \frac{1}{K_{eq}} = \frac{k_-}{k_+} \ominus \frac{\text{time}^{-1}}{\text{conc}^{-1}\,\text{time}^{-1}} \, , \tag{4.22}$$

and has physical dimensions of concentration.

Conserved Quantities

Eq. 4.21 may appear to be an underdetermined equation, because there are three unknown concentrations ($[L]_{ss}$, $[R]_{ss}$ and $[LR]_{ss}$). However, the ODE system Eq. 4.20 has two conserved quantities, the total concentration of ligand ($\ell_T \overset{\text{def}}{:=} [L] + [LR]$) and the total concentration of receptor ($r_T \overset{\text{def}}{:=} [R] + [LR]$). At steady state,

$$\ell_T = [L]_{ss} + [LR]_{ss} \tag{4.23}$$
$$r_T = [R]_{ss} + [LR]_{ss} \, . \tag{4.24}$$

The parameters k_+, k_-, ℓ_T, and r_T are given quantities, so Eqs. 4.21, 4.23 and 4.24 may be solved simultaneously for $[L]_{ss}$, $[R]_{ss}$, $[LR]_{ss}$ (Problem 4.14).

Equilibrium Relations

The most important aspect of the steady state concentrations ($[L]_{ss}$, $[R]_{ss}$ and $[LR]_{ss}$) is the functional relationship between the ligand-bound receptor ($[LR]_{ss}$) and free ligand ($[L]_{ss}$). This is found by writing Eq. 4.21 as

$$[LR]_{ss}K = [L]_{ss}[R]_{ss} \, ,$$

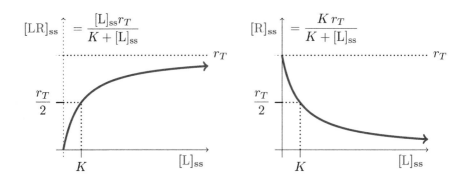

Figure 4.6   Equilibrium relations (binding curves) for bimolecular association of ligand and receptor (Eq. 4.19). Left: $[LR]_{ss}$ is an increasing hyperbolic function of $[L]_{ss}$ (Eq. 4.25). Right: $[R]_{ss}$ is a decreasing hyperbolic function of $[L]_{ss}$ (Eq. 4.26).

where $K = k_-/k_+$ is the dissociation constant (Eq. 4.22). Substituting $[R]_{ss} = r_T - [LR]_{ss}$ (Eq. 4.24), and solving for $[LR]_{ss}$, we obtain

$$K[LR]_{ss} = [L]_{ss} (r_T - [LR]_{ss})$$
$$K[LR]_{ss} + [L]_{ss}[LR]_{ss} = [L]_{ss} r_T$$
$$(K + [L]_{ss}) [LR]_{ss} = [L]_{ss} r_T ,$$

and finally

$$[LR]_{ss} = \frac{[L]_{ss} r_T}{K + [L]_{ss}} . \tag{4.25}$$

A similar calculation gives

$$[R]_{ss} = \frac{K r_T}{K + [L]_{ss}} . \tag{4.26}$$

Fig. 4.6 plots these **equilibrium relations** (also called **binding curves**). Observe that the concentration of ligand-bound receptor ($[LR]_{ss}$) is an increasing **hyperbolic**[4] function of free ligand ($[L]_{ss}$). When $[L]_{ss}$ is low, most receptor is in the ligand-free form (R). When $[L]_{ss}$ is high, most receptor is in the ligand-bound form (LR). When the free ligand concentration is equal to the dissociation constant ($[L]_{ss} = K$), the bound ligand concentration is half maximal ($[LR]_{ss} = r_T/2$). To see this, substitute K for $[L]_{ss}$ in Eq. 4.25,

$$[LR]_{ss} = \frac{K r_T}{K + K} = \frac{K r_T}{2K} = \frac{r_T}{2} . \tag{4.27}$$

In quantitative pharmacology, the term EC_{50} refers to the concentration of a drug that gives half maximal response. Assuming that cellular response is proportional to the fraction of ligand-bound receptors,

$$\text{response} \propto \frac{[LR]_{ss}}{r_T} = \frac{[L]_{ss}}{K + [L]_{ss}} , \tag{4.28}$$

our model suggests that the EC_{50} will be the ligand-receptor dissociation constant K, because $[L]_{ss}/(K + [L]_{ss}) \to 1$ as $[L]_{ss} \to \infty$ and evaluates to $1/2$ at $[L]_{ss} = K$.

4.6 Sequential Binding

With practice it is possible to calculate the equilibrium relations predicted by more complex binding schemes. For example, consider the following two bimolecular association reactions,

$$L + R \underset{k_a^-}{\overset{k_a^+}{\rightleftharpoons}} LR \qquad L + LR \underset{k_b^-}{\overset{k_b^+}{\rightleftharpoons}} L_2R, \tag{4.29}$$

where LR and L_2R are receptor bound to one or two ligands, respectively, and k_a^+, k_a^-, k_b^+ and k_b^- are rate constants. This reaction may also be written in a manner that highlights the sequential ligand binding,

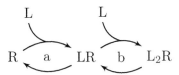

To analyze this binding scheme, write the conservation law for the coupled bimolecular association reactions shown in Eq. 4.29,

$$\frac{d[L]}{dt} = -j_a^+ + j_a^- - j_b^+ + j_b^- \tag{4.30a}$$

$$\frac{d[R]}{dt} = -j_a^+ + j_a^- \tag{4.30b}$$

$$\frac{d[LR]}{dt} = +j_a^+ - j_a^- - j_b^+ + j_b^- \tag{4.30c}$$

$$\frac{d[L_2R]}{dt} = +j_b^+ - j_b^- . \tag{4.30d}$$

Assuming mass action kinetics, $j_a^+ = k_a^+[L][R]$, $j_a^- = k_a^-[LR]$, $j_b^+ = k_b^+[L][LR]$ and $j_b^- = k_b^-[L_2R]$. Substituting gives a system of four ODEs,

$$\frac{d[L]}{dt} = -k_a^+[L][R] + k_a^-[LR] - k_b^+[L][LR] + k_b^-[L_2R] \tag{4.31a}$$

$$\frac{d[R]}{dt} = -k_a^+[L][R] + k_a^-[LR] \tag{4.31b}$$

$$\frac{d[LR]}{dt} = k_a^+[L][R] - k_a^-[LR] - k_b^+[L][LR] + k_b^-[L_2R] \tag{4.31c}$$

$$\frac{d[L_2R]}{dt} = k_b^+[L][LR] - k_b^-[L_2R] . \tag{4.31d}$$

Steady State Concentrations and Equilibrium Relations

The steady state concentrations ($[L]_{ss}$, $[R]_{ss}$, $[LR]_{ss}$, $[L_2R]_{ss}$) are found by setting the rates of change to zero. The steady state is a chemical equilibrium, because $d[R]/dt = 0 \Rightarrow j_a^+ = j_a^-$ and $d[L_2R]/dt = 0 \Rightarrow j_b^+ = j_b^-$. These provide two (of four) algebraic equations solved by the steady state concentrations,

$$k_a^+[L]_{ss}[R]_{ss} = k_a^-[LR]_{ss} \tag{4.32}$$

$$k_b^+[L]_{ss}[LR]_{ss} = k_b^-[L_2R]_{ss}. \tag{4.33}$$

The remaining two algebraic relationships are provided by conserved quantities, namely, the total ligand concentration (ℓ_T) and total receptor concentration (r_T),

$$r_T \overset{\text{def}}{=} [R]_{ss} + [LR]_{ss} + [L_2R]_{ss} \tag{4.34}$$

$$\ell_T \overset{\text{def}}{=} [L]_{ss} + [LR]_{ss} + 2[L_2R]_{ss}, \tag{4.35}$$

where the factor of 2 multiplying $[L_2R]_{ss}$ is needed because the species of L_2R has two ligand molecules bound. Solving Eqs. 4.32–4.35 simultaneously leads to the following equilibrium relationships,

$$[R]_{ss} = \frac{K_a K_b \, r_T}{K_a K_b + [L]_{ss} K_b + [L]_{ss}^2} \tag{4.36a}$$

$$[LR]_{ss} = \frac{[L]_{ss} K_b \, r_T}{K_a K_b + [L]_{ss} K_b + [L]_{ss}^2} \tag{4.36b}$$

$$[L_2R]_{ss} = \frac{[L]_{ss}^2 \, r_T}{K_a K_b + [L]_{ss} K_b + [L]_{ss}^2} \tag{4.36c}$$

where $K_a = k_a^-/k_a^+$, $K_b = k_b^-/k_b^+$ are two dissociation constants.

Fig. 4.7 plots these steady state concentrations as a function of free ligand concentration. The concentration of free receptor ($[R]_{ss}$) is a decreasing function of the ligand concentration ($[L]_{ss}$), while the concentration of doubly bound receptor ($[L_2R]_{ss}$) increases. The binding curve for singly bound receptor ($[LR]_{ss}$) is **biphasic**,[5] that is, it increases to a maximum and then decreases.

Problem 4.6   Show that ℓ_T and r_T are conserved quantities by differentiating Eqs. 4.34 and 4.35 to obtain $d\ell_T/dt = 0$ and $dr_T/dt = 0$.

4.7   Sigmoidal Binding Curves

Fig. 4.7 shows that the binding curve for L_2R is **sigmoidal**, that is, the slope of the binding curve is nearly zero when $[L]_{ss}$ is small and there is an inflection point near $[L]_{ss} = \sqrt{K_a K_b}$. To see this, use Eq. 4.36c and differentiate $[L_2R]_{ss}$ with respect to $[L]_{ss}$,

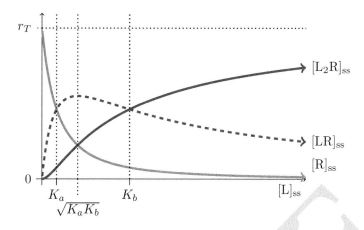

Figure 4.7 The steady state concentration of the three different forms of receptor ($[R]_{ss}$, $[LR]_{ss}$ and $[L_2R]_{ss}$) as a function of free ligand concentration ($[L]_{ss}$) according to the equilibrium relations of Eq. 4.36.

$$
\begin{aligned}
[L_2R]'_{ss} &= \left(\frac{[L]_{ss}^2 r_T}{K_a K_b + [L]_{ss} K_b + [L]_{ss}^2} \right)' \\[2mm]
&= \left(\frac{[L]_{ss}^2}{K_a K_b + [L]_{ss} K_b + [L]_{ss}^2} \right)' r_T \\[2mm]
\vdots \\[2mm]
&= \frac{K_b [L]_{ss} \left(2K_a + [L]_{ss} \right) r_T}{\left(K_a K_b + [L]_{ss} K_b + [L]_{ss}^2 \right)^2} > 0 .
\end{aligned}
\tag{4.37}
$$

The positive derivative shows that $[L_2R]_{ss}$ is an increasing function of $[L]_{ss}$. Observe that for $[L]_{ss} > 0$ the derivative asymptotically approaches zero,

$$
[L_2R]'_{ss} \to 0 \quad \text{as} \quad [L]_{ss} \to 0,
$$

consistent with Fig. 4.8 (right column). This sigmoidal binding curve differs from the hyperbolic curve observed of a receptor that binds only one ligand (Fig. 4.6, Eq. 4.25) where the derivative of $[LR]_{ss}$ with respect to $[L]_{ss}$ is also positive,

$$
[LR]'_{ss} = \left[\frac{[L]_{ss} r_T}{K + [L]_{ss}} \right]' = \frac{K r_T}{(K + [L]_{ss})^2} > 0 ,
\tag{4.38}
$$

but the limiting derivative is not zero; rather, the limiting derivative is positive,

$$
[LR]'_{ss} \to \left. \frac{K r_T}{(K + [L]_{ss})^2} \right|_{[L]_{ss}=0} = \frac{K r_T}{K^2} = \frac{r_T}{K} \quad \text{as} \quad [L]_{ss} \to 0 .
\tag{4.39}
$$

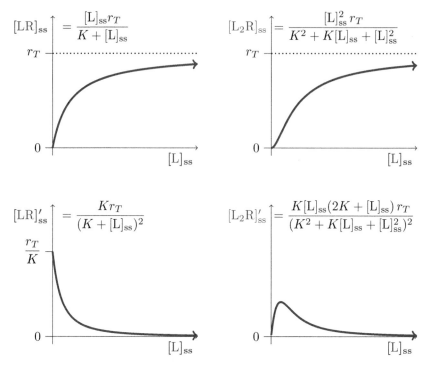

$$[LR]_{ss} = \frac{[L]_{ss} r_T}{K + [L]_{ss}}$$

$$[L_2R]_{ss} = \frac{[L]_{ss}^2 r_T}{K^2 + K[L]_{ss} + [L]_{ss}^2}$$

$$[LR]'_{ss} = \frac{K r_T}{(K + [L]_{ss})^2}$$

$$[L_2R]'_{ss} = \frac{K[L]_{ss}(2K + [L]_{ss}) r_T}{(K^2 + K[L]_{ss} + [L]_{ss}^2)^2}$$

Figure 4.8 Comparison of the binding curves for a bimolecular association reaction between ligand and receptor (Eq. 4.19) and the sequential binding kinetic scheme (Eq. 4.29) with $K_a = K_b = K$.

4.8   Binding Curves and Hill Functions

The sequential binding model simplifies when the affinity of the second ligand binding reaction (b) is much higher than the first (a). In terms of dissociation constants (the reciprocal of affinity) this corresponds to $K_b \ll K_a$. Evaluating Eq. 4.36 in the limit as $K_b \to 0$ shows that (all other things being equal) a high affinity second binding reaction (b) causes the doubly bound state to be highly favored ($[L_2R]_{ss} \to 1$ while $[LR]_{ss} \to 0$ and $[R]_{ss} \to 0$). What is more interesting is to consider the limit $K_b \to 0$ with fixed $K_a K_b$. Rewrite Eq. 4.36 in terms of K_b and the quantity that is not changing ($K_*^2 = K_a K_b$) and, subsequently, taking the limit as $K_b \to 0$, we find

$$\frac{[R]_{ss}}{r_T} = \frac{K_*^2}{K_*^2 + [L]_{ss} K_b + [L]_{ss}^2} \to \frac{K_*^2}{K_*^2 + [L]_{ss}^2} \tag{4.40a}$$

$$\frac{[LR]_{ss}}{r_T} = \frac{[L]_{ss} K_b}{K_*^2 + [L]_{ss} K_b + [L]_{ss}^2} \to 0 \tag{4.40b}$$

$$\frac{[L_2R]_{ss}}{r_T} = \frac{[L]_{ss}^2}{K_*^2 + [L]_{ss} K_b + [L]_{ss}^2} \to \frac{[L]_{ss}^2}{K_*^2 + [L]_{ss}^2}. \tag{4.40c}$$

This shows that when the (geometric) average $K_* = \sqrt{K_b K_a}$ is not particularly small, but K_b / K_a is small, there is negligible concentration of receptors with only one ligand bound ($[LR]_{ss}$). Furthermore, the expressions for the unbound ($[R]_{ss}$) and doubly bound ($[L_2R]_{ss}$) receptor have simplified.

The binding curves of Eq. 4.40 are referred to as **Hill functions**, after the English physiologist A. V. Hill (1886–1977). An increasing[6] Hill function takes the form

$$h(x) = \frac{v\, x^n}{\kappa^n + x^n} \qquad v, \kappa > 0,\tag{4.41}$$

where the **Hill coefficient** $n \geq 1$ is often an integer and, evidently, $\kappa \Leftrightarrow x$. For $n = 1$ the Hill function is hyperbolic, and when $n \geq 2$ it is sigmoidal. To see this, calculate the derivative of $h(x)$ with respect to x,

$$h'(x) = v\left(\frac{x^n}{\kappa^n + x^n}\right)' = v\,\frac{\left(x^n\right)' \cdot \left(\kappa^n + x^n\right) - \left(\kappa^n + x^n\right)' \cdot x^n}{(\kappa^n + x^n)^2}$$

$$= v\,\frac{n x^{n-1}\left(\kappa^n + x^n\right) - n x^{n-1} x^n}{(\kappa^n + x^n)^2} = \frac{v n x^{n-1} \kappa^n}{(\kappa^n + x^n)^2}.$$

Observe that for $n = 1$, the derivative of the Hill function,

$$h'(x) = \frac{v \kappa}{(\kappa + x)^2},$$

is similar to Eq. 4.38. For $n = 2$, the derivative is similar to Eq. 4.37:

$$h'(x) = \frac{v\, 2 x \kappa^2}{\left(\kappa^2 + x^2\right)^2}.$$

Fig. 4.9 plots hyperbolic ($n = 1$) and sigmoidal ($n = 2$ and 4) Hill functions using standard and logarithmic x-axes, revealing that the derivative at the half maximum ($x = \kappa, h = v/2$) is an increasing function of n. In fact, for large n the Hill function

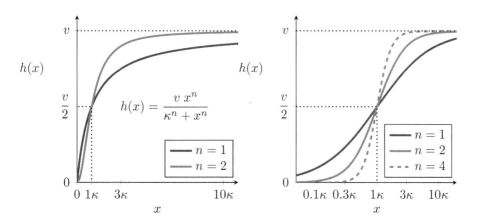

Figure 4.9 The Hill function $h(x) = v x^n/(\kappa^n + x^n)$ for various values of n. The semilog plot (right) highlights the sigmoidal shape (for $n \geq 2$).

resembles the Heaviside step function (defined on p. 29). To see this, evaluate the derivative of the increasing Hill function for general n (Eq. 4.42) at $x = \kappa$,

$$h'(x = \kappa) = \frac{vn\kappa^{n-1}\kappa^n}{(\kappa^n + \kappa^n)^2} = \frac{vn\kappa^{2n-1}}{4\kappa^{2n}} = \frac{vn}{4\kappa}.$$

This result makes it clear that the Hill coefficient n may be chosen to fit the observed cooperativity of a response. When $n \geq 1$ is an integer, the Hill coefficient may be interpreted as the number of sequential ligand binding steps in a phenomenological receptor model.

4.9   Discussion

Segel and Edelstein-Keshet (2013, Chapter 2) is a gentle introduction to biochemical kinetics. Basic principles of chemical kinetics and derivation of steady states are covered in Cornish-Bowden (2012, Chapters 1 and 4). Lauffenburger and Linderman (1996) is a highly recommended monograph on modeling of ligand-receptor interactions. Segel (1993) is a classic monograph on enzyme kinetics with numerous examples of steady state analysis.

Finding Conserved Quantities

Conserved quantities may be found by checking various linear combinations of chemical species concentrations until one or more are found for which the sum has zero rate of change. For example, the right sides of Eq. 4.20a and Eq. 4.20c sum to zero, so $[L] + [LR]$ is a conserved quantity in this ODE system. In Section 4.6, the quantity $\chi = [A] + [B] + 2[C]$ is conserved. This result may be derived by assuming a conserved quantity $\chi = a[A] + b[B] + c[C]$, where it remains to specify a, b and c. This is achieved by differentiating and collecting terms,

$$
\begin{aligned}
\frac{d\chi}{dt} &= a\frac{d[A]}{dt} + b\frac{d[B]}{dt} + c\frac{d[C]}{dt} \\
&= a[-j_1 + j_2] + b[-j_1 + 3j_2] + c[j_1 - 2j_2] \\
&= [-a - b + c]j_1 + [a + 3b - 2c]j_2,
\end{aligned}
$$

where the first step uses Eq. 4.5. Because $j_1 = k_1[A]^2[B]$ and $j_2 = k_2[C]^2$ are linearly independent, $d\chi/dt = 0$ requires that a, b and c solve

$$-a - b + c = 0 \tag{4.42a}$$

$$a + 3b - 2c = 0. \tag{4.42b}$$

This linear algebraic system is underdetermined (2 equations, 3 unknowns). We may assume $a = 1$, to obtain

$$-n_b + n_c = 1$$

$$3n_b - 2n_c = -1,$$

and solve this linear system to find $n_b = 1$ and $n_c = 2.$[7]

Supplemental Problems

Problem 4.7   For each reaction scheme below, write ODEs for chemical species concentrations under the assumption of mass action kinetics. Determine the physical dimensions of each rate constant.

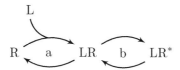

(a) $A + B \underset{k_2}{\overset{k_1}{\rightleftharpoons}} C + D$ (b) $2A + B \underset{k_4}{\overset{k_3}{\rightleftharpoons}} C + 2B$ (c) $A \underset{k_{ba}}{\overset{k_{ab}}{\rightleftharpoons}} B \underset{k_{cb}}{\overset{k_{bc}}{\rightleftharpoons}} C$

Problem 4.8   Write ODEs for the receptor model below, in which LR and LR* are two forms of ligand-bound receptor. Write the equilibrium constants as conc $\ominus$ $K_a^{eq} = k_a^+ / k_a^-$ and 1 $\ominus$ $K_b^{eq} = k_b^+ / k_b^-$ (dimensionless). Determine the physical dimensions of each rate constant. Find the equilibrium relation for $[LR^*]_{ss}$ as a function of $[L]_{ss}$ and the dissociation constant $K_a = 1/K_a^{eq}$.

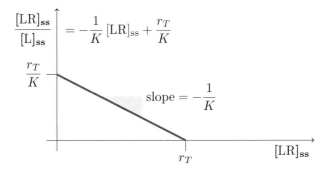

Problem 4.9   In an isomerization reaction for which the reactant-to-product rate constant is smaller than the product-to-reactant rate constant, is the free energy difference $\Delta G = G_{product} - G_{reactant}$ positive or negative?

Problem 4.10   What are the physical dimensions of ΔG, K_{eq}, α and β in Eq. 4.18?

Problem 4.11   Show that the total concentrations ℓ_T and r_T defined in Eqs. 4.23 and 4.24 are conserved.

Problem 4.12   Use Eqs. 4.21 and 4.24 to derive Eq. 4.26.

Problem 4.13   The **Scatchard plot** below shows the ratio of bound to unbound ligand ($[LR]_{ss}/[L]_{ss}$) as a function of the bound ligand ($[LR]_{ss}$) for the bimolecular association reaction of Eq. 4.19.

$$\frac{[LR]_{ss}}{[L]_{ss}} = -\frac{1}{K}[LR]_{ss} + \frac{r_T}{K}$$

with axes labeled $\dfrac{r_T}{K}$ (vertical intercept), slope $= -\dfrac{1}{K}$, r_T and $[LR]_{ss}$ (horizontal axis).

Beginning with the equilibrium relations (Eqs. 4.25 and 4.26) show that $[LR]_{ss}/[L]_{ss}$ is a linear function of $[LR]_{ss}$ (as the above plot suggests).

Problem 4.14   Solve Eqs. 4.21–4.24 simultaneously to obtain expressions for $[L]_{ss}$, $[R]_{ss}$ and $[LR]_{ss}$ in terms of K, r_T and ℓ_T. Hint: You will need to solve a quadratic polynomial of the form $ax^2 + bx + c = 0$ using $x = (-b \pm \sqrt{b^2 - 4ac})/(2a)$. Which of the two solutions is physical?

Problem 4.15   Solve Eqs. 4.32–4.34 simultaneously to obtain Eq. 4.36.

Problem 4.16   In Fig. 4.7, the curves for free (R) and singly bound (LR) receptor intersect as though $[R]_{ss} = [LR]_{ss}$ when $[L]_{ss} = K_a$. Confirm this using Eq. 4.36.

Problem 4.17   Using Eq. 4.36, show that (a) $[LR]_{ss} = [L_2R]_{ss}$ when $[L]_{ss} = K_b$, (b) $[R]_{ss} = [L_2R]_{ss}$ when $[L]_{ss} = \sqrt{K_aK_b}$, (c) $[R]_{ss}$ is a decreasing function of $[L]_{ss}$, and (d) $[L_2R]_{ss}$ is an increasing function of $[L]_{ss}$.

Problem 4.18   Take the time derivative of both sides of Eqs. 4.34 and 4.35 to show that $d\ell_T/dt = 0$ and $dr_T/dt = 0$.

Problem 4.19   If a bimolecular association rate constant is given by $k_+ = 3.7\,\mathrm{M^{-1}\,s^{-1}}$, what is its value in units of $\mu\mathrm{M^{-1}\,ms^{-1}}$? In units of $\mathrm{M^{-1}\,s^{-1}}$?

Problem 4.20   Consider this ODE for the dynamics of a cell signaling molecule,

$$\frac{d[A]}{dt} = k_1[A] - k_2[A]^2 \quad \text{where} \quad k_1, k_2 > 0. \tag{4.43}$$

(a)   If time is in seconds and concentration is in micromolar, what are the units of k_2?
(b)   Find the steady state concentration of A in terms of k_1 and k_2.
(c)   Determine the stability of the steady state by analytically checking the sign of the appropriate derivative.
(d)   Draw a phase diagram for the model. Label both axes and indicate (i) the direction of flow on the horizontal axis and (ii) the stability of steady states.
(e)   If $[A]_{ss} = 5\ \mu\mathrm{M}$ and $k_1 = 10\ \mathrm{ms^{-1}}$, what is k_2?
(f)   Approximate the time constant for relaxation when $[A] \approx 5\ \mu\mathrm{M}$.

Problem 4.21   The activation of receptors in the absence of stimulating ligands and **inverse agonism** can be understood using a four-state receptor model:

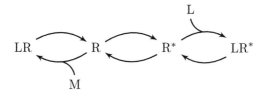

In the above scheme, R and R* are inactive and active forms of the receptor. L is the agonist ligand that binds to the active form of the receptor. M is an *inverse* agonist that binds to the inactive form. The equilibrium relations are

$$\frac{[M]_{ss}[R]_{ss}}{[MR]_{ss}} = K_M \qquad \frac{[R^*]_{ss}}{[R]_{ss}} = K_R \qquad \frac{[L]_{ss}[R^*]_{ss}}{[LR^*]_{ss}} = K_L, \tag{4.44}$$

where K_R is dimensionless and $K_M \doteq K_L \doteq$ conc. The steady state fraction of activated receptor is,

$$1 \doteq \rho^*_{ss} = \frac{[R^*]_{ss} + [LR^*]_{ss}}{[MR]_{ss} + [R]_{ss} + [R^*]_{ss} + [LR^*]_{ss}} \doteq \frac{conc}{conc}.$$

Using Eq. 4.44 and $r_T \overset{\text{def}}{=} [MR] + [R] + [R^*] + [LR^*]$, find the fraction of activated receptor (ρ^*_{ss}) as a function of $[M]_{ss}$ and $[L]_{ss}$.

Problem 4.22 The reaction scheme for Michaelis-Menton enzyme kinetics is

$$S + E \underset{k_1^-}{\overset{k_1^+}{\rightleftharpoons}} ES \overset{k_2}{\rightarrow} E + P,$$

where S is the substrate, E is the enzyme, ES is the enzyme-substrate complex, and P is the product. (a) Assuming mass action kinetics, write the ODEs for $d[S]/dt$, $d[E]/dt$, $d[C]/dt$ and $d[P]/dt$. (b) Find a conserved sum of concentrations. (c) Assuming rapid equilibrium between enzyme and substrate, $k_1^+[S][C] = k_1^-[ES]$, we may write

$$[S][E]/[ES] = K_1 \quad \text{where} \quad K_1 = k_1^-/k_1^+ \tag{4.45}$$

is the dissociation constant. Use your answer from (b) to express [ES] as a function of a conserved quantity, K_1, and [S]. Your answer will be a hyperbolic Hill function $(n - 1)$.

Solutions

Problem 4.1 In $j_{out} = kc$, $k \doteq$ time^{-1} is a first-order rate constant. The influx rate constant $j \doteq$ conc/time is zeroth order.

Problem 4.4 $K_{eq} = 1$.

Problem 4.5 $k_+ \doteq$ conc^{-1}time^{-1} (second-order) and $k_- \doteq$ time^{-1} (first-order). The units would be $\mu M^{-1} s^{-1}$ (per micromolar per second) and s^{-1}.

Problem 4.8 The physical dimensions of the association rate constants is $k_a^+ \doteq$ conc^{-1} time^{-1}. The rate constants k_a^-, k_b^+ and k_b^- are unimolecular (physical dimensions of time^{-1}). The equilibrium relation is

$$[LR^*]_{ss} = \frac{(K_b^{eq}[L]_{ss}/K_a)r_T}{1 + [L]_{ss}/K_a + K_b^{eq}[L]_{ss}/K_a}$$

where $r_T = [L] + [LR] + [LR^*]$ is conserved.

Problem 4.9 Reactant is favored, $G_{reactant} < G_{product}$ and $\Delta G > 0$.

Problem 4.10 $\alpha \doteq \beta \doteq$ time^{-1}, $K_{eq} \doteq 1$ (dimensionless) and $\Delta G \doteq k_B T \doteq$ energy.

Problem 4.11 Differentiating gives $d\ell_T/dt = d[L]/dt + d[LR]/dt = -j_+ + j_- + j_+ - j_- = 0$ and similarly for dr_T/dt.

Problem 4.16 Setting $[L]_{ss} = K_a$ in Eq. 4.36a,

$$[R]_{ss} \overset{?}{=} [LR]_{ss}$$

$$\frac{K_a K_b}{K_a K_b + [L]_{ss} K_b + [L]_{ss}^2} \overset{?}{=} \frac{[L]_{ss} K_b}{K_a K_b + [L]_{ss} K_b + [L]_{ss}^2}$$

$$\frac{K_a K_b}{K_a K_b + K_a K_b + K_a^2} \overset{\text{yes}}{=} \frac{K_a K_b}{K_a K_b + K_a K_b + K_a^2} \cdot \checkmark$$

Problem 4.21 Using Eq. 4.44, write r_T as a function of $[R^*]$,

$$r_T = [M]_{ss}[R^*]_{ss}/(K_M K_R) + [R^*]_{ss}/K_R + [R^*]_{ss} + [L]_{ss}[R^*]_{ss}/K_L ,$$

we obtain

$$[R^*] = r_T/ \left([M]_{ss}/(K_M K_R) + 1/K_R + 1 + [L]_{ss}/K_L\right) .$$

Similar calculations for $[MR]_{ss}$, $[R^*]_{ss}$ and $[LR^*]_{ss}$ lead to

$$\rho_{ss}^* = \frac{1 + [L]_{ss}/K_L}{\left(1 + [M]_{ss}/K_M\right)/K_R + 1 + [L]_{ss}/K_L}. \tag{4.46}$$

Observe that a high concentration of agonist (L) increases the activated fraction of receptor (eventually to 100%), that is, $\rho_{ss}^* \to 1$ as $[L]_{ss} \to \infty$. High concentration of inverse agonist (M) drives the activated fraction of receptor to zero, $\rho_{ss}^* \to 0$ as $[M]_{ss} \to \infty$. The concentration of inverse agonist M changes the EC_{50} for the agonist L, which is the value of $[L]_{ss}$ leading to $\rho_{ss}^* = 1/2$ and solving

$$[L]_{ss}/K_L + 1 = 1/K_R \left(1 + [M]_{ss}/K_M\right) .$$

Problem 4.22 (a) The governing ODEs are:

$$\frac{d[S]}{dt} = -k_1^+ [S][E] + k_1^- [ES] \tag{4.47a}$$

$$\frac{d[E]}{dt} = -k_1^+ [S][E] + k_1^- [ES] \tag{4.47b}$$

$$\frac{d[ES]}{dt} = k_1^+ [S][E] - k_1^- [ES] - k_2[ES] \tag{4.47c}$$

$$\frac{d[P]}{dt} = k_2[ES] . \tag{4.47d}$$

(b) The one conserved quantity is

$$[E]_T \overset{\text{def}}{=} [E] + [ES] = \text{constant}. \tag{4.48}$$

The reader may confirm that $d[E]_T/dt = 0$ by summing Eqs. 4.47b and 4.47c. (c) Solving Eqs. 4.45 and 4.48 simultaneously, we obtain

$$[ES] = \frac{[S][E]_T}{[S] + K_1} .$$

Compare Eq. 4.41 using $n = 1$, $x \leftrightarrow [S]$, $\kappa \leftrightarrow K_1$ and $v \leftrightarrow k_2[E]_T$. Under this rapid equilibrium assumption,[8] the rate of formation of product is

$$\frac{d[P]}{dt} = k_2[ES] = \frac{k_2[S][E]_T}{[S] + K_1} .$$

Notes

1. A chemical "equation" such as A + B → C is *elementary* when the molecules of chemical species A and B may interact to produce the chemical species C. This may occur either through association and binding, yielding an A-B complex denoted by C, or through a chemical change that occurs when A and B collide in solution. In either case, the reaction is elementary and mass action kinetics is a good candidate for modeling the chemical flux (rate of loss of A and B and rate of gain of C). Chemical equations may also represent the *overall stoichiometry* of a chemical process that may involve multiple steps, e.g., for photosynthesis,

   $$6CO_2 + 6H_2O \rightarrow C_6H_{12}O_6 + 6O_2 .$$

   Chemical equations of this type are not meant to suggest anything about how the net reaction rate depends on reactant concentration.

2. In Eqs. 4.6a–4.6c, $k_1 \ominus conc^{-2} time^{-1}$ is a third-order rate constant,

   $$\frac{conc}{time} \ominus \frac{d[A]}{dt} \ominus j_1 = k_1 [A]^2[B] \ominus \frac{1}{conc^2 \, time} \cdot conc^2 \cdot conc .$$

3. Steady states of chemical reactions do not always correspond to equilibrium.

4. The term hyperbolic is discussed further on p. 82.

5. See p. 84 for another biphasic function.

6. The Hill function $h(x) = vx^n/(\kappa^n + x^n)$ is increasing; $g(x) = v\kappa^n/(\kappa^n + x^n) = 1 - h(x)$ is a *decreasing* Hill function.

7. Students of linear algebra will appreciate that Eqs. 4.6 and 4.42 can be written

   $$\frac{d}{dt}\begin{pmatrix} [A] \\ [B] \\ [C] \end{pmatrix} = \underbrace{\begin{pmatrix} -1 & 1 \\ -1 & 3 \\ 1 & -2 \end{pmatrix}}_{S} \begin{pmatrix} j_1 \\ j_2 \end{pmatrix} \quad \text{and} \quad \underbrace{\begin{pmatrix} -1 & -1 & 1 \\ 1 & 3 & -2 \end{pmatrix}}_{S^T} \begin{pmatrix} a \\ b \\ c \end{pmatrix} = \begin{pmatrix} 0 \\ 0 \end{pmatrix} .$$

   The stoichiometric coefficient vector $(a\ b\ c)^T$ is in the *nullspace* of S^T, the transpose of the stoichiometric matrix S. Elementary row operations yield

   $$\begin{pmatrix} 1 & 1 & -1 \\ 1 & 3 & -2 \end{pmatrix} \rightarrow \begin{pmatrix} 1 & 1 & -1 \\ 0 & 2 & -1 \end{pmatrix} \rightarrow \begin{pmatrix} 1 & 1 & -1 \\ 0 & 1 & -1/2 \end{pmatrix} \rightarrow \begin{pmatrix} 1 & 0 & -1/2 \\ 0 & 1 & -1/2 \end{pmatrix} ,$$

   so a and b are leading variables and c is a free variable that we may set to an arbitrary value (say, $c = \theta$). The second row gives $b - c/2 = 0 \Rightarrow b = c/2 = \theta/2$; the first row means $a - c/2 = 0 \Rightarrow a = c/2 = \theta/2$. Thus, any vector of the form $(a\ b\ c)^T = (1/2\ 1/2\ 1)^T\theta$ is in the nullspace of S^T. $\theta = 2$ leads to the integer values $a = 1$, $b = 1$ and $c = 2$.

8. The rate of formation of product may also be analyzed via a quasi-static assumption ($d[ES]/dt \approx 0$). This gives a hyperbolic Hill function with a different half maximum: $k_2[S][E]_T/([S] + K_m)$ where $K_m = (k_1^- + k_2)/k_1^+ = K_1 + k_2/k_1^+$. These approximations agree when $k_2 \ll k_1^+$.

5 Function Families and Characteristic Times

Cellular responses to antagonist drugs (blockers) and the membrane voltage response of neurons stimulated by the excitatory neurotransmitter glutamate provide case studies for qualitative analysis of functions. We also introduce an important concept in dynamics: the characteristic time for changes in a function.

5.1 Functions and Relations

A real-valued function $y = f(x)$ must associate exactly one output y to each input x in its domain. The set of points (x, y) in the Cartesian plane that satisfy $x^2 + xy + y^2 = 1$ is an algebraic curve (an ellipse, below left),[1] but not a function, because, e.g., both $y = 1$ and $y = -1$ are associated to $x = 0$.

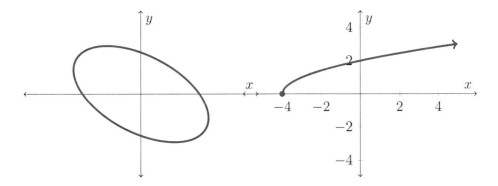

On the other hand, $y = \sqrt{x + 4}$ (above right) is a function with natural domain $-4 \leq x < \infty$ and range $0 \leq y < \infty$, because, when $x + 4$ is nonnegative, $\sqrt{x + 4}$ has a unique nonnegative principal square root. The **natural domain** of a real-valued function is the subset of the real number line for which the function is defined.

5.2   Scaling and Shifting of Functions

Fig. 5.1 shows that the equation $y = x/(x+2)$ defines a hyperbola whose asymptotes are the lines $x = -2$ and $y = 1$. With the exception of $x = -2$, where the function is not defined, every input x results in a single output $y = f(x) = x/(x+2)$. The gray region of Fig. 5.1 (left) shows how the positive branch of the hyperbola may be used to define a function with **domain of definition** given by $0 \le x < \infty$ and range $0 \le y < 1$.

Fig. 5.2 shows several hyperbolic functions that are quantitatively different with regard to their horizontal asymptotes, $\lim_{x\to\infty} f(x)$, and slopes near the origin, $\lim_{x\downarrow 0} f'(x)$. On the other hand, all eight functions are monotone increasing and concave.[2] Each can be put in the form

$$y = f(x; v, \kappa) = \frac{vx}{x+\kappa} \qquad x \ge 0 \qquad v, \kappa > 0 \tag{5.1}$$

with an appropriate choice for v and κ. For example, by dividing the numerator and denominator of $y = 10x/(2x+1)$ by 2 we obtain $y = 5x/(x+1/2)$, so in this case $v = 5$

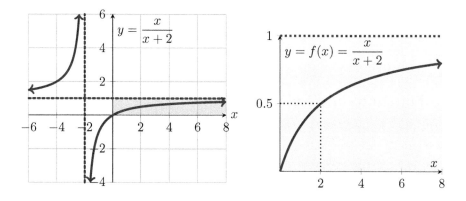

Figure 5.1   A plot of $y = x/(x+2)$ in the Cartesian plane is a hyperbola that may be used to define a function $y = f(x)$ on the domain $0 \le x < \infty$.

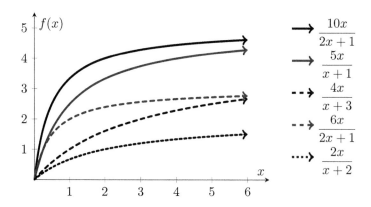

Figure 5.2   Hyperbolic functions with domain of definition $0 \le x \le \infty$ that are equivalent up to scaling of the coordinate axes.

and $\kappa = 1/2$. Because we are free to choose both v and κ, Eq. 5.1 is a *two-parameter family* of functions.

The hyperbolic functions of Fig. 5.2 are all **equivalent up to scaling** of the coordinate axes. To see this, divide both sides of Eq. 5.1 by v and, subsequently, divide the numerator and denominator on the right hand side by κ,

$$y = \frac{vx}{x+\kappa} \iff y/v = \frac{x}{x+\kappa} \iff y/v = \frac{x/\kappa}{x/\kappa + 1}.$$

The rightmost equation makes it clear that we may transform the vertical and horizontal axes using $\hat{y} = y/v$ and $\hat{x} = x/\kappa$ to obtain

$$\hat{y} = \frac{\hat{x}}{\hat{x}+1}. \tag{5.2}$$

For this reason, the parameters v and κ are referred to as **scale parameters**.

Problem 5.1 The binding curve $[LR]_{ss} = [L]_{ss}r_T/(K + [L]_{ss})$ shown in Fig. 4.6 (left) is an example of a hyperbolic function (input = $[L]_{ss}$, output = $[LR]_{ss}$). What is this function's domain of definition and range?

Problem 5.2 Show that the three-parameter family of functions

$$y = f(x; \alpha, \beta, \gamma) = \frac{\alpha x}{\beta x + \gamma} \qquad x \geq 0 \qquad \alpha, \beta, \gamma > 0 \tag{5.3}$$

is equivalent to the two-parameter family given by Eq. 5.1.

Fig. 5.3 shows two Gaussian (bell-shaped) curves with natural domain given by the entire real number line. The curves are Gaussians of the form $y = \exp[-(x - \beta)^2]$ where exp denotes the exponential function. The parameter β is positive for the blue curve (rightward shift) and negative for the black curve (leftward shift). The two

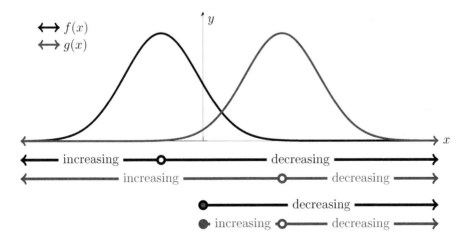

Figure 5.3 Two shifted Gaussian functions that are qualitatively similar on their natural domain (the entire real number line). On the domain of definition $0 \leq x < \infty$, the functions are qualitatively different: $f(x)$ is decreasing, $g(x)$ is biphasic (increasing then decreasing).

curves are not equivalent up to scaling of the coordinate axes, but we may shift the x-axis using $\hat{x} = x - \beta$ to obtain a Gaussian with no free parameters: $y = \exp(-\hat{x}^2)$. The two curves are **equivalent up to shifting** of the horizontal axis.[3] On the other hand, these Gaussian curves may be used to define functions on a restricted semi-infinite domain $0 \leq x < \infty$. On this domain of definition, the two functions are qualitatively different; for example, $f(x)$ is a decreasing function for $x \geq 0$, while $g(x)$ is biphasic, first increasing and then decreasing (Fig. 5.3).

5.3   Qualitative Analysis of Functions

Functions that are both continuous and differentiable may be characterized as increasing, decreasing, nondecreasing, nonincreasing, convex, concave, and so on. The presence or absence of inflection points is another important qualitative feature of a function, as are the number of local maxima and minima. Let us agree that *functions are qualitatively similar on a specified domain when the functions' sign (negative, zero, positive) and the sign of the functions' first and second derivatives agree on the sequence of open intervals whose boundaries are given by sign changes in f, f' and f''.*[4] That is a mouthful, but the idea is straightforward to apply in practice, as we will illustrate in several different physiological contexts.

Antagonist Drug Response

The blue and black curves of Fig. 5.4 represent typical *decreasing* cellular responses, as might be observed when a membrane receptor is inhibited by an **antagonist** drug

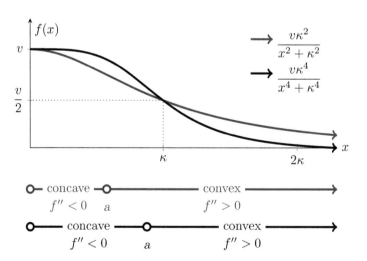

Figure 5.4   In pharmacological research, antagonist inhibition of a membrane receptor yields decreasing responses (x = antagonist concentration, $f(x)$ = magnitude of response). The parameter κ is the half maximal inhibitory concentration (IC$_{50}$), a measure of antagonist drug potency.

(a blocker). Denoting the antagonist concentration as x and the magnitude of response as $y = f(x)$, the two functions take the form

$$f(x; v, \kappa, n) = \frac{v\kappa^n}{x^n + \kappa^n} \qquad x \geq 0 \tag{5.4}$$

with parameters $v, \kappa > 0$ and $n = 2$ (blue) or 4 (black). Both functions achieve their maximum value v in the absence of antagonist ($x = 0$). The parameter κ is the half maximal inhibitory concentration (IC_{50}), that is, the dose leading to 50% inhibition: $f(x = \kappa) = v/2$. IC_{50} is commonly used as a measure of antagonist drug potency in pharmacological research. The previously discussed half maximal effective concentration (EC_{50}) is the analogous concept for agonist drugs (see Eq. 4.28 and nearby text).

The blue function $f(x; n = 2) = v\kappa^2/(x^2 + \kappa^2)$ and the black function $f(x; n = 4) = v\kappa^4/(x^4 + \kappa^4)$ cannot be made to coincide by scaling or shifting the coordinate axes, and yet the two functions are qualitatively similar in many ways, e.g., the functions are both positive ($f > 0$) and both functions achieve their maximum at $x = 0$. Taking the first derivative shows that both functions are also monotone decreasing ($f' < 0$ for $x > 0$).

The situation with regard to the second derivative is slightly more complicated. Both functions have a single inflection point where the second derivative is zero. Although the inflection points have different quantitative locations, the second derivative of both functions is negative ($f'' < 0$) for $0 < x < a$ where a is the inflection point that solves $f''(x = a) = 0$ (the functions are concave on this interval). For $x > a$ the second derivatives are positive ($f'' > 0$) and the functions are convex. One may summarize the qualitative agreement of these functions using a table:

$$
\begin{array}{cccc}
[0] & (0,a) & [a] & (a,\infty) \\
f \quad + & + & + & + \\
f' \quad 0 & - & - & - \\
f'' \quad - & - & 0 & + \\
\end{array}
\tag{5.5}
$$

where the first row uses **interval notation** (see Table 5.1) for connected subsets of the real number line: $[0] = \{0\}$, $(0,a) = \{x \in \mathbb{R} : 0 < x < a\}$, $[a] = \{a\}$ and $(a,\infty) = \{x \in \mathbb{R} : a < x < \infty\}$.[5] The location of the inflection point (denoted by a) is a different value for each function, and the derivatives at $[0]$ are evaluated from the right (one-sided limits with x decreasing to 0).

If one ignores the singleton intervals $[0]$ and $[a]$ that contain one point and have no length,[6] the qualitative analysis can easily be done by eye. We might write an abbreviated table:

$$
\begin{array}{ccc}
& (0,a) & (a,\infty) \\
f & + & + \\
f' & - & - \\
f'' & - & + \\
\end{array}
$$

In words: "The functions $f(x; n = 2) = v\kappa^2/(x^2 + \kappa^2)$ and $f(x; n = 4) = v\kappa^4/(x^4 + \kappa^4)$ are qualitatively similar up to the second derivative. Both functions are positive and decreasing. In the first phase, both functions are concave. In the second phase, both are convex."

Table 5.1   Intervals are connected subsets of the real numbers $\mathbb{R}$

Interval	Set builder notation	Number line
$[a]$	$\{a\},\ a \in \mathbb{R}$	
$[a, b]$	$\{x \in \mathbb{R} : a \leq x \leq b\}$	
$[a, b)$	$\{x \in \mathbb{R} : a \leq x < b\}$	
$(a, b]$	$\{x \in \mathbb{R} : a < x \leq b\}$	
(a, b)	$\{x \in \mathbb{R} : a < x < b\}$	
$(-\infty, b]$	$\{x \in \mathbb{R} : -\infty < x \leq b\}$	
$(-\infty, b)$	$\{x \in \mathbb{R} : -\infty < x < b\}$	
$[a, \infty)$	$\{x \in \mathbb{R} : a \leq x < \infty\}$	
(a, ∞)	$\{x \in \mathbb{R} : a < x < \infty\}$	
$(-\infty, \infty)$	$\mathbb{R}$	

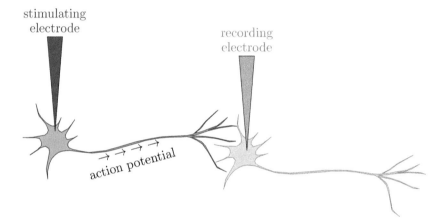

Figure 5.5   Diagram of an experiment involving a stimulating electrode and a recording electrode positioned so that they penetrate the soma of two synaptically connected neurons.

Ionotropic and Metabotropic Responses

Fig. 5.5 shows the configuration of an experiment involving a stimulating electrode and a recording electrode that penetrate the soma of two different neurons. The neurons are synaptically connected (see Fig. 4.4), so when the presynaptic neuron is stimulated, glutamate is released into synaptic cleft(s) and binds ligand-gated receptors. Activated postsynaptic receptors may recruit ionic currents of the postsynaptic membrane leading to a cellular response known as an **excitatory postsynaptic potential** (EPSP).[7] The EPSPs may be fast (10 ms) or slow (50–200 ms) depending on whether the postsynaptic glutamate receptors are ionotropic or metabotropic (see Fig. 5.6).

   Ionotropic neurotransmitter receptors are ligand-gated ion channels that open in response to the binding of neurotransmitter. The resulting membrane current causes a fast transient increase in membrane voltage (about 10 ms in duration). The blue curve of Fig. 5.7 shows a fast postsynaptic response characteristic of AMPA-type[8] postsynaptic glutamate ionotropic receptors activated by binding of ligand (glutamate).

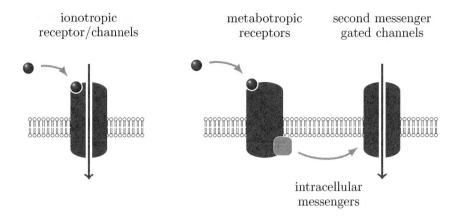

ionotropic
receptor/channels

metabotropic
receptors

second messenger
gated channels

intracellular
messengers

Figure 5.6   Postsynaptic receptors can be divided into two broad classes. Ionotropic receptors are ligand-gated receptors and also ion channels. Metabotropic receptors are not ion channels, but are coupled to their associated channels by a second messenger cascade.

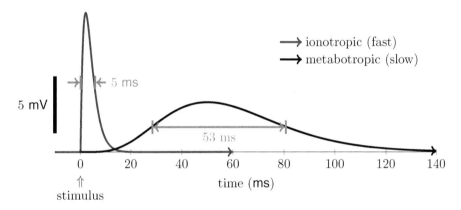

Figure 5.7   The full width at half maximum (FWHM) is an intuitive measure of the duration of a response.

Metabotropic receptors are membrane proteins, but not ion channels. Similar to other members of the group C family of G-protein-coupled receptors (GPCRs), metabotropic glutamate receptors (mGluRs) act through intracellular second messengers to regulate ion channels. The postsynaptic responses resulting from activation of metabotropic receptors (e.g., mGluR$_1$ or mGluR$_5$)[9] can have duration of hundreds of milliseconds or more (Fig. 5.7, black curve) and there may be considerable delay between activation of mGluRs and changes in membrane voltage.

The qualitative differences between ionotropic and metabotropic EPSPs are highlighted in Fig. 5.8 by rescaling the coordinate axes so that the maxima coincide. Both functions are positive and biphasic (increasing then decreasing) with a single local maximum. However, the ionotropic response begins with a positive derivative while the metabotropic response does not.

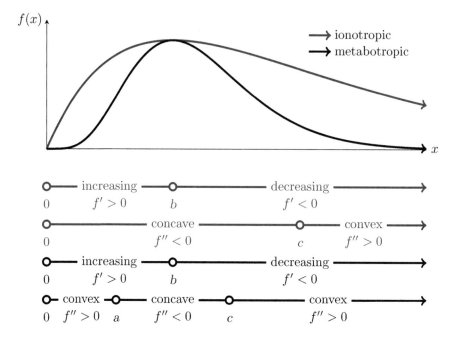

Figure 5.8   Responses of Fig. 5.6 with coordinate axes scaled so that the maxima coincide. The metabotropic response (black) has two inflection points, while the ionotropic response (blue) has only one.

	metabotropic response				ionotropic response		
	$(0,a)$	(a,b)	(b,c)	(c,∞)	$(0,b)$	(b,c)	(c,∞)
f	+	+	+	+	+	+	+
f'	+	+	−	−	+	−	−
f''	+	−	−	+	−	−	+

The above analysis shows that the responses are qualitatively similar during the final two intervals, (b,c) and (c,∞), but distinct during the first few intervals, $(0,a)$ and (a,b) versus $(0,b)$.[10] The first nonsingleton interval of the metabotropic response is convex ($f'' > 0$); conversely, the first nonsingleton interval of the ionotropic response is concave ($f'' < 0$). The metabotropic response also has one more inflection point than the ionotropic response. For these reasons we conclude that the ionotropic and metabotropic postsynaptic responses are *not* qualitatively similar.

5.4   Characteristic Times

The temporal scale of physiological responses can be understood through the concept of **characteristic time**.[11] To analytically calculate (or visually estimate) the characteristic time of a signal $s(t)$ on a domain of interest, one identifies (1) the difference between the minimum and maximum[12] of the signal on this interval, $\max s - \min s$, and (2) the absolute rate of change of the signal, $\max |s'|$. The characteristic time (denoted τ_{char}) is the ratio of these two quantities,

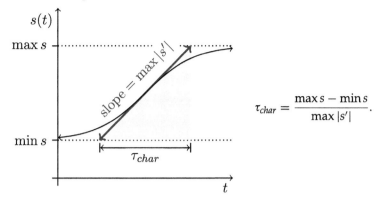

$$\tau_{char} = \frac{\max s - \min s}{\max |s'|}.$$

Thus, the characteristic time of a signal is the amount of time required to traverse the range of the signal (on a domain of interest) when moving at the maximum rate of change.

Fig. 5.9 (top) reproduces the ionotropic response and illustrates how one would visually estimate its characteristic time. The maximum rate of change is $\lim_{t \downarrow 0} s'(t)$

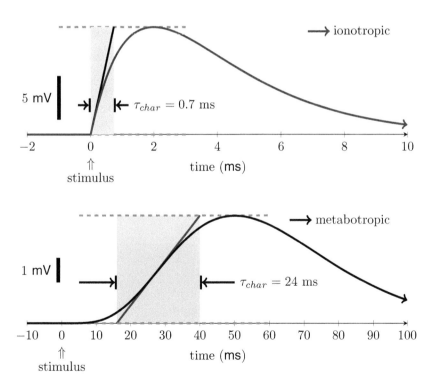

Figure 5.9 Visual estimation of the characteristic time – a measure of the time scale on which changes of a response occur – for the ionotropic and metabotropic responses (top and bottom, respectively).

(black line). The characteristic time of 0.7 ms is smaller than the FWHM of 5 ms (Fig. 5.7), because the former is focused on the fast rising phase of the ionotropic response. For the metabotropic response (Fig. 5.9, bottom), the maximum rate of change occurs at the first inflection point (intersection of blue line and black curve). As expected, the characteristic time of 24 ms for the metabotropic response is much larger than the 7 ms characteristic time of the ionotropic response.

> **Problem 5.3**   The characteristic time is well defined for any function that is continuous, differentiable and bounded. What are the properties required for a function to have a well-defined full width at half maximum (FWHM)?

5.5   Discussion

Time Scales of Cellular Biophysics and Neuroscience

The dynamic phenomena of cellular biophysics and neuroscience occur on widely different time scales. Similarly, the length scales of relevant anatomical structures occupy several different orders of magnitude. Fig. 5.10 represents a range of physiological processes and their characteristic times. Is 10^{-4} to 10^6 seconds the full range of time scales relevant to neuroscience? Sketch an analogous diagram for the length scales relevant to neuroscience.

The Characteristic Time is not the Period

Fig. 5.11 shows a schematic electrocardiogram, a noninvasive measurement of the electrical activity of the heart using electrodes attached to the surface of the chest. One cycle of a typical electrocardiogram (ECG) consists of a P wave, a QRS complex, a T wave, and a U wave (not shown). The characteristic time for changes in the ECG is about half of the width of the QRS complex (60–120 ms), much shorter than the period of the cardiac cycle, which is 0.5–1.0 seconds corresponding to a heart rate of

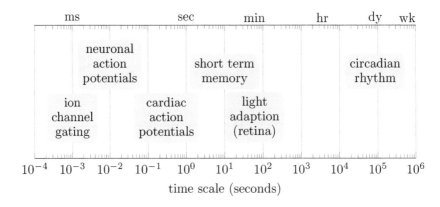

Figure 5.10   Time scales of cellular biophysics and neuroscience.

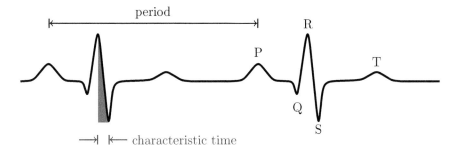

Figure 5.11   The characteristic time for changes in an electrocardiogram is about half the width of the QRS complex, much shorter than the period of the cardiac cycle. *Question*: What are the physical dimensions of this plot?

60–120 beats per minute. The time scale on which a signal is repeating may also be characterized by the natural frequency

$$[\text{natural frequency}] = \frac{1}{[\text{period}]}.$$

The characteristic time of a repeating signal is not to be confused with the signal's period.

An important technical aspect of recording biological signals is the **sampling rate** given by the reciprocal of the time between successive measurements,

$$[\text{sampling rate}] = \frac{1}{[\text{time interval between measurements}]}.$$

If the goal of an experimental recording is to generate information regarding the important features of a biological signal, should the time between observations be shorter or longer than τ_{char}? What is a reasonable sampling rate for the electrocardiogram of Fig. 5.11? What is a reasonable sampling rate for blood samples measuring hormone concentration changes associated with the circadian rhythm?

Estimation of Characteristic Times

Fig. 5.12 may be used to practice visually estimating characteristic times. Remember that the characteristic time is sensitive to the most rapidly changing epochs of the signal; consequently, it need not correspond to a measure of response duration such as FWHM.

The signals plotted in Fig. 5.12 are **rational functions**—the quotient of two polynomials—that take the form

$$y = \frac{p(x)}{q(x)} = \frac{p_3 x^3 + p_2 x^2 + p_1 x + p_0}{x^4 + q_3 x^3 + q_2 x^2 + q_1 x + q_0},$$

where $q_4 = 1$ so that $q(x)$ is a polynomial of higher degree than $p(x)$. What qualitative property of these functions is ensured by this choice? What is required of $q(x)$ for the rational function $f(x)$ to be continuous and differentiable?

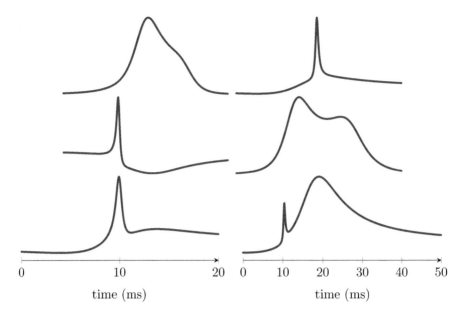

Figure 5.12   Estimate the characteristic time for changes of these functions.

How to Determine Equivalence up to Scaling

How does one determine if two functions are equivalent up to scaling of the coordinate axes? If scale parameters are not known in advance, one approach is to define a scaling of the coordinate axes such as $\hat{x} = x/\phi$ and $\hat{y} = y/\psi$ where ϕ and ψ are positive unspecified constants. Rescaling $g(x)$ to obtain $f(x)$ is possible if there are values for ϕ and ψ such that $y = f(x)$ is the same function as $\hat{y} = g(\hat{x})$. Because

$$\hat{y} = g(\hat{x}) \;\Leftrightarrow\; y/\psi = g(x/\phi) \;\Leftrightarrow\; y = \psi g(x/\phi),$$

$y = f(x)$ is the same function as $\hat{y} = g(\hat{x})$ if and only if ϕ and ψ can be chosen so that

$$f(x) = \psi g(x/\phi).$$

For example, suppose we wished to check if $y = f(x) = 3x^2 - 5$ is equivalent up to scaling of the coordinate axes to $y = g(x) = x^2 - 1$. If so, we should be able to choose ϕ and ψ such that

$$\underbrace{3x^2 - 5}_{f(x)} \overset{?}{=} \underbrace{\psi\left[(x/\phi)^2 - 1\right]}_{\psi g(x/\phi)} = (\psi/\phi^2)x^2 - \psi.$$

For the two quadratics to agree for all x, they must agree term by term.[13] Equating the coefficients gives

$$3 = \psi/\phi^2 \quad -5 = -\psi.$$

Simultaneously solving the above equations for ϕ and ψ gives $\psi = 5$ and

$$3 = 5/\phi^2 \;\Leftrightarrow\; \phi^2 = 5/3 \;\Leftrightarrow\; \phi = \sqrt{5/3}.$$

Thus, we find that the scaling $\phi = \sqrt{5/3}$ and $\psi = 5$ transforms $y = f(x)$ to $\hat{y} = g(\hat{x})$. To confirm this, begin with $y = f(x) = 3x^2 - 5$ and rescale using $\hat{x} = x/\sqrt{5/3}$ and $\hat{y} = y/5$ to find

$$5\hat{y} = 3\left(\sqrt{\frac{5}{3}}\hat{x}\right)^2 - 5 \Rightarrow \hat{y} = \frac{3}{5}\left(\sqrt{\frac{5}{3}}\hat{x}\right)^2 - 1 \Rightarrow \hat{y} = \frac{3}{5}\frac{5}{3}\hat{x}^2 - 1$$

where the last expression implies $\hat{y} = g(\hat{x}) = \hat{x}^2 - 1$.

Supplemental Problems

Problem 5.4 For each of the functions below, (i) sketch a graph, (ii) indicate the location of parameters on the s or t axis as appropriate, (iii) check your sketch using software, (iv) analytically calculate the characteristic time of the function, and (v) indicate τ_{char} on the graph. You may assume α and β are positive ($\alpha, \beta > 0$).

(a) $s(t; \alpha, \beta) = ve^{-kt}$ $t \geq 0$ (b) $s(t; \alpha, \beta) = \dfrac{v}{\sqrt{2\pi k}}e^{-t^2/2k^2}$

Problem 5.5 Show that $y = 8x^2 + 2x - 7$ is equivalent up to scaling and translating to $y = x^2 + x + 1$.

Problem 5.6 Functions that are symmetric about the y-axis are called even functions and have the property $f(-x) = f(x)$. Conversely, odd functions satisfy $f(-x) = -f(x)$ and are symmetric about the origin (reflected across both x- and y-axes). Assuming a and b are nonzero, classify the following functions as even, odd or neither.

(a) $y = x$ (d) $y = ax^2 + b$ (g) $y = \dfrac{1}{x^4 + 1}$
(b) $y = x + a$ (e) $y = x^{-2}$ (h) $y = ax^3 - bx$
(c) $y = 1/x$ (f) $y = x^3$ (i) $y = ax^3 - bx^2$

Problem 5.7 Consider the oscillatory function $x(t) = a\sin(\omega t)$ where $\omega = 2\pi f$ is angular frequency, f is natural frequency, and $T = 1/f$ is the period.

(a)   Show that $\tau_{char} = 2/\omega$.
(b)   Why does τ_{char} not depend on a?
(c)   Why is τ_{char} inversely proportional to ω?
(d)   Show that τ_{char} is proportional to oscillation period T.

Problem 5.8 Perform qualitative analysis of the following functions. Write your answer in tabular form, using numerical values for interval endpoints. Plot each function. Mark up the plot indicating when the function is positive versus negative, increasing versus decreasing, convex versus concave.

(a) $y = x^2 - 9x + 14$ **(b)** $y = x^3 - 12x^2 + 41x - 42$

Problem 5.9 The functions shown in Fig. 5.8 are equivalent up to scaling to $f(x) = xe^{-x}$ and $g(x) = x^5 e^{-x}$. Analytically calculate the first and second derivatives of $f(x)$ and $g(x)$ and plot these functions. Check your result by comparing to the qualitative analysis of Eq. 5.3.

Problem 5.10 Perform qualitative analysis of the binding curves for the sequential binding kinetic scheme (Eq. 4.29, Fig. 4.8).

Problem 5.11 Show that the Hill function of Eq. 4.41 is antisymmetric around the half maximum κ when plotted with a logarithmic x-axis (see Fig. 4.9, right).

Solutions

Figure 5.11 voltage and time.

Problem 5.1 The domain of definition is $[L]_{ss} \geq 0$ (because concentrations must be nonnegative) and the range is $0 \leq [LR]_{ss} \leq r_T$.

Problem 5.2 In general, Eq. 5.1 is obtained from Eq. 5.3 as follows:

$$f(x; \alpha, \beta, \gamma) = \frac{\alpha x}{\beta x + \gamma} = \frac{\alpha/\beta \cdot x}{x + \gamma/\beta} = \frac{vx}{x + \kappa} = f(x; v, \kappa),$$

where $v = \alpha/\beta$ and $\kappa = \gamma/\beta$.

Problem 5.6

(a) odd	**(d)** even	**(g)** even
(b) neither	**(e)** even	**(h)** odd
(c) odd	**(f)** odd	**(i)** neither

Problem 5.8

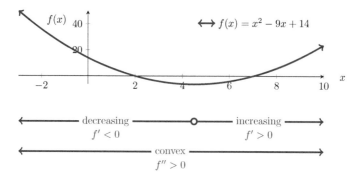

The function $f(x) = x^2 - 9x + 14 = (x - 2)(x - 7)$ has roots $x \in \{2,7\}$, derivative $f'(x) = 2x - 9$. The parabola is convex with minimum at $x = 4.5$.

	$(-\infty, 2)$	$(2, 4.5)$	$(4.5, 7)$	$(7, \infty)$
f	$+$	$-$	$-$	$+$
f'	$-$	$-$	$+$	$+$
f''	$+$	$+$	$+$	$+$

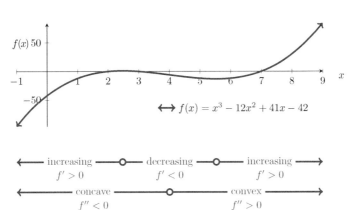

$$\longleftrightarrow f(x) = x^3 - 12x^2 + 41x - 42$$

increasing —○— decreasing —○— increasing
$f' > 0$ $f' < 0$ $f' > 0$

concave —————○————— convex
$f'' < 0$ $f'' > 0$

The cubic function $f(x) - x^3 - 12x^2 + 41x - 42 - (x - 7)(x - 3)(x - 2)$ has roots $x \in \{2,3,7\}$, derivative $f'(x) = 3x^2 - 24x + 41$ with roots $4 - \sqrt{21}/3 \approx 2.47$ and $4 + \sqrt{21}/3 \approx 5.53$. The second derivative $f''(x) = 6x - 24$ implies an inflection point at $x = 4$.

	$(-\infty, 2)$	$(2, 2.47)$	$(2.47, 3)$	$(3, 4)$	$(4, 5.53)$	$(5.53, 7)$	$(7, \infty)$
f	$-$	$+$	$+$	$-$	$-$	$-$	$+$
f'	$+$	$+$	$-$	$-$	$-$	$+$	$+$
f''	$-$	$-$	$-$	$-$	$+$	$+$	$+$

Problem 5.9

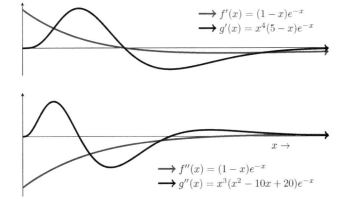

$$\longrightarrow f'(x) = (1 - x)e^{-x}$$
$$\longrightarrow g'(x) = x^4(5 - x)e^{-x}$$

$x \to$

$$\longrightarrow f''(x) = (1 - x)e^{-x}$$
$$\longrightarrow g''(x) = x^3(x^2 - 10x + 20)e^{-x}$$

Problem 5.10   The abbreviated qualitative analyses are:

	$(0, \infty)$		$(0, \cdot)$	$(\cdot, \infty)$
[LR]	$+$	[L$_2$R]	$+$	$+$
[LR]$'$	$-$	[L$_2$R]$'$	$+$	$+$
[LR]$''$	$-$	[L$_2$R]$''$	$+$	$-$

Problem 5.11   The domain of the Hill function is $0 \leq x < \infty$ and the range is $0 \leq h < v$, with $h(0) = 0$ and $h \to v$ as $x \to \infty$. The half maximum of the Hill function defined by $h(x) = v/2$ occurs when $x = \kappa$. The difference of h and $v/2$ on the linear axis is $\hat{h} = h - v/2$ with $-v/2 \leq \hat{h} < v/2$. The difference between x and κ (when measured as a linear distance on a semilog x plot) is $\hat{x} = \ln x - \ln \kappa$ where $x > 0$ and $-\infty < \hat{x} < \infty$. (Note that $\hat{x}$ is not defined when $x = 0$, corresponding to the fact that the $x = 0$ cannot appear on a semilog x plot.) Using $x = \kappa e^{\hat{x}}$ and $\hat{h} = h + v/2$ we write

$$\hat{h}(\hat{x}) = v\left(\frac{e^{2\hat{x}}}{1 + e^{2\hat{x}}} - \frac{1}{2}\right).$$

To show that the Hill function Eq. 4.41 is antisymmetric around the half maximum κ when plotted with a logarithmic x-axis, we must show that $\hat{h}(\hat{x})$ is symmetric about the origin (an odd function), that is,

$$\hat{h}(-\hat{x}) = -\hat{h}(\hat{x}).$$

To check this, we write,

$$v\left(\frac{e^{-2\hat{x}}}{1 + e^{-2\hat{x}}} - \frac{1}{2}\right) \stackrel{?}{=} -v\left(\frac{e^{2\hat{x}}}{1 + e^{2\hat{x}}} - \frac{1}{2}\right)$$

$$\frac{e^{-2\hat{x}}}{1 + e^{-2\hat{x}}} - \frac{1}{2} \stackrel{?}{=} \frac{1}{2} - \frac{e^{2\hat{x}}}{1 + e^{2\hat{x}}}$$

$$\frac{e^{-2\hat{x}}}{1 + e^{-2\hat{x}}} + \frac{e^{2\hat{x}}}{1 + e^{2\hat{x}}} \stackrel{?}{=} 1$$

$$\frac{e^{-2\hat{x}}\left(1 + e^{2\hat{x}}\right) + e^{-2\hat{x}}\left(1 + e^{-2\hat{x}}\right)}{\left(1 + e^{-2\hat{x}}\right)\left(1 + e^{2\hat{x}}\right)} \stackrel{?}{=} 1$$

$$\frac{e^{-2\hat{x}} + 1 + e^{2\hat{x}} + 1}{1 + e^{2\hat{x}} + e^{-2\hat{x}} + 1} \stackrel{\text{yes}}{=} 1. \checkmark$$

Notes

1.   This ellipse has the rational parameterization,

$$x = \frac{1 - t^2}{1 + t + t^2} \qquad y = \frac{t(t + 2)}{1 + t + t^2}.$$

2.   A differentiable function f is concave on an interval if its derivative function f' is monotone nonincreasing on that interval, i.e., a concave

function has a nonincreasing slope. A function f is convex on an interval if f' is monotone nondecreasing on that interval, i.e., a convex function has a nondecreasing slope.

3. Equivalent up to a horizontal translation.

4. We terminate the qualitative analysis at the second derivative so that it is straightforward to perform "by eye." We allow differences on singleton sets so that the method is easily applied to continuous piecewise differentiable functions.

5. For a review of set intervals and point set topology see Sutherland (2009).

6. One property of an interval $[a, b]$ is its length,

$$\text{len}([a, b]) \overset{\text{def}}{=} b - a.$$

This definition implies that a singleton interval $[a]$ has length zero,

$$\text{len}([a]) = \text{len}([a, a]) = a - a = 0.$$

For the length of disjoint intervals such as $[a, b)$ and $[b]$ to sum to $b - a$, we require $\text{len}((a, b)) = \text{len}([a, b)) = \text{len}((a, b]) = \text{len}([a, b])$.

7. The causes of the electronegativity of neurons and modeling of neuronal voltage are the subject of Chapters 7 and 8.

8. AMPA is the compound α-amino-3-hydroxy-5-methyl-4-isoxazolepropionic acid, a specific agonist for this ionotropic membrane receptor ion channel.

9. mGluR$_1$ and mGluR$_5$ are Group I metabotropic glutamate receptors. The compound 3,5-dihydroxyphenylglycine (DHPG) is a potent agonist of Group I mGluRs, but not Group II (mGluR$_2$ and mGluR$_3$) or Group III (mGluR$_4$, mGluR$_6$, mGluR$_7$ and mGluR$_8$) mGluRs.

10. We do not tabulate the sign of f, f', f'' on singleton intervals $[0], [a], [b], [c]$.

11. This definition of characteristic time follows Segel (1984) (pp. 56–57).

12. Strictly speaking, the definition of the characteristic time uses the supremum and infimum, as opposed to maximum and minimum, to allow the asymptotic values $\lim_{t \to \pm\infty} s(t)$ to be included in determining the full range of the signal. For example, the sigmoidal function of Eq. 5.6 has no maximum or minimum value, but does have a greatest lower bound (infimum) and least upper bound (supremum).

13. Because x^2 and 1 are linearly independent.

6 Bifurcation Diagrams of Scalar ODEs

Bifurcation theory is the mathematical analysis of how the number and/or stability of steady state solutions of an ordinary differential equation depends on one or more parameters.

6.1 A Single-Parameter Family of ODEs

Consider the following scalar ordinary differential equation,

$$\frac{dy}{dt} = f(y; \mu) = y^2 - 2y + \mu \qquad y(0) = y_0, \tag{6.1}$$

where μ is a constant but unspecified parameter. Eq. 6.1 is not *one* ordinary differential equation initial value problem, but rather a *single-parameter family* of ODE IVPs. The semicolon in $f(y; \mu)$ calls attention to the fact that μ is neither the dependent nor the independent variable of the ODE.[1]

Fig. 6.1 shows how the phase diagram $dy/dt = f(y; \mu)$ changes with the parameter μ. For any value of μ, the plot of $f(y; \mu)$ as a function of y is a convex parabola with a minimum that occurs at $y = 1$. To see this, take the derivative of the right side of Eq. 6.1 with respect to y to obtain

$$f'(y) = 2y - 2, \tag{6.2}$$

then set $f'(y)$ to zero to obtain $0 = 2y - 2 \Rightarrow y = 1$. The minimum value of $f(y; \mu) = y^2 - 2y + \mu$ depends on μ. It is found by substituting $y = 1$:

$$\min_y f(y; \mu) = f(y = 1; \mu) = 1^2 - 2 \cdot 1 + \mu = -1 + \mu. \tag{6.3}$$

Why leave the parameter μ unspecified? One reason is that analysis of Eq. 6.1 can proceed without choosing a particular value. For example, the steady states of

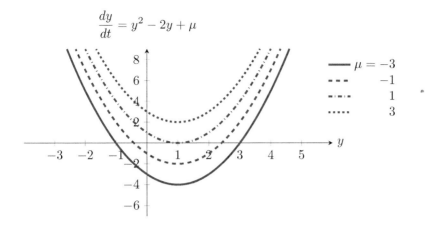

Figure 6.1   Phase diagrams for a single-parameter family of ODEs (Eq. 6.1). *Question*: What is the critical value for the bifurcation parameter μ?

Eq. 6.1 are found by setting the left side to zero and solving for y_{ss}. Using the quadratic formula,[2] we find

$$0 = f(y_{ss}; \mu) = y_{ss}^2 - 2y_{ss} + \mu \Rightarrow y_{ss} = \frac{2 \pm \sqrt{4 - 4\mu}}{2} = 1 \pm \sqrt{1 - \mu}. \qquad (6.4)$$

We interpret $y_{ss} = 1 \pm \sqrt{1 - \mu}$ as follows:

$\mu < 1$:   two real-valued solutions $y_{ss} = 1 - \sqrt{1 - \mu}$ or $y_{ss} = 1 + \sqrt{1 - \mu}$
$\mu = 1$:   one real-valued solution $y_{ss} = 1$
$\mu > 1$:   no real-valued solution.

The stability analysis can be performed by considering each case. When $\mu > 1$, there are no steady states. When $\mu = 1$, the steady state at $y_{ss} = 1$ is unstable because $dy/dt > 0$ for $y \neq 1$. When $\mu < 1$, the lower steady state $y_{ss} = 1 - \sqrt{1 - \mu}$ is stable, because the intersection with the horizontal axis has negative slope. The upper steady state $y_{ss} = 1 + \sqrt{1 - \mu}$ is unstable, because the intersection with the horizontal axis has positive slope. This can also be shown by evaluating $f'(y)$ (Eq. 6.2) at $y_{ss} = 1 \pm \sqrt{1 - \mu}$,

$$f'(1 + \sqrt{1 - \mu}) = 2\sqrt{1 - \mu} > 0 \qquad \text{(unstable)}$$
$$f'(1 - \sqrt{1 - \mu}) = -2\sqrt{1 - \mu} < 0 \qquad \text{(stable)},$$

where we have used $\sqrt{1 - \mu} > 0$ when $\mu < 1$.

Fig. 6.2 shows how the phase line changes with the bifurcation parameter μ. The **critical value** of the bifurcation parameter is $\mu = 1$, because the number of steady states changes at this value of μ.

6.2   Fold Bifurcation

Fig. 6.3 (right) is a **bifurcation diagram** for Eq. 6.1 that plots the relationship between the parameter μ and the steady states y_{ss}. Stable steady states (attractors) are solid lines

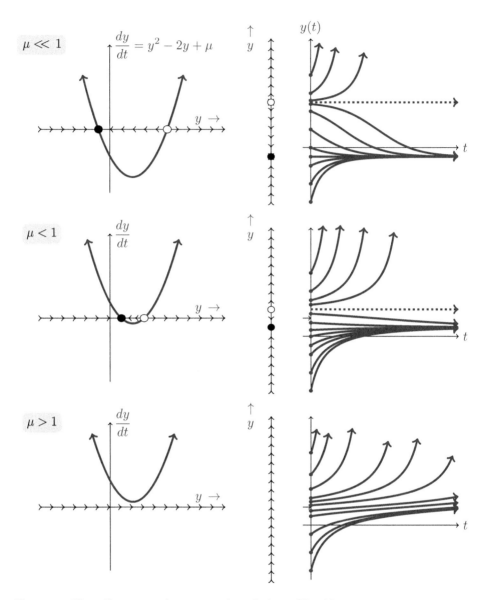

Figure 6.2   Phase diagrams and representative solutions of Eq. 6.1.

and unstable steady states (repellors) are broken lines. As shown in Fig. 6.3 (left), the direction of flow can be easily discerned for any value of the bifurcation parameter. In fact, one may imagine constructing the bifurcation diagram from vertically oriented phase lines of Fig. 6.2, each located at the appropriate value of μ (horizontal axis).

The phenomenon summarized in Fig. 6.3 is a **fold bifurcation** (also called a saddle-node bifurcation).[3] The term fold bifurcation is apt, because in Fig. 6.3 the unstable branch of steady states is folded over the stable branch of steady states. In Figs. 6.2 and 6.3 the distance between the stable and unstable steady states decreases

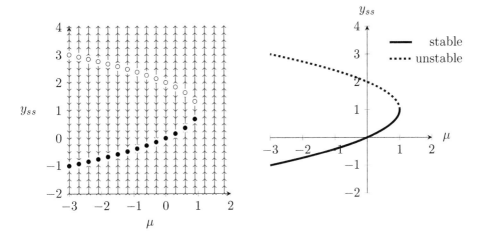

Figure 6.3   Left: Collection of phase lines (blue) for Eq. 6.1 with arrows indicating direction of flow for a given value of μ. When $\mu < 1$ there are two steady states (one stable and the other unstable), when $\mu = 1$ there is one unstable steady state, and when $\mu > 1$ there are no steady states. Right: Bifurcation diagram shows a fold at the critical value $\mu = 1$. *Question:* Which branch of the bifurcation diagram is an attractor?

as the bifurcation parameter μ increases. The steady states coalesce and disappear at the critical value $\mu = 1$ where the fold bifurcation occurs.

6.3   Transcritical Bifurcation

Changes in the number and/or stability of steady states of a scalar ODE may occur as a fold bifurcation or by other means. For example, a **transcritical bifurcation** may be illustrated through the analysis of the following single-parameter family of ODEs,

$$\frac{dy}{dt} = f(y; \mu) = \mu y - y^2 . \tag{6.5}$$

Fig. 6.4 shows several phase diagrams for Eq. 6.5, each using a different value of μ. The single-parameter family of parabolas, $f(y; \mu) = \mu y - y^2$, intersects with the horizontal axis in three different ways depending on the sign of μ. To see this, set the left side of Eq. 6.5 to zero and solve for steady states:

$$0 = f(y_{ss}; \mu) = \mu y_{ss} - y_{ss}^2 = y_{ss}(\mu - y_{ss}) \Rightarrow y_{ss} = 0 \text{ or } y_{ss} = \mu . \tag{6.6}$$

Note that the number of steady states changes from 2 to 1 when the bifurcation parameter μ is at the critical value $\mu = 0$,

$$\mu < 0: \quad y_{ss} = \mu \text{ or } y_{ss} = 0 \quad \text{two solutions}$$
$$\mu = 0: \quad y_{ss} = 0 \quad \text{one solution}$$
$$\mu > 0: \quad y_{ss} = 0 \text{ or } y_{ss} = \mu \quad \text{two solutions.}$$

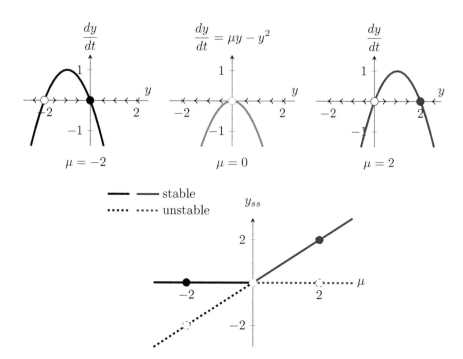

Figure 6.4   Phase diagrams (top) and bifurcation diagram (bottom) for Eq. 6.5 that exhibits a transcritical bifurcation at $\mu = 0$.

To perform stability analysis, differentiate the right side of Eq. 6.5,

$$f'(y) = \mu - 2y,$$
(6.7)

and consider how the sign of the derivatives $f'(y_{ss} = 0) = \mu$ and $f'(y_{ss} = \mu) = \mu - 2\mu = -\mu$ depends on the bifurcation parameter μ,

$$\mu < 0: \quad f'(\mu) > 0 \text{ (unstable) and } f'(0) < 0 \text{ (stable)}$$
$$\mu = 0: \quad f'(0) = 0 \text{ and } f''(0) < 0 \text{ (unstable)}$$
$$\mu > 0: \quad f'(0) > 0 \text{ (unstable) and } f'(\mu) < 0 \text{ (stable)}.$$

We conclude (i) the steady state at $y_{ss} = 0$ is stable when $\mu < 0$ and unstable when $\mu \geq 0$ and (ii) the steady state at $y_{ss} = \mu \neq 0$ is unstable when $\mu < 0$ and stable when $\mu > 0$. The transcritical bifurcation diagram plots the steady states y_{ss} as a function of μ (Fig. 6.4).

6.4   Pitchfork Bifurcations

The **pitchfork bifurcation** is our third example of how the number and stability of steady states in a scalar ODE may change as a parameter is varied. The pitchfork bifurcation will be illustrated through the analysis of

$$\frac{dy}{dt} = f(y; \mu) = \mu y - y^3.$$
(6.8)

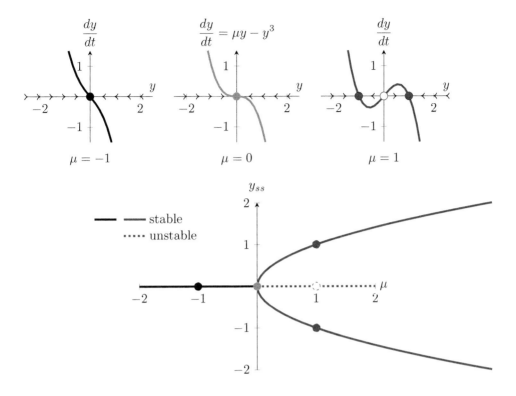

Figure 6.5   Phase diagrams (top) and bifurcation diagram (bottom) for Eq. 6.8 that exhibits a supercritical pitchfork bifurcation at $\mu = 0$.

Fig. 6.5 shows phase diagrams for Eq. 6.8 for different values of μ that cause the single-parameter family of cubics, $f(y; \mu) = \mu y - y^3$, to intersect with the horizontal axis in one of three ways (depending on the sign of μ). The steady states are given by

$$0 = f(y_{ss}; \mu) = \mu y_{ss} - y_{ss}^3 = y_{ss}(\mu - y_{ss}^2) \Rightarrow y_{ss} = 0 \text{ or } y_{ss} = \pm\sqrt{\mu},$$

that is,

$$
\begin{array}{lll}
\mu < 0: & y_{ss} = 0 & \text{one solution} \\
\mu = 0: & y_{ss} = 0 & \text{one solution} \\
\mu > 0: & y_{ss} = -\sqrt{\mu}, 0, \text{ or } \sqrt{\mu} & \text{three solutions.}
\end{array}
$$

For stability analysis we differentiate the right side of Eq. 6.8,

$$f'(y) = \mu - 3y^2, \tag{6.9}$$

and consider the sign of $f'(y_{ss})$ in each case,

$$f'(y_{ss} = 0) = \mu$$
$$f'(y_{ss} = \pm\sqrt{\mu}) = \mu - 3\mu = -2\mu.$$

We conclude (i) the steady state at $y_{ss} = 0$ is stable for $\mu < 0$ and unstable for $\mu > 0$ and (ii) the steady states at $y_{ss} = \pm\sqrt{\mu}$ for $\mu > 0$ are stable. At $y_{ss} = 0$ we have

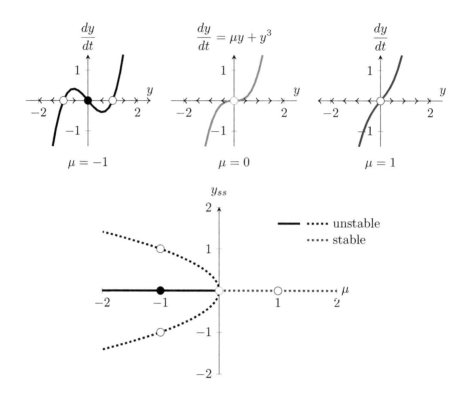

Figure 6.6   Phase diagrams (top) and bifurcation diagram (bottom) for Eq. 6.10 that exhibits a subcritical pitchfork bifurcation at $\mu = 0$.

$f'(y_{ss}) = 0$, which is neither positive (indicating instability) nor negative (indicating stability). However, the second derivative $f''(0) < 0$ and the phase diagram indicate that for $\mu = 0$ the steady state at $y_{ss} = 0$ is stable.

Subcritical Versus Supercritical Pitchfork Bifurcations

Fig. 6.6 shows that the following ODE,

$$\frac{dy}{dt} = f(y; \mu) = \mu y + y^3,\qquad\qquad(6.10)$$

has a pitchfork bifurcation at $\mu = 0$. This version of the pitchfork bifurcation is *subcritical*, while the pitchfork bifurcation of Fig. 6.5 is *supercritical*. The distinction between a supercritical and subcritical bifurcation is not referring to the left-right orientation of the pitchfork; rather, the terms indicate that the multistability occurs before (subcritical) or after (supercritical) the central branch of steady states transitions from stable to unstable as μ increases.

Problem 6.1   Carry out an analytical stability analysis of Eq. 6.10 and compare with the analysis of Eq. 6.8 (above).

6.5 Bifurcation Types and Symmetry

The similarities and differences between Figs. 6.5 and 6.6 can be understood as a consequence of symmetry in the corresponding ODEs (Eqs. 6.8 and 6.10). If one begins with the ODE for the supercritical pitchfork bifurcation ($dy/dt = \mu y - y^3$) and makes the replacement $\hat{\mu} = -\mu$, the ($\hat{\mu}, y_{ss}$)-bifurcation diagram for $dy/dt = -\hat{\mu}y - y^3$ is a reflection across the vertical axis of the (μ, y_{ss})-bifurcation diagram (Fig. 6.7, top row). Though the ODE is modified by redefining the bifurcation parameter, the bifurcation type (supercritical pitchfork) does not change.[4] If one begins with the ODE for the subcritical pitchfork bifurcation ($dy/dt = y^3 - \mu y$) and makes the replacement $\hat{\mu} = -\mu$, a similar reflection occurs (Fig. 6.7, bottom row).

Continuing this reasoning, one might ask what replacement would transform a supercritical pitchfork into a subcritical pitchfork or vice versa (Figs. 6.5 and 6.6). The answer may be surprising. This transformation is achieved by changing the sign of the independent variable of the ODE. This has the effect of reversing the direction of flow and may be interpreted physically as reversing the direction of time. To see this, begin with Eq. 6.8 and make the replacement $\hat{t} = -t$,

$$\frac{dy}{dt} = \mu y - y^3 \;\Rightarrow\; -\frac{dy}{d\hat{t}} = \mu y - y^3 \;\Rightarrow\; \frac{dy}{d\hat{t}} = -\mu y + y^3 .$$

This transformation changes the sign of the right side of the ODE and yields an ODE equivalent to Eq. 6.10. The supercritical and subcritical pitchfork bifurcations are time-reversed versions of one another. When time runs backwards, attractors become repellors and repellors become attractors.

There is no supercritical/subcritical distinction for fold and transcritical bifurcations, because reversing time (and changing stability of the various branches in the bifurcation diagram) is equivalent to a horizontal or vertical reflection (Fig. 6.8). Scaling, translating or reflecting a bifurcation does not change its type.

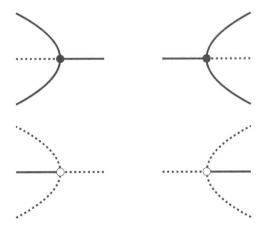

Figure 6.7 Four pitchfork bifurcations. *Question:* Which of these bifurcations are subcritical? supercritical?

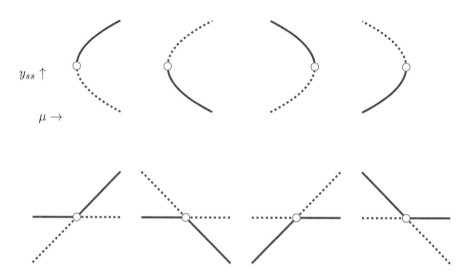

y_{ss} ↑

$\mu \rightarrow$

Figure 6.8   The bifurcation diagrams for $dy/dt = f(y; \mu)$ can be reflected across the vertical or horizontal axis with the replacement $\hat{y} = -y$ or $\hat{\mu} = -\mu$.

6.6   Structural Stability

The fold (saddle-node) is the most important type of bifurcation discussed in this chapter, because the fold bifurcation is **structurally stable**, while the transcritical and pitchfork bifurcations are not.

The structural stability of a bifurcation refers to its robustness to perturbations (small changes in the governing ODE). For example, consider how the bifurcation analysis of Eq. 6.5 (transcritical bifurcation) changes when the bifurcation parameter λ is introduced as an additive constant,

$$\frac{dy}{dt} = f(y; \mu, \lambda) = \lambda + \mu y - y^2. \tag{6.11}$$

When $\lambda = 0$, Eq. 6.11 is identical to Eq. 6.5 and, as discussed above, the ODE has a transcritical bifurcation at $\mu = 0$. However, when $\lambda \neq 0$, the transcritical bifurcation is no longer present. To see this, set the left side of Eq. 6.11 to zero and solve for steady states:

$$0 = f(y_{ss}; \mu, \lambda) = \lambda + \mu y_{ss} - y_{ss}^2 \;\Rightarrow\; y_{ss} = \frac{\mu \pm \sqrt{\mu^2 + 4\lambda}}{2}. \tag{6.12}$$

Considering the square root, we see that real-valued y_{ss} are obtained only when $\mu^2 + 4\lambda \geq 0$, that is, $\mu^2 > -4\lambda$. If λ is positive, then for all μ there are two steady states y_{ss}, one positive and one negative (Fig. 6.9, right). There is no critical value for μ leading to a change in the number or stability of steady states. If λ is negative, then -4λ is positive, there are zero steady states for $\mu = 0$, two steady states for $\mu > \sqrt{-4\lambda}$ or $\mu < -\sqrt{-4\lambda}$, and one steady state for $\mu = \pm\sqrt{-4\lambda}$. As shown in Fig. 6.9 (left), this corresponds to two fold bifurcations. Because the transcritical bifurcation only occurs when $\lambda = 0$, this bifurcation is *structurally unstable*.

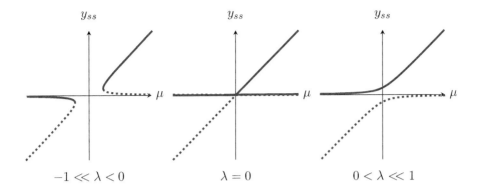

Figure 6.9   The transcritical bifurcation of Eq. 6.11 is structurally unstable.

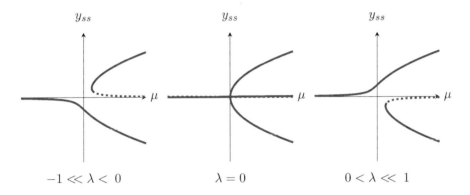

Figure 6.10   The supercritical pitchfork bifurcation of Eq. 6.13 is structurally unstable.

Fig. 6.10 shows that the supercritical pitchfork bifurcation is also structurally unstable. These plots were constructed by introducing a second bifurcation parameter into Eq. 6.8,

$$\frac{dy}{dt} = f(y; \mu, \lambda) = \lambda + \mu y - y^3 \tag{6.13}$$

and noting that steady states satisfy,

$$0 = y_{ss}^3 - \mu y_{ss} - \lambda. \tag{6.14}$$

When $\lambda = 0$, Eq. 6.13 is identical to Eq. 6.8 and a supercritical pitchfork bifurcation is observed at $\mu = 0$ (Fig. 6.10, middle). When $\lambda \neq 0$, there is no pitchfork bifurcation; rather, a fold bifurcation is observed (Fig. 6.10, left and right). We conclude that the supercritical pitchfork bifurcation is structurally unstable.

Problem 6.2   An analytical expression for y_{ss} in terms of λ and μ is unwieldly, because Eq. 6.14 is a cubic polynomial. However, $0 = y_{ss}^3 - \mu y_{ss} - \lambda \Rightarrow \mu y_{ss} = y_{ss}^3 - \lambda$, which implies

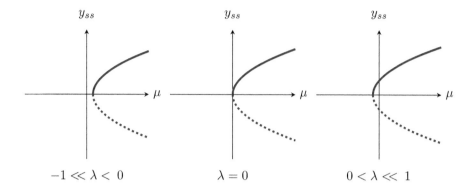

Figure 6.11   The fold bifurcation of Eq. 6.16 is structurally stable.

$$\mu = \frac{y_{ss}^3 - \lambda}{y_{ss}} = y_{ss}^2 - \frac{\lambda}{y_{ss}} \quad \text{for} \quad y_{ss} \neq 0.$$

Fig. 6.10 was created by plotting μ as a function of y_{ss} for three values of λ. Apply this technique to the subcritical pitchfork bifurcation (Eq. 6.10). Your answer will be a series of three diagrams analogous to Fig. 6.10.

As mentioned at the beginning of this section, the fold (saddle-node) bifurcation is more robust to perturbations than the transcritical and pitchfork bifurcations. Consider the following ODE,

$$\frac{dy}{dt} = f(y; \mu) = -y^2 + \mu, \tag{6.15}$$

with steady states $y_{ss} = \pm\sqrt{\mu}$ (two steady states when $\mu > 0$, one steady state when $\mu = 0$, and no steady state when $\mu < 0$). This ODE exhibits a fold bifurcation at $\mu = 0$ (Fig. 6.11, middle). When a second bifurcation parameter is introduced in an additive fashion,

$$\frac{dy}{dt} = f(y; \mu, \lambda) = -y^2 + \mu + \lambda, \tag{6.16}$$

the steady states become $y_{ss} = \pm\sqrt{\mu + \lambda}$ and the critical value of the fold bifurcation changes to $\mu = -\lambda$, but the type of bifurcation does not change (Fig. 6.11, left and right).

6.7   Further Reading

For further reading on bifurcations in scalar ODEs, physical and physiological models that exhibit these bifurcations, see Kaplan and Glass (2012, Chapter 4), Strogatz (2014, Chapter 3) and Tyson et al. (2003).

Supplemental Problems

Problem 6.3   Draw the phase line for the following equations using representative values of the parameter μ. Choose at least one value of μ for every possible qualitatively different phase line. You do not have to specify a numerical value for each, it is sufficient to specify a range, e.g., "For $0 < \mu < 1$, the phase line is..."

(a) $\dfrac{dy}{dt} = \mu - y^2$

(c) $\dfrac{dy}{dt} = \mu - y + y^3$

(b) $\dfrac{dy}{dt} = -\mu y + y^3$

(d) $\dfrac{dy}{dt} = \mu + y - y^3$

Problem 6.4   The logistic equation is a scalar ODE model of population growth given by

$$\frac{du}{dt} = r\left(1 - \frac{u}{K}\right)u \tag{6.17}$$

where $r > 0$ is a growth rate and $K > 0$ is the carrying capacity. If u is a mass density of bacteria, what are the physical dimensions of r and K? Rescale the dependent variable using $\hat{u} = u/K$ to obtain

$$\frac{d\hat{u}}{dt} = r\left(1 - \hat{u}\right)\hat{u}.$$

Next, rescale the independent variable (time) using $\hat{t} = rt$ to obtain

$$\frac{d\hat{u}}{d\hat{t}} = \left(1 - \hat{u}\right)\hat{u}. \tag{6.18}$$

What are the physical dimensions of $\hat{u}$ and $\hat{t}$? Draw the phase diagram for Eq. 6.18. Find the steady states $\hat{u}_{ss}$ and analyze their stability. What are the steady state values of the unscaled dependent variable u? Show that

$$\hat{u}(\hat{t}) = \frac{u_0}{u_0 + (1 - u_0)\exp\left(-\hat{t}\right)}$$

solves Eq. 6.18 with the initial condition $u(0) = u_0$. Rescale this solution to find the solution $u(t)$ of Eq. 6.17 and check that the ODE and initial value are satisfied. For more information on modeling the growth of microorganisms see Edelstein-Keshet (2005, Chapter 4).

Problem 6.5   The analytical solution of Eq. 6.1 can be derived by separation of variables. Consider this simpler example:

$$\frac{dy}{dt} = y^2 + \alpha^2$$

$$\frac{dy}{y^2 + \alpha^2} = dt$$

$$\int_{y_0}^{y} \frac{d\hat{y}}{\hat{y}^2 + \alpha^2} = \int_{0}^{t} d\hat{t}$$

$$\tan^{-1}(y/\alpha)\Big|_{y_0}^{y} = \hat{t}\Big|_{0}^{t}$$

$$\vdots$$

$$y = \alpha \left(\frac{\alpha \tan \alpha t + y_0}{\alpha - y_0 \tan \alpha t} \right).$$

Notice that when $\alpha = 0$ there is a steady state at $y_{ss} = 0$. Is this steady state stable or unstable? For $\alpha \neq 0$, $y^2 + \alpha^2 > 0$ so there is no steady state satisfying $0 = y_{ss}^2 + \alpha^2$. Using this example as a guide, derive an analytical solution to Eq. 6.1.

Solutions

Figure 6.1   The critical value is $\mu = 1$.

Figure 6.3   The stable branch is an attractor. The unstable branch is a repellor.

Figure 6.7   The top row are supercritical pitchfork bifurcations. The bottom row are subcritical.

Problem 6.2   Adding a parameter to the subcritical pitchfork bifurcation of Eq. 6.10 gives $dy/dt = \lambda + \mu y + y^3$ with steady states solving $-\mu = y_{ss}^2 + \lambda/y_{ss}$ for $y_{ss} \neq 0$, as follows:

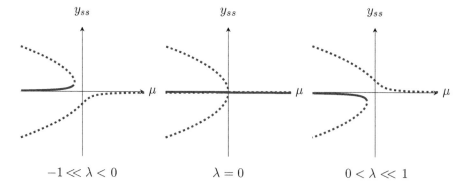

Notes

1. One also may write $f_\mu(y)$ instead of $f(y; \mu)$, or even $f(y)$ when the importance of the parameter μ is understood from context.
2. Solutions of the quadratic polynomial equation $0 = ax^2 + bx + c$ are given by

$$x = \frac{-b \pm \sqrt{b^2 - 4ac}}{2a}.$$

Here we use the replacements $x = y_{ss}$, $a = 1$, $b = -2$, $c = \mu$.
3. The significance of the term saddle-node bifurcation will not be clear until we have considered bifurcations in systems of ODEs.
4. One may also scale and translate the bifurcation diagram's horizontal axis with the substitution $\hat{\mu} = \alpha\mu - \beta$ ($\alpha \neq 0$) without changing the bifurcation type.

PART II
Passive Membranes

7 The Nernst Equilibrium Potential

The focus of this chapter is *the electrical potential difference between the two sides of the cell plasma membrane* (intracellular versus extracellular, inside versus outside). This transmembrane potential is essential to neuronal signaling and is the starting point of our discussion of electrophysiology.

7.1 Cellular Compartments and Electrical Potentials

Membrane-delimited compartments within living cells have different ionic concentrations and **electrical potentials**. Ionic concentration gradients are maintained by transporters and pumps, and the electrical potential within various compartments is related to the selective permeability of biological membranes to different ions.

The subject of this chapter is the electrical potential difference between the interior and exterior of "quiescent" or "resting" neurons, that is, neurons that are not depolarized through chemical neurotransmission or current injection. By definition, a resting excitable cell is not spiking (i.e., repetitively firing action potentials) and is not experiencing depolarization-induced calcium influx (recall Fig. 2.6) or stimulus-induced intracellular signaling (e.g., sperm-mediated activation of an oocyte).

The electrical potential difference between the interior and exterior of resting neurons is primarily due to the potassium concentration difference inside and out and the plasma membrane's selective permeability to potassium. The resting membrane's permeabilities to sodium, chloride, calcium and other ions have second-order effects that will be considered in the sequel (e.g., p. 125).

We will write ϕ_{in} for the electrical potential of the neuron's interior (cytosol),[1] and ϕ_{out} for the electrical potential of the extracellular space (see Fig. 7.1). The **membrane potential**, i.e., the electrical potential difference across the cell plasma membrane, is given by

$$\Delta\phi_{pm} = \phi_{in} - \phi_{out} \approx -65\,\text{mV}. \tag{7.1}$$

Because the electrical potential within a cell is typically lower than the electrical potential outside the cell ($\phi_{in} < \phi_{out}$), the membrane potential is usually negative. Although

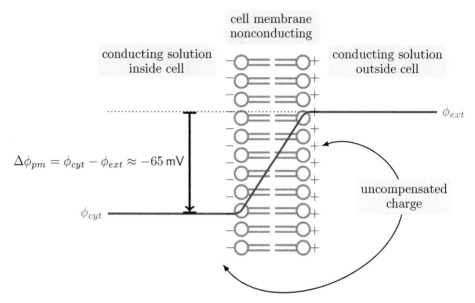

Figure 7.1 The cell plasma membrane is composed of a nonconducting lipid bilayer that separates uncompensated charges inside and outside the cell. Adapted from *Physicochemical and Environmental Plant Physiology*, 4th edn, Nobel, PS, Figure 3–1, Copyright 2009, with permission from Elsevier.

the membrane potential of resting neurons is highly variable, $\Delta\phi_{pm}$ is usually about -65 ± 15 millivolts (mV).

Fig. 7.2 shows electrical potentials typical for the intracellular space (cytosol), the lumen of the endoplasmic reticulum (ER), and the cellular compartment within the inner mitochondrial membrane (i.e., the mitochondrial matrix), each expressed relative to the electrical potential of the extracellular space.[2] For clarity, numerous cellular organelles are not shown, including the cell nucleus, liposomes, the Golgi apparatus, and so on. Indeed, this diagram is highly schematic and does not accurately represent the number of mitochondria and spatial organization of the ER membrane. In Fig. 7.2, $\phi_{out} = 0$ mV is the reference potential, that is, the extracellular space is "ground" (as in electric circuit theory). Consistent with Eq. 7.1, the cytosolic potential is $\phi_{in} = -65$ mV. The electrical potential of the ER (ϕ_{er}) is equal to the cytosol (ϕ_{cyt}) so there is no electrical potential difference across the ER membrane ($\Delta\phi_{er} = \phi_{er} - \phi_{cyt} = 0$ mV). The electrical potential difference across the inner mitochondrial membrane is $\Delta\phi_{mito} = \phi_{mito} - \phi_{cyt} = -250$ mV $- (-65$ mV$) = -185$ mV. Because electrical potential differences are important to cell physiology, one ought to commit their approximate values to memory.

7.2   Nernst Equilibrium Potential

The lipid bilayer of the cell plasma membrane is a barrier to the movement of dissolved physiological ions such as potassium, sodium, calcium and chloride. This allows various transporters and pumps to regulate the concentration of intracellular

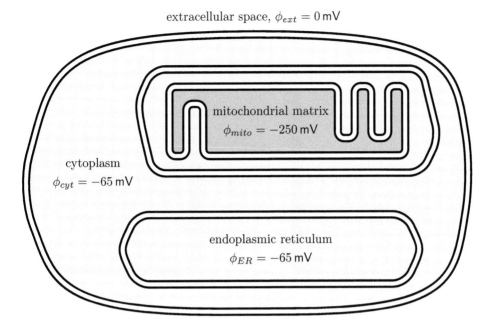

extracellular space, $\phi_{ext} = 0\,$mV

mitochondrial matrix
$\phi_{mito} = -250\,$mV

cytoplasm
$\phi_{cyt} = -65\,$mV

endoplasmic reticulum
$\phi_{ER} = -65\,$mV

Figure 7.2   The electrical potentials of the cytosol, ER and mitochondria compared to the extracellular space.

and extracellular physiological ions by catalyzing their transport across the plasma membrane. For example, the sodium-potassium ATPase is a transmembrane protein that uses energy from adenosine triphosphate to pump sodium out of (and potassium into) neurons, both against their concentration gradients. The sodium-potassium ATPase and other highly regulated membrane transport processes generate characteristic ionic concentration differences between the cell interior and exterior (see Table 7.2 below).

Ionic channels are transmembrane proteins with an aqueous pore that may allow dissolved ions to move very rapidly down an electrochemical (concentration and electrical potential) gradient from one side of the plasma membrane to the other. The ionic channels of excitable membranes are highly regulated and many are **selectively permeable** to a particular ion (e.g., N-type Ca^{2+} channels from Chapter 2). In resting neurons, the vast majority of ion channels that are "open" – i.e., in a permissive state that allows dissolved ions to move through their aqueous pore – are highly selective for potassium. The potassium concentration difference between the inside and outside of neurons and the permeability of the resting plasma membrane to potassium generates the characteristic electrical potential difference between the cell interior and exterior (Fig. 7.2).

Fig. 7.3 illustrates a thought experiment that explains how the selective permeability of a membrane may lead to an electronegative cell interior. The upper left panel of Fig. 7.3 shows a water-filled chamber divided by a membrane (vertical solid line) that is entirely impermeable to ions in solution. Within the left subcompartment of the

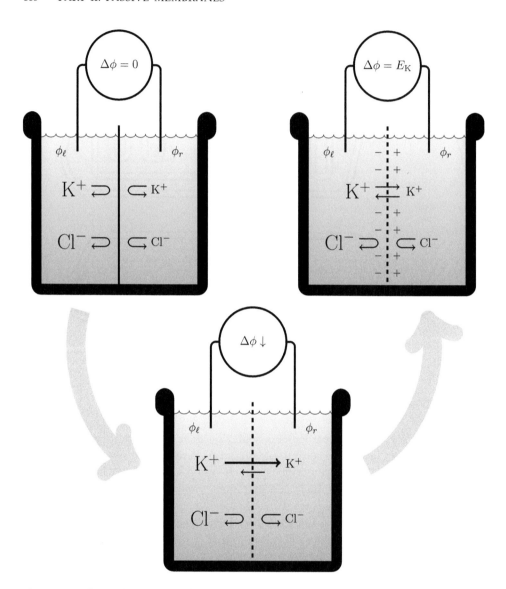

Figure 7.3   Electrical potential difference across a membrane selectively permeable to potassium ions. Font size indicates concentration of potassium and chloride, while + and − indicate uncompensated charges. Adapted from *Ionic Channels of Excitable Membranes* by Hille (2001) Fig. 1.4, p. 13. By permission of Oxford University Press.

divided chamber, a large amount of potassium chloride $(KCl)^3$ has been dissolved. As a result, there is a high concentration of dissociated potassium and chloride ions in solution on the left side of the membrane. In the right subcompartment of the divided chamber, a lesser amount of KCl was added, resulting in a lower concentration of both potassium and chloride on the right side of the membrane. Because the (positively charged) cation and (negatively charged) anion concentrations are exactly balanced, both sides of the chamber are electrically neutral. A voltmeter makes electrical connection with the membrane-separated compartments of the chamber. The voltmeter indicates no electrical potential difference between the left and right subcompartments ($\Delta\phi = \phi_r - \phi_\ell = 0\,\text{mV}$).

In Fig. 7.3 (lower middle) we imagine that the membrane instantly becomes permeable to potassium, but not chloride. Potassium is now able to diffuse through the membrane from high to low concentration (down its concentration gradient). However, because chloride ions (Cl^-) cannot cross the membrane along with potassium ions (K^+), the net movement of potassium ions from left to right increases the electrical potential on the right relative to the left ($\phi_\ell < \phi_r$).

Fig. 7.3 (top right) shows the equilibrium situation, in which the electrical potential difference between the subcompartments is high enough that the tendency for potassium ions to move down their concentration gradient (left to right) is balanced by the tendency of positively charged potassium ions to move from high to low electrical potential (right to left). The electrical potential difference leading to this balance is referred to as the *Nernst equilibrium potential* for potassium[4] and is denoted E_K.

The thought experiment of Fig. 7.3 provides important qualitative information about the Nernst equilibrium potential for potassium. At equilibrium there will be more positive charge associated with the right chamber and, consequently, the electrical potential on the right side of the compartment is greater than that on the left ($\phi_r > \phi_\ell$). Thus, the *Nernst equilibrium potential* for potassium is negative ($\Delta\phi = \phi_\ell - \phi_r = E_K < 0$). The next section shows how the quantitative value of Nernst equilibrium potential is related to the potassium chloride concentrations initially chosen for the two sides of the chamber.

7.3   Derivation of the Nernst Equation

To determine the quantitative relationship between ionic concentrations on opposite sides of a semipermeable membrane and the resulting Nernst equilibrium potential, we apply the Boltzmann-Gibbs distribution from statistical mechanics,

$$\frac{p_r}{p_\ell} = \exp\left\{-\frac{u_r - u_\ell}{k_B T}\right\}. \tag{7.2}$$

This equation relates the equilibrium probabilities (p_ℓ and p_r) of finding a particle in one of two states to the energy difference between these states, $u_r - u_\ell$, in units of joules (J). In Eq. 7.2, T is the absolute temperature and $k_B = 1.3807\,\text{J}\,\text{K}^{-1}$ is Boltzmann's constant (joules per degree Kelvin). Take a moment to convince yourself that Eq. 7.2 is consistent with the lower energy state having higher probability, e.g., $u_r < u_\ell$ implies $p_r > p_\ell$. As required, the argument of the exponential function is dimensionless, because $u_r - u_\ell \ominus$ energy and $k_B T \ominus$ (energy/temp) · temp = energy.

Eq. 7.2 may also be expressed in terms of ionic concentrations and molar energies, as follows

$$\frac{[K^+]_r}{[K^+]_\ell} = \exp\left\{-\frac{U_r - U_\ell}{RT}\right\}, \tag{7.3}$$

Table 7.1 Physical constants that are relevant to cellular biophysics

Name	Symbol	Value
Avogadro's number	N	$6.02 \times 10^{23} \, \text{mol}^{-1}$
Boltzmann's constant	k_B	$1.3807 \times 10^{-23} \, \text{J} \, \text{K}^{-1}$
Gas constant	$R = N k_B$	$8.3145 \, \text{J} \, \text{mol}^{-1} \, \text{K}^{-1}$
Elementary charge	q	$1.6022 \times 10^{-19} \, \text{C}$
Faraday's constant	$F = Nq$	$9.6485 \times 10^4 \, \text{C} \, \text{mol}^{-1}$

where $[K^+]_r / [K^+]_\ell$ is the ratio of potassium concentrations, $U_r - U_\ell$ is the molar energy difference of the two sides (units of joules per mole), and $R = 8.314 \, \text{J} \, \text{mol}^{-1} \, \text{K}^{-1}$ is the gas constant given by $R = k_B N_A$ where $N_A = 6.022 \times 10^{23} \, \text{mol}^{-1}$ is Avogadro's number (see Table 7.1).

The molar energies U_r and U_ℓ depend on the ionic charge of potassium and the electrical potentials, ϕ_r and ϕ_ℓ. For example, the molar energy on the right side is

$$\frac{\text{energy}}{\text{amount}} \ominus \quad U_r = z_K F \phi_r \quad \ominus \frac{\text{charge}}{\text{amount}} \cdot \text{voltage} \tag{7.4}$$

and similarly $U_\ell = z_K F \phi_\ell$ where $z_K = +1$ is the (dimensionless) valence of potassium and $F = 9.6487 \times 10^4 \, \text{C} \, \text{mol}^{-1}$ is Faraday's constant, the absolute value of the charge of one mole of electrons (see Problem 7.2). In Eq. 7.4, we have written the physical dimension of ϕ_r as voltage = electrical potential, preferring to use a single word rather than two. (Voltage and electrical potential are synonyms and we use the terms interchangeably.) The physical dimensions of Eq. 7.4 balance because charge·voltage = energy (see Discussion).[5]

The Nernst equilibrium potential for potassium, denoted by E_K, is the electrical potential difference

$$E_K = \phi_\ell - \phi_r$$

that satisfies Eq. 7.3 (Fig. 7.3, top right). To determine its relationship to the ionic concentrations $[K^+]_r$ and $[K^+]_\ell$, substitute the molar energies $U_r = z_K F \phi_r$ and $U_\ell = z_K F \phi_\ell$ into Eq. 7.3 to obtain

$$\frac{[K^+]_r}{[K^+]_\ell} = \exp\left\{ -\frac{z_K F (\phi_r - \phi_\ell)}{RT} \right\} = \exp\left\{ \frac{z_K F E_K}{RT} \right\}.$$

Taking the logarithm of both sides gives

$$\ln \frac{[K^+]_r}{[K^+]_\ell} = \frac{z_K F E_K}{RT} \quad \Rightarrow \quad E_K = \frac{RT}{z_K F} \ln \frac{[K^+]_r}{[K^+]_\ell}.$$

This is the sought-after relationship between the Nernst equilibrium potential E_K and the ionic concentrations $[K^+]_\ell$ and $[K^+]_r$. This equation is used to calculate numerical values for the Nernst equilibrium potentials E_K, E_{Cl}, E_{Na} and E_{Ca} (see Section 7.4 and Fig. 7.4).

Table 7.2   Representative ionic concentrations and Nernst equilibrium potentials, adapted from Hille (2001)

Ion	$[S]_o$	$[S]_i$	E_S (mV)	
S	(mM)	(mM)	21°C	37°C
K^+	4	155	-93.0	-98.1
Na^+	145	12	$+63.4$	$+66.8$
Ca^{2+}	1.5	10^{-4}	$+112$	$+118$
Cl^-	123	4.2	-85.9	-90.1

7.4   Calculating Nernst Equilibrium Potentials

Let S denote potassium (K^+), chloride (Cl^-), sodium (Na^+) or calcium (Ca^{2+}). Following Eq. 7.3, the Nernst equilibrium potential for S is

$$E_S = \frac{RT}{z_S F} \ln \frac{[S]_o}{[S]_i} , \tag{7.5}$$

where $[S]_o$ (o = outside) and $[S]_i$ (i = inside) denote extracellular and intracellular concentrations of S (see Table 7.2 for typical values).

RT/F is About 25 millivolts

Both the valence z_S and the ratio $[S]_o/[S]_i$ of Eq. 7.5 are dimensionless; consequently, the physical dimensions of RT/F must be the same as $E_K \ominus$ voltage. To see this, write

$$\frac{RT}{F} \ominus \frac{\frac{energy}{amount \cdot temp} \cdot temp}{charge/amount} = \frac{energy}{charge} = voltage .$$

If we take room temperature to be 21°C, the quantity RT/F is

$$\frac{RT}{F} = \frac{8.3145 \, J \, mol^{-1} \, K^{-1} \cdot (273.15 \, K + 21 \, K)}{9.6485 \times 10^4 \, C \, mol^{-1}} = 25.43 \, mV , \tag{7.6}$$

where we have used energy/charge = voltage (J/C = V). Because the quantity RT/F often occurs in biophysical formulas, it is helpful to memorize the approximate value

$$\frac{RT}{F} \approx 25 \, mV . \tag{7.7}$$

Approximate Values of E_K, E_{Cl}, E_{Na} and E_{Ca}

Eq. 7.5 may be used to calculate numerical values for Nernst equilibrium potentials. For potassium, we obtain an easy-to-remember approximate value using $z_K = +1$, $RT/F \approx 25 \, mV$ and $[K^+]_o/[K^+]_i \approx 1/40$,

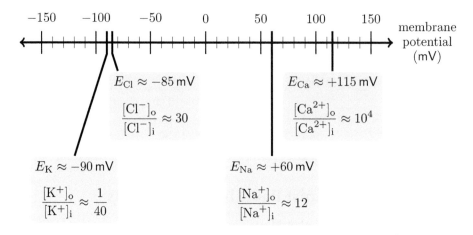

Figure 7.4 Ionic ratios and Nernst equilibrium potentials (Eqs. 7.8–7.11).

$$E_K = \frac{RT}{z_K F} \ln \frac{[K^+]_o}{[K^+]_i} = 25 \, mV \cdot \underbrace{\ln \frac{1}{40}}_{-3.69} \approx -90 \, mV \, . \tag{7.8}$$

Similar calculations for sodium, chloride and calcium are

$$E_{Na} = \frac{RT}{z_{Na} F} \ln \frac{[Na^+]_o}{[Na^+]_i} = 25 \, mV \cdot \underbrace{\ln 12}_{2.48} \approx +60 \, mV \tag{7.9}$$

$$E_{Cl} = \frac{RT}{z_{Cl} F} \ln \frac{[Cl^-]_o}{[Cl^-]_i} = -25 \, mV \cdot \underbrace{\ln 30}_{3.40} \approx -85 \, mV \tag{7.10}$$

$$E_{Ca} = \frac{RT}{z_{Ca} F} \ln \frac{[Ca^{2+}]_o}{[Ca^{2+}]_i} = 12.5 \, mV \cdot \underbrace{\ln 10^4}_{9.21} \approx +115 \, mV \tag{7.11}$$

where $RT/(z_{Cl}F) \approx -25$ mV (note sign) and $RT/(z_{Ca}F) \approx +12.5$ mV (decreased by a factor of two because $z_{Ca} = 2$). Fig. 7.4 shows these four Nernst equilibrium potentials on a horizontal axis with units of mV. These values should be committed to memory; most importantly, the sign of each and their order,

$$E_K < E_{Cl} < 0 < E_{Na} < E_{Ca} \, .$$

7.5   Chemical Potential

An alternative derivation of the Nernst equilibrium potential formula begins with the chemical potential μ in an ideal solution

$$\mu = \bar{\mu} + RT \ln c$$

where c is the ionic concentration and $\bar{\mu}$ is the standard state potential for this ion.[6] The *electrochemical potential* $\hat{\mu}$ is given by the sum of the chemical potential and $zF\phi$, that is,

$$\hat{\mu} = \mu + zF\phi$$

where μ is the chemical potential, z is the valence (i.e., charge number) of the ion, F is Faraday's constant, and ϕ is the electrical potential. Combining these expressions gives

$$\hat{\mu} = \bar{\mu} + RT \ln c + zF\phi.$$

Note that increasing the ion concentration increases the electrochemical potential, and an increasing electrical potential ϕ can either increase or decrease the electrochemical potential depending on whether the ion is positively ($z > 0$) or negatively ($z < 0$) charged.

At equilibrium the intracellular and extracellular electrochemical potentials will be equal, $\hat{\mu}_i = \hat{\mu}_o$, that is,

$$\bar{\mu} + RT \ln c_i + zF\phi_i = \bar{\mu} + RT \ln c_o + zF\phi_o, \tag{7.12}$$

as illustrated in Fig. 7.5. Let us denote the Nernst equilibrium potential – i.e., the electrical potential difference leading to zero net flux – as $E = \phi_i - \phi_o$. Subtracting the standard state chemical potential $\bar{\mu}$ from both sides and then rearranging this expression, we find

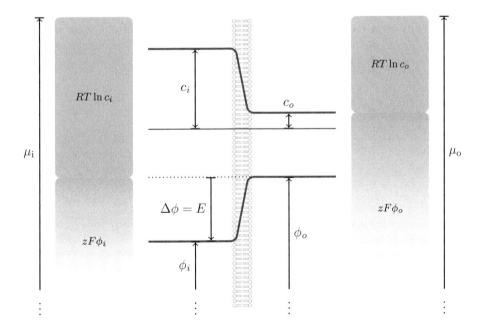

Figure 7.5 Equilibrium across a semipermeable cell membrane occurs when the (electro)chemical potentials are equal, that is, $\hat{\mu}_i = \hat{\mu}_o$ (see Eq. 7.12). For a monovalent cation (such as potassium) $RT \ln c_i > RT \ln c_o$ and $zF\phi_i < zF\phi_o$. Adapted from *Physicochemical and Environmental Plant Physiology*, 4th edn, Nobel, PS, Figure 3–2, Copyright 2009, with permission from Elsevier.

$$RT \ln c_i + zF\phi_i = RT \ln c_o + zF\phi_o$$
$$zF\phi_i - zF\phi_o = RT \ln c_o - RT \ln c_i$$
$$\underbrace{zF\,(\phi_i - \phi_o)}_{E} = RT \ln \frac{c_o}{c_i}.$$

Thus, the Nernst equilibrium potential is given by

$$\boxed{E = \frac{RT}{zF} \ln \frac{c_o}{c_i}}.$$

Fig. 7.5 illustrates the equilibrium across a semipermeable cell membrane that occurs when the (electro)chemical potentials are equal ($\hat{\mu}_i = \hat{\mu}_o$).

7.6   Discussion

The Number of Uncompensated Charges in Resting Cells

In our discussion of Nernst equilibrium potentials, we have assumed that intracellular and extracellular ionic concentrations are fixed quantities (at least approximately so). This might seem dubious, given that the development of the Nernst equilibrium potential involves movement of ions across the plasma membrane (Fig. 7.3). The assumption is justified because the number of uncompensated charges required to achieve equilibrium is small compared to the total number of ions within a cell.

In the case of potassium, for example, the number of uncompensated charges needed to establish the Nernst equilibrium potential E_K is given by

$$n = \frac{Q}{q} = \frac{C_{mem}\,|E_K|}{q},$$

q is the elementary charge and the total charge $Q = C_{mem}\,|E_K|$ is the product of the capacitance of the plasma membrane C_{mem} and the magnitude of the electrical potential accumulated from the net outward movement of potassium ions.[7] Assuming a spherical cell of radius $10\,\mu m$, the cell surface area is $S_{mem} = 4\pi\,(10\,\mu m)^2 = 1257\,\mu m^2 = 1.257 \times 10^{-5}\,cm^2$ where we have used $1\,cm = 10^4\,\mu m$. Assuming a specific capacitance of $1\,\mu F/cm^2$ for the cell membrane, the total cell capacitance is $C_{mem} = 1.257 \times 10^{-5}\,cm^2 \cdot 1\,\mu F/cm^2 = 1.257 \times 10^{-11}\,F$ (farad).[8] Using $|E_K| = 0.09\,V$ and $C = F\,V$ (a coulomb is a farad-volt), the number of uncompensated charges is

$$n = \frac{\overbrace{1.257 \times 10^{-11}\,F \cdot 0.09\,V}^{1.131\times 10^{-12}\,C}}{1.602 \times 10^{-19}\,C} = 7.016 \times 10^6,$$

that is, about 7 million potassium ions. For comparison, the total number of potassium ions in the cell interior can be estimated as the product of $[K^+]_i \approx 155\,mM$ and the volume of the cell, $v = \frac{4}{3}\pi\,(10\,\mu m)^3 = 4189\,\mu m^3$. Using $1\,\mu m^3 = 10^{-15}$ liter, we find

$$n_T = 4189 \,\mu m^3 \cdot 155 \,mM = 4.189 \times 10^{-12} \,liter \cdot 0.155 \,mol/liter$$
$$= 4.189 \times 10^{-12} \cdot 0.155 \cdot 6.02 \times 10^{23} = 3.90 \times 10^{11},$$

that is, 390 billion potassium ions. The ratio

$$\frac{n}{n_T} = \frac{7.016 \times 10^6}{3.90 \times 10^{11}} = 1.80 \times 10^{-5}$$

shows that the uncompensated charges may be provided by about two thousandths of a percent of the total number of intracellular potassium ions.

The Goldman-Hodgkin-Katz Voltage Equation

The electronegativity of the cell cytosol compared to extracellular space is predominantly due to the selective permeability of the plasma membrane to potassium. However, there are lesser contributions due to permeability of the resting membrane to other monovalent cations (e.g., sodium, $z_{Na} = +1$), monovalent anions (e.g., chloride, $z_{Cl} = -1$), and divalent cations (e.g., calcium, $z_{Ca} = +2$). The Goldman-Hodgkin-Katz voltage equation,

$$E_{ghk} = \frac{RT}{F} \ln \frac{\sum_j P_{S_j^+}[S_j^+]_o + \sum_j P_{S_j^-}[S_j^-]_i}{\sum_j P_{S_j^+}[S_j^+]_i + \sum_j P_{S_j^-}[S_j^-]_o}, \tag{7.13}$$

yields an **effective equilibrium potential** (E_{ghk}) that captures the contributions of multiple *monovalent* ions (e.g., potassium, sodium and chloride). In Eq. 7.13, j is an index over monovalent ions, and $P_{S_j^+}$ and $P_{S_j^-}$ are membrane permeabilities for the cations (S_j^+) and anions (S_j^-), respectively. For a membrane permeable to potassium and sodium, but impermeable to other ions, the GHK voltage equation (Eq. 7.13) becomes

$$E_{ghk} = \frac{RT}{F} \ln \frac{P_{Na}[Na^+]_o + P_K[K^+]_o}{P_{Na}[Na^+]_i + P_K[K^+]_i}. \tag{7.14}$$

Assuming the resting membrane is $\gamma \approx 20$ times more permeable to potassium than sodium, $P_K = \gamma P_{Na}$, we have

$$E_{ghk} = \frac{RT}{F} \ln \frac{[Na^+]_o + \gamma [K^+]_o}{[Na^+]_i + \gamma [K^+]_i}. \tag{7.15}$$

Using $\gamma = 20$ and the concentrations from Table 7.2 we find

$$E_{ghk} = 25 \,mV \cdot \ln \frac{145 \,mM + 20 \cdot 4 \,mM}{12 \,mM + 20 \cdot 155 \,mM} = -65.7 \,mV. \tag{7.16}$$

This value is in the range of voltages (-80 to $-50 \,mV$) commonly observed in resting neurons. Note that this effective equilibrium potential satisfies (see Fig. 7.6)

$$-90 \,mV \approx E_K < E_{ghk} < E_{Na} \approx +60 \,mV. \tag{7.17}$$

Furthermore, E_{ghk} is nearer to E_K than E_{Na}, because Eq. 7.16 assumes the potassium permeability is 20 times greater than the sodium permeability.

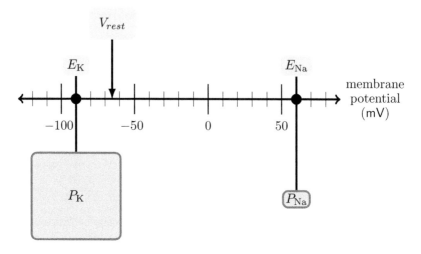

Figure 7.6   According to the Goldman-Hodgkin-Katz voltage equation, the resting membrane potential (V_{rest}) will be closer to the Nernst reversal potential for potassium (E_K) than for sodium (E_{Na}) when the permeability for potassium is greater than that for sodium ($P_K > P_{Na}$, see Eq. 7.21).

Energy and Potential

What is voltage? The concept of *electrical potential* (voltage) is related to the concept of *electrical potential energy*. Let us recall the meaning of potential energy in a more familiar context (gravity).

Gravitational Potential Energy and Gravitational Potential

Some time ago my cat Nellie (a tabby) ran up a tree to escape Harvey, my dog (a shetland mix). The effort required for Nellie to climb the tree is due to the gravitational attraction between Nellie and the earth. The *gravitational potential energy* difference between Nellie on the ground and Nellie on the branch, $\Delta \mathcal{E} = \mathcal{E}_{branch} - \mathcal{E}_{ground}$, is the work (energy = force · length) required to raise Nellie to the height of her perch ($\Delta h = h_{branch} - h_{ground}$).[9] The gravitational force Nellie overcame as she climbed the tree is the product of her mass (m) and the *standard acceleration due to gravity* (g). Thus, Nellie's gravitational potential energy increase upon climbing the tree is

$$\Delta \mathcal{E} = m g \, \Delta h. \tag{7.18}$$

The physical dimensions and SI units of this expression are

$$
\begin{bmatrix} \text{gravitational} \\ \text{potential} \\ \text{energy} \end{bmatrix}
=
\begin{bmatrix} \text{mass} \\ \text{of} \\ \text{cat} \end{bmatrix}
\cdot
\begin{bmatrix} \text{gravitational} \\ \text{field} \\ \text{strength} \end{bmatrix}
\cdot
\begin{bmatrix} \text{height} \\ \text{of} \\ \text{branch} \end{bmatrix}
$$

$$
\begin{array}{cccc}
\Updownarrow & \Updownarrow & \Updownarrow & \Updownarrow \\
\text{energy} & \text{mass} & \text{acceleration} = \text{force/mass} & \text{length} \\
\text{J} & \text{kg} & \text{m/s}^2 = \text{N/kg} & \text{m}
\end{array}
$$

$$\underbrace{\qquad\qquad\qquad\qquad\qquad}_{\text{force (N)}}.$$

Using $g \approx 9.81\,\text{m/s}^2$, a branch height of $\Delta h = 4\,\text{m}$, and cat mass of $m = 5\,\text{kg}$, the gravitational potential energy difference $\Delta \mathcal{E} = mg\Delta h$ is

$$\Delta \mathcal{E} = 5.00\,\text{kg} \cdot 9.81\,\text{m/s}^2 \cdot 4.00\,\text{m} = 196\,\text{J}$$

where $J = \text{kg} \cdot \text{m}^2/\text{s}^2$ (joule) is the SI unit of energy and work.

Gravitational potential is the quotient of gravitational potential energy and mass. The gravitational potential difference between branch and ground is $\Delta \phi = \Delta \mathcal{E}/m$. Dividing Eq. 7.18 by m we obtain

$$\Delta \phi = g\,\Delta h \tag{7.19}$$

with dimensions and units,

$$\begin{bmatrix} \text{gravitational} \\ \text{potential} \end{bmatrix} = \begin{bmatrix} \text{gravitational} \\ \text{field} \\ \text{strength} \end{bmatrix} \cdot \begin{bmatrix} \text{height} \\ \text{of branch} \end{bmatrix}$$

$$\Updownarrow \qquad\qquad \Updownarrow \qquad\qquad\qquad \Updownarrow$$

energy/mass   acceleration = force/mass   length
J/kg $\text{m/s}^2 = \text{N/kg}$ m

$$\underbrace{\qquad\qquad\qquad\qquad\qquad}_{\text{energy/mass}}.$$

The gravitational potential is the work per unit mass required to raise not just 5 kg of cat, but any object at all, to the height of the branch. Substituting $g \approx 9.81\,\text{m/s}^2$ and $\Delta h = 4\,\text{m}$ into Eq. 7.19 gives the gravitational potential difference between branch and ground,

$$\Delta \phi = g\,\Delta h = 9.81\,\text{m/s}^2 \cdot 4.00\,\text{m} = 39.2\,\text{J/kg},$$

where $\text{m}^2/\text{s}^2 = \text{J/kg}$. Together, Eqs. 7.18 and 7.19 imply that

$$\Delta \mathcal{E} = \Delta \phi \cdot m \tag{7.20}$$

$$\begin{bmatrix} \text{gravitational} \\ \text{potential} \\ \text{energy} \end{bmatrix} = \begin{bmatrix} \text{gravitational} \\ \text{potential} \end{bmatrix} \cdot \begin{bmatrix} \text{mass} \\ \text{of whatever} \end{bmatrix}$$

$$\Updownarrow \qquad\qquad \Updownarrow \qquad\qquad \Updownarrow$$

energy   energy/mass   mass
J   J/kg   kg.

While gravitational potential energy $\Delta \mathcal{E}$ is a property of Nellie the cat, the gravitational potential $\Delta \phi$ is a property of location (height above ground). Compared to ground, when located on the branch, the gravitational potential energy of cat, squirrel, bowling ball, etc., is $\Delta \phi = 39.2\,\text{J/kg}$ times the object's mass (Eq. 7.20).

Electric Potential Energy and Electric Potential
Electrical potential energy, electrical potential (voltage)[10] and electrical field strength are analogous to gravitational potential energy, gravitational potential and gravitational field strength. The product of charge and electrical field strength yields a force, much like the product of mass and gravitational field strength. The physical dimensions and units are

$$\begin{bmatrix} \text{electrical} \\ \text{potential} \\ \text{energy} \end{bmatrix} = \begin{bmatrix} \text{charge} \\ \text{of ion} \end{bmatrix} \cdot \begin{bmatrix} \text{electric field} \\ \text{strength} \end{bmatrix} \cdot [\text{ displacement }]$$

$$\Updownarrow \qquad\qquad \Updownarrow \qquad\qquad\qquad \Updownarrow \qquad\qquad\qquad \Updownarrow$$

energy   charge $\dfrac{\text{mass} \cdot \text{length}}{\text{charge} \cdot \text{time}^2} = \dfrac{\text{force}}{\text{charge}}$ length

J   C $\dfrac{\text{kg} \cdot \text{m}}{\text{C} \cdot \text{s}^2} = \dfrac{\text{N}}{\text{C}}$ m

$$\underbrace{\qquad\qquad\qquad\qquad\qquad}_{\text{force (N).}}$$

While gravitational field strength has physical dimensions force/mass $\ominus$ N/kg = m/s^2 = acceleration, the dimensions of electric field strength are force/charge $\ominus$ N/C. Electrical potential is electric potential energy divided by charge,

$$\begin{bmatrix} \text{electric} \\ \text{potential} \end{bmatrix} = \begin{bmatrix} \text{electric} \\ \text{field strength} \end{bmatrix} \cdot [\text{ displacement }]$$

$$\Updownarrow \qquad\qquad\qquad \Updownarrow \qquad\qquad\qquad \Updownarrow$$

$\dfrac{\text{energy}}{\text{charge}}$ $\dfrac{\text{mass} \cdot \text{length}}{\text{charge} \cdot \text{time}^2} = \dfrac{\text{force}}{\text{charge}}$ length

$V = \dfrac{J}{C}$ $\dfrac{\text{kg} \cdot \text{m}}{\text{C} \cdot \text{s}^2} = \dfrac{\text{N}}{\text{C}} = \dfrac{\text{V}}{\text{m}}$ m

$$\underbrace{\qquad\qquad\qquad\qquad\qquad}_{\text{energy/charge.}}$$

The physical dimensions of electrical potential (voltage) are energy/charge (V = J/C). Thinking about it the other way, a joule (J) is the work required to move an electric charge of one coulomb (C) through an electrical potential difference of one volt (V). The electric field strength can be expressed using the unit volt, because force/charge $\ominus$ N/C = V/m. Electrical potential (voltage) is the proportionality constant between electrical potential energy and charge,

$$\begin{bmatrix} \text{electric} \\ \text{potential} \\ \text{energy} \end{bmatrix} = \begin{bmatrix} \text{electric} \\ \text{potential} \end{bmatrix} \cdot \begin{bmatrix} \text{charge} \\ \text{of ion} \end{bmatrix}$$

$$\Updownarrow \qquad\qquad \Updownarrow \qquad\qquad \Updownarrow$$

energy   energy/charge   charge

J   J/C = V   C.

When the electric field strength is constant, the electrical potential is a linear function of position.

One caveat to this comparison of electrical and gravitational potentials is that charge is a signed quantity.[11] The electrical potential energy of an ion depends on the strength and orientation of the electric field, the direction of displacement, and the sign of its charge (negative or positive). The electrical potential energy of a cation such as sodium decreases upon moving from outside to inside ($z_{Na} = +1$ so $\phi_i < \phi_o \Rightarrow z_{Na}\phi_i < z_{Na}\phi_o$). Conversely, the electrical potential energy of an anion such as chloride increases upon moving from outside to inside ($z_{Cl} = -1$ so $\phi_i < \phi_o \Rightarrow z_{Cl}\phi_i > z_{Cl}\phi_o$).

Supplemental Problems

Problem 7.1 According to Eq. 7.6, RT/F is approximately 25 mV at room temperature (21°C). What is RT/F at $T = 37°C$ (body temperature)? What is the relative error of using room temperature instead of body temperature in a calculation of RT/F?

Problem 7.2 Show that Faraday's constant $F = 9.6487 \times 10^4$ C mol^{-1} is the product of the elementary positive charge $q = 1.602 \times 10^{19}$ C (the electric charge carried by a single proton) and Avogadro's number ($N_A = 6.02 \times 10^{23}$ mol^{-1}).

Problem 7.3 Use Eq. 7.5 to show that increasing the extracellular potassium concentration $[K^+]_o$ causes E_K to increase (become more positive, closer to zero).

Problem 7.4 Calculate the electric field strength (units of V/m) within the cell plasma membrane assuming a membrane thickness of 40 nm.

Problem 7.5 Intracellular potassium concentration is about forty times higher than extracellular (see Table 7.2). With so much more positively charged potassium ions inside versus outside, why is the interior of the cell not positively charged compared to the exterior?

Problem 7.6 Using your knowledge of intracellular and extracellular ionic concentrations, sketch a version of Fig. 7.5 that illustrates the balance of (electro)chemical potential for a membrane selectively permeable to sodium. Then make two more: one for calcium and one for chloride. Which two of these four diagrams are most similar?

Problem 7.7 Estimate the percentage increase in the sodium concentration inside a spherical cell of radius 5 μm as the result of a single action potential. Model the action potential as a 1 ms duration square pulse of 1 mA/cm^2 inward sodium current.

Problem 7.8 Show that Eq. 7.14 can be written as follows,

$$E_{ghk} = \frac{RT}{F} \ln \frac{r_{Na}[Na^+]_o + r_K[K^+]_o}{r_{Na}[Na^+]_i + r_K[K^+]_i} \tag{7.21}$$

where $r_{Na} = P_{Na}/(P_{Na} + P_K)$, $r_K = P_K/(P_{Na} + P_K)$, and $r_{Na} + r_K = 1$. Use Eq. 7.21 to show that the Goldman-Hodgkin-Katz voltage satisfies Eq. 7.17.

Problem 7.9 Using plotting software of your choosing, graph E_{ghk} as a function of the relative permeability for sodium $0 \le r_{Na} \le 1$ where

$$r_{Na} = \frac{P_{Na}}{P_{Na} + P_K},$$

the total permeability $P_T = P_{Na} + P_K$ is constant, and $r_K + r_{Na} = 1$ (see Eq. 7.21). Compare this plot to the weighted average of the Nernst equilibrium potentials $(r_K E_K + r_{Na} E_{Na})$.

Solutions

Problem 7.1 A calculation similar to Eq. 7.6 gives 26.7 mV when $T = 37°C$. This is a relative error of approximately 5%, because

$$\frac{|25.43 \, mV - 26.7 \, mV|}{26.7 \, mV} = 0.048 \,.$$

Problem 7.4 The electric field strength is

$$\frac{|-65 \, mV|}{40 \, nm} = \frac{65 \times 10^{-3} \, V}{40 \times 10^{-9} \, m} = 1.625 \times 10^6 \, V/m$$

or about 2 megavolts per meter. For comparison, the generation of an electric spark in air (e.g., by the flame ignitor of a gas stove burner) requires about 3 volts per meter. A much higher field strength is required for dielectric breakdown of lipid.

Problem 7.5 The principle of electroneutrality of solutions requires that nearly every cation in the cytosol has a nearby anion. This is true in the cytosol of a cell as well as in extracellular space. The Nernst equilibrium potential E_K is due to the diffusion driven charge separation that occurs in the context of a membrane with selective permeability to potassium.

Notes

1. The aqueous component of the cell cytoplasm.
2. The properties of the inter-membrane space between the outer and inner mitochondrial membranes will not concern us.
3. An odorless white chemical compound similar to table salt.
4. The Nernst equilibrium potential is sometimes referred to as a Donnan potential or diffusion potential.
5. The product of coulomb and volt is joule ($C \cdot V = J$). The correspondence voltage = electrical potential is easy to remember because voltage recalls the SI unit of electrical potential (the volt, denoted V).
6. We are ignoring the possibility of an activity coefficient and assuming constant temperature and pressure.
7. C_{mem} or C for membrane capacitance is not to be confused with the SI unit coulomb (C).

8. The SI unit of capacitance, the farad (F), is not to be confused with Faraday's constant $F = 9.6485 \times 10^4\,\text{C}\,\text{mol}^{-1}$.

9. Both h_{branch} and h_{ground} are elevations measured from some reference (e.g., mean sea level). However, the chosen reference elevation does not affect the value of Δh, because it is included in both h_{branch} and h_{ground}.

10. The electrical potential is sometimes called the *electrical field potential* or the *electrostatic potential*.

11. Mass is not a signed quantity; it is nonnegative.

8 The Current Balance Equation

Although resting neurons are typically polarized to $-65 \pm 15\,\mathrm{mV}$, the transmembrane potential in neurons is a dynamic quantity. Cell membranes may be depolarized or hyperpolarized by ionic currents mediated by cations and/or anions permeating the aqueous pore of channels. Membrane potential may also be manipulated by injecting current into the cell interior using a stimulating electrode.

8.1 Membrane Voltage

For simplicity, let the symbol V and phrase *membrane voltage* denote the difference in the electrical potential of the cell interior compared to the cell exterior,

$$V = \Delta\phi = \phi_i - \phi_o\,.$$

Membrane voltage is the difference – inside versus outside a cell – in electrical potential energy per unit charge. *Voltage* is less wordy than transmembrane "electrical potential difference" or "electrical potential energy difference per unit charge." Its usage implies the comparison of electrical potentials (it is the difference $\Delta\phi$ that matters) as well as quantification in the SI unit of volt ($\mathsf{V} = \mathsf{J}/\mathsf{C}$, joule per coulomb).

8.2 Ionic Fluxes and Currents

Electric current is defined as positive in the direction of movement of positive charge.[1] For this reason, physiologists refer to the movement of cations (positive charge) across a cell membrane from the extracellular to intracellular space as an *inward* ionic current.

Assuming the cell interior is electronegative, the effect of inward ionic current is a decrease in the magnitude of the electrical polarization typical of the cell interior, a phenomenon described as **depolarization**. Movement of cations from the intracellular to extracellular space (the opposite direction) is an *outward* ionic current. Outward ionic currents intensify a cell's electrical polarization, a phenomenon physiologists refer to as **hyperpolarization**.

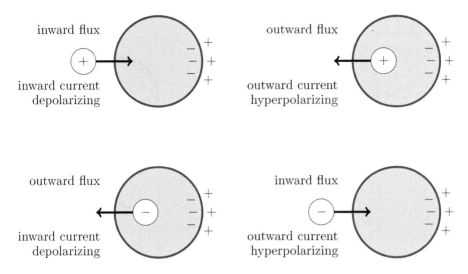

Figure 8.1   Fluxes of cations and anions and resulting inward and outward ionic membrane currents.

For anions (negative charge), an inward material flux – movement of an anion across a cell membrane from the extracellular to intracellular space – corresponds to an *outward* ionic current (hyperpolarizing). An outward anionic flux results in a depolarizing, inward ionic current (see Fig. 8.1).

8.3   Ionic Currents and Voltage

The standard method of visually displaying the dynamics of membrane potential is to graph $V(t)$ with voltage on the vertical axis. Because $V \approx -65$ mV for resting neurons, -100 mV and 0 mV are reasonable lower and upper limits for the V-axis. Fig. 8.2 (left column) shows typical depolarizing and hyperpolarizing voltage responses to step changes in ionic membrane current. These plots begin with $V = -65$ mV and the process of depolarization is recorded as an increase of $V(t)$ (upward). Hyperpolarization is recorded as a decrease of $V(t)$ (downward):

depolarization	increasing $V(t)$	$dV/dt > 0$
hyperpolarization	decreasing $V(t)$	$dV/dt < 0$

When physiologists visually display the dynamics of ionic membrane current, the convention is for outward ionic current to be positive ($I_{ion} > 0$) and inward ionic current to be negative ($I_{ion} < 0$). This choice is arbitrary, but consistent with the outward pointing arrow labelled I_{ion} in Fig. 8.2.[2] Fig. 8.2 (left column) shows that outward ionic currents, defined as positive ($I_{ion} > 0$), are hyperpolarizing ($dV/dt < 0$), while inward ionic currents, defined as negative ($I_{ion} < 0$), are depolarizing ($dV/dt > 0$).

Beware! Fig. 8.2 gives no information about the charge carrier. If a positive ionic membrane current (I_{ion}, blue curves) leading to hyperpolarization is mediated by

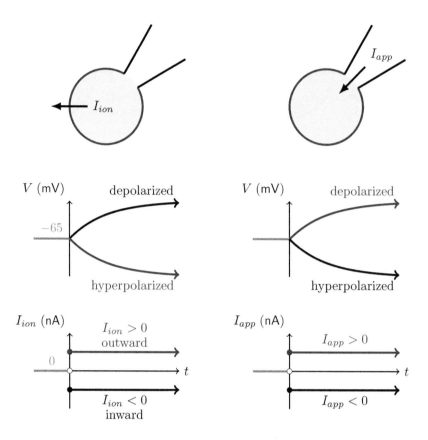

Figure 8.2   Positive (outward) ionic membrane currents are hyperpolarizing ($dV/dt < 0$); negative (inward) ionic currents are depolarizing ($dV/dt > 0$). Positive applied currents are depolarizing; negative applied currents are hyperpolarizing.

cations (e.g., through channels selective for potassium), then the ionic flux is directed as the ionic current. In this case the ionic flux is outward because positive ionic membrane current flows in the direction of the arrow labelled I_{ion}. However, positive (outward) ionic membrane currents can also be mediated by inward flux of anions (e.g., via chloride channels). Regardless of whether the charge carrier is a cation or anion:

	ionic current		effect on voltage
outward	$I_{ion} > 0$	positive	hyperpolarizing
inward	$I_{ion} < 0$	negative	depolarizing

8.4   Applied Currents and Voltage

Experimentalists manipulate the membrane potential of neurons in many ways. One method is to inject current using a penetrating glass microelectrode. Inward applied current corresponds to positive charge entering the cell and, consequently, inward applied current is depolarizing. For the visual display of changes in applied current, the convention is that inward (depolarizing) applied current is positive ($I_{app} > 0$). This is the opposite of the sign convention for ionic membrane current, and there are good reasons for the choice (see Chapter 12). However, it may be confusing at first.

Fig. 8.2 (right column) shows typical depolarizing and hyperpolarizing voltage responses to step changes in applied current. Note that a rapid increase in the current applied through an electrode (I_{app} changing from zero to a positive value) is indicated on a plot of $I_{app}(t)$ as an upward step (Fig. 8.2, bottom right). The model diagram (top right) indicates that positive applied current corresponds to positive charge entering the cell; consequently, the upward step in I_{app} results in depolarization ($dV/dt > 0$). Decreasing the applied current from zero to a negative value leads to hyperpolarization ($dV/dt < 0$):

applied current			effect on voltage
inward	$I_{app} > 0$	positive	depolarizing
outward	$I_{app} < 0$	negative	hyperpolarizing

8.5   The Current Balance Equation

Fig. 8.3 shows a schematic cell membrane punctured by an electrode with arrows indicating the direction of positive current (i.e., the direction of movement of positive charge) that corresponds to $I_{cap} > 0$, $I_{ion} > 0$ and $I_{app} > 0$. This diagram highlights the fact that the total membrane current (I_{mem}) is composed of two parts: (1) the *ionic current*[3] that is mediated by ions able to permeate the membrane via the aqueous pore of channels (described above); and (2) capacitative current (I_{cap}) that flows as positive and negative charge is separated across the plasma membrane.

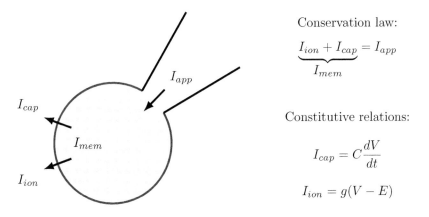

Conservation law:

$$\underbrace{I_{ion} + I_{cap}}_{I_{mem}} = I_{app}$$

Constitutive relations:

$$I_{cap} = C\frac{dV}{dt}$$

$$I_{ion} = g(V - E)$$

Figure 8.3   Conservation of charge requires a balance of currents entering and leaving the cell (Kirchoff's current law).

According to **Kirchoff's current law** (conservation of electric charge), the current entering the cell through the electrode (I_{app}) must equal the current leaving the cell through the ionic and capacitative membrane currents ($I_{mem} = I_{ion} + I_{cap}$). The following **current balance equation**,

$$\underbrace{I_{ion} + I_{cap}}_{I_{mem}} = I_{app}, \tag{8.1}$$

is a mathematical statement of Kirchoff's law. The convention that positive applied current is inward, while positive membrane current is outward, explains why I_{app} and I_{mem} are on opposite sides of this equation.

Because the applied current is under experimental control, we do not require a constitutive relation for I_{app}; it will be zero, some constant, or a prescribed function of time chosen by the experimentalist or modeler. But we are required to choose constitutive relations for the ionic and capacitative membrane currents. It is helpful to begin with the constitutive relations for I_{cap}. The properties of capacitors are likely unfamiliar to the reader, so please trust (for the moment) that an adequate constitutive relation is,

$$\text{current} \ominus I_{cap} = C\frac{dV}{dt} \ominus \text{capacitance} \cdot \frac{\text{voltage}}{\text{time}}, \tag{8.2}$$

where C, a nonnegative constant, is the capacitance of the cell membrane (see Discussion). Substituting Eq. 8.2 into Eq. 8.1 gives

$$I_{ion} + C\frac{dV}{dt} = I_{app}. \tag{8.3}$$

Rearranging this expression and making explicit the possible dependence of applied membrane current on time, and ionic membrane current on time and voltage, we arrive at the skeleton of a first-order differential equation for membrane voltage,

$$\frac{dV}{dt} = \frac{1}{C}\left[I_{app}(t) - I_{ion}(V,t)\right]. \tag{8.4}$$

The relationships between the terms of this current balance equation are a consequence of biophysicists' sign conventions (Sections 8.2–8.4). If you have trouble remembering these conventions, recall Eq. 8.4 and consider limiting cases.

- *Are negative applied currents hyperpolarizing or depolarizing?*
  To answer this, consider the possibility that $I_{app} < 0$ whilst $I_{ion} = 0$. Because $C > 0$, current balance (Eq. 8.4) requires $dV/dt < 0$. We conclude that negative applied current is hyperpolarizing.
- *Is negative ionic membrane current depolarizing or hyperpolarizing?*
  Well, if $I_{app} = 0$ and $I_{ion} < 0$, Eq. 8.4 says $dV/dt > 0$. So negative ionic membrane current is depolarizing.
- *Is negative ionic membrane current inward or outward?*
  We just convinced ourselves that negative ionic membrane current is depolarizing. Depolarization occurs when cations (positive charge) move from extracellular to intracellular. The direction of positive current corresponds to the direction of movement of positive charge. So negative ionic currents are inward membrane currents.

And so on. With practice it gets easier.

8.6   Constitutive Relation for Ionic Membrane Current

What is an appropriate constitutive relation for ionic membrane current? In the middle of the twentieth century, cell physiologists began to *measure* ionic membrane currents using a technique now called voltage-clamp recording (Chapter 12). Measurements of ionic membrane currents are visualized as plots of $I_{ion}(V)$ known as **current-voltage relations**.

8.6.1   Ohm's Law

As you may have learned in introductory physics, the constitutive relation for an Ohmic resistor in an electrical circuit is

$$\text{voltage} \ominus \ V_{res} = I_{res}R \ominus \text{current} \cdot \text{resistance} . \tag{8.5}$$

This is usually interpreted to mean that the voltage drop across the resistor (V_{res}) is directly proportional to the current passing through the resistor (I_{res}). The proportionality constant is the resistance (R). Voltage plotted as a function of current is a line through the origin with slope R $\ominus$ resistance = voltage/current, i.e., rise over run (Fig. 8.4, left). The constitutive relation Eq. 8.5 is known as **Ohm's law**. If the unit of voltage is millivolts (mV) and the unit of current is nanoamps (nA), then the unit of resistance is megaohms (MΩ = mV/nA), because Ω = V/A and $10^6 = 10^{-3}/10^{-9}$.

In physiological contexts, it is helpful to rearrange Eq. 8.4 as follows,

$$\text{current} \ominus \ I_{res} = \frac{1}{R}V_{res} = gV_{res} \ominus \text{conductance} \cdot \text{voltage} , \tag{8.6}$$

where the resistor is characterized by conductance, the reciprocal of the resistance,

$$\text{conductance} \ominus \ g = \frac{1}{R} \ominus \frac{1}{\text{resistance}} .$$

Fig. 8.4 (right) plots the version of Ohm's law given by Eq. 8.6. The current through the resistor (I_{res}) is directly proportional to the voltage drop across the resistor (V_{res}). The proportionality constant is the conductance (g). Current plotted as a function of voltage is a line through the origin with slope g $\ominus$ conductance = current/voltage (Fig. 8.4, right). The unit of conductance on this plot is the reciprocal of megaohms,

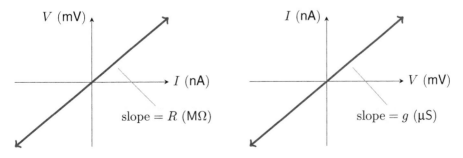

Figure 8.4   Graphical representations of Ohm's law.

that is, microsiemen ($\mu S = 1/M\Omega = nA/mV$), because $S = 1/\Omega = A/V$ and $10^{-6} = 1/10^6 = 10^{-9}/10^{-3}$.

8.6.2   Ionic Current, Conductance and Reversal Potential

The functional relationship between membrane voltage and ionic membrane currents can be complex. However, when the experimental situation is simplified by working with nonexcitable cells or, in the case of neurons, by pharmacologically blocking the ion channels responsible for action potentials, ionic membrane currents are sometimes Ohmic, that is, their current-voltage relation is well approximated by

$$\text{current} \ominus I_{ion} = g\,(V - E) \ominus \text{conductance} \cdot \text{voltage}. \qquad (8.7)$$

Fig. 8.5 shows that an Ohmic membrane current is an increasing function of voltage with constant slope, a line in the (V, I_{ion})-plane. The slope of this line, denoted g, has physical dimensions of conductance $=$ current/voltage (rise over run). On this plot, the unit of voltage is millivolts (mV) and the unit of current is nanoamps (nA), so the unit of conductance is microsiemen ($\mu S = nA/mV$, see Table 8.1).

The ionic membrane current plotted in Fig. 8.5 is zero when the membrane voltage is $E = -65\,mV$. Cell physiologists would describe this ionic membrane current as *reversing* at $V = E$, because I_{ion} is inward (negative) when $V < E$ and outward (positive) when $V > E$. The parameter E in the ionic membrane current (Eq. 8.7) is the *reversal potential* for I_{ion}, i.e., the voltage at which $I_{ion}(V)$ changes sign. The quantity $(V - E)$ in Eq. 8.7, the deviation of the membrane voltage from E, is referred to as the *driving force* for the ionic membrane current. Eq. 8.7 is fundamental because ionic

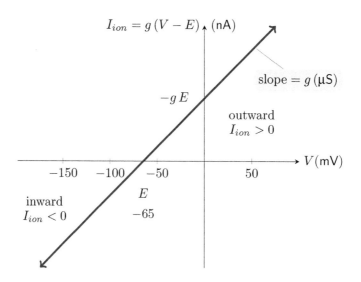

Figure 8.5   Current-voltage relation for an ionic membrane current of the form $I_{ion} = g(V - E)$ (Eq. 8.7). The horizontal axis intercept is $E = -65\,mV$ and the vertical axis intercept is $-gE$ (positive because $g > 0$ and $E < 0$).

Table 8.1 Physical dimensions and units for ionic membranes

Physical dimension	Unit	Abbreviation
$\text{conductance} = \dfrac{1}{\text{resistance}}$	$\text{siemen} = \dfrac{1}{\text{ohm}}$	$\text{S} = \dfrac{1}{\Omega}$
$\text{current} = \dfrac{\text{charge}}{\text{time}}$	$\text{ampere} = \dfrac{\text{coulomb}}{\text{second}}$	$\text{A} = \dfrac{\text{C}}{\text{s}}$
$\text{capacitance} = \dfrac{\text{charge}}{\text{voltage}}$	$\text{farad} = \dfrac{\text{coulomb}}{\text{volt}}$	$\text{F} = \dfrac{\text{C}}{\text{V}}$
$\text{resistance} = \dfrac{\text{voltage}}{\text{current}}$	$\text{ohm} = \dfrac{\text{volt}}{\text{ampere}}$	$\Omega = \dfrac{\text{V}}{\text{A}}$
$\text{conductance} = \dfrac{\text{current}}{\text{voltage}}$	$\text{siemen} = \dfrac{\text{ampere}}{\text{volt}}$	$\text{S} = \dfrac{\text{A}}{\text{V}}$

membrane currents that are not Ohmic are often conceptualized as the product of a linear driving force and a voltage-dependent conductance (see Chapter 10).

8.7 The Phase Diagram for Voltage of Passive Membranes

Assuming the voltage dependence of ionic membrane current has been measured, knowledge of the functional form of $I_{ion}(V)$ allows us to put flesh on the skeleton ODE for membrane voltage introduced in Section 8.5. Substituting the ionic membrane current of Eq. 8.7 into the current balance equation (Eq. 8.4), we obtain a linear first-order differential equation for membrane voltage,

$$\frac{dV}{dt} = \frac{1}{C}\left[I_{app} - \underbrace{g\,(V-E)}_{I_{ion}}\right]. \tag{8.8}$$

This ODE model for a passive neuronal membrane is one of the most important formulas in cellular biophysics. Assuming constant applied current, Eq. 8.8 is an autonomous first-order ODE that can be analyzed using the phase diagram technique (Chapter 3).

Fig. 8.6 shows the phase diagram for Eq. 8.8. Putting the ODE in the familiar form $dV/dt = f(V) = \jmath - kV$,

$$\frac{dV}{dt} = \underbrace{\frac{I_{app}+gE}{C}}_{\jmath} - \underbrace{\frac{g}{C}}_{k} V, \tag{8.9}$$

shows that $f(V)$ is a negatively sloped line (slope $= -k = -g/C < 0$) with vertical axis intercept $\jmath = (I_{app}+gE)/C$ (as shown in Fig. 8.6). The horizontal axis intercept is the

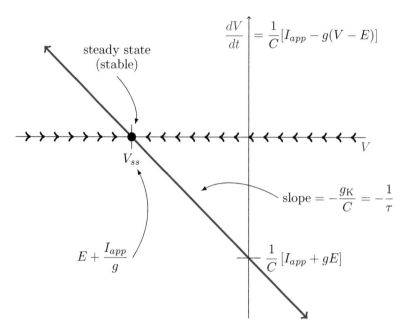

Figure 8.6   The phase diagram for a passive membrane (Eq. 8.9).

steady state membrane voltage $V_{ss} = J/k = E + I_{app}/g$. This steady state can also be found by setting the left side of Eq. 8.8 to zero, multiplying by C, and solving for V_{ss},

$$I_{app} - g\ (V_{ss} - E) \Rightarrow 0 = I_{app} - gV_{ss} + gE \Rightarrow gV_{ss} = I_{app} + gE,$$

which implies

$$\text{voltage} \ominus V_{ss} = E + \frac{I_{app}}{g} \ominus \frac{\text{current}}{\text{conductance}}. \tag{8.10}$$

Fig. 8.6 indicates that when the voltage is more hyperpolarized than this steady state $V < V_{ss}$, the rate of change of voltage is positive $dV/dt > 0$ (increasing, see rightward pointing arrows on the phase line). Conversely, when the voltage is more depolarized than this steady state $V > V_{ss}$, the rate of change of voltage is negative $dV/dt < 0$ (decreasing, leftward pointing arrows). We conclude that the steady state $V = V_{ss}$ is stable.

8.8   Exponential Time Constant for Membrane Voltage

Given the form of Eq. 8.9, it should be no surprise that solutions to the ODE with initial condition $V(0) = V_0$ are exponential relaxations,

$$V(t) = V_{ss} + (V_0 - V_{ss}) e^{-t/\tau}. \tag{8.11}$$

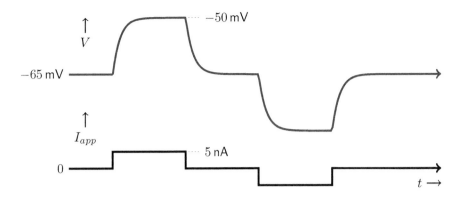

Figure 8.7 Passive membrane model response (V, blue) to positive and negative steps in applied current ($I_{app}(t)$ in Eq. 8.9). *Question*: Assuming the pulses are 50 ms in duration, what is the membrane time constant?

The exponential time constant $\tau = 1/k = C/g$ is referred to as the *membrane time constant*,

$$\text{time} \; \leftrightarrow \; \tau = \frac{C}{g} \; \leftrightarrow \; \frac{\text{capacitance}}{\text{conductance}} .$$

From Eq. 8.11 we understand that the voltage response to a step in applied current will be fast when τ is small (e.g., a high conductance membrane), and slow when τ is large (low conductance). Conductance is the reciprocal of resistance ($R = 1/g$), so the membrane time constant can also be written as $\tau = RC$, a formula you may recall from introductory physics (electric circuit theory). Membrane time constants of neurons are on the order of 5 to 50 ms.

Fig. 8.7 shows the membrane voltage of the ODE model (Eq. 8.9) when the applied current (I_{app}) steps from zero to a positive value for a duration of 50 ms (V depolarizes in response). After returning to zero for 50 ms, I_{app} steps down to a negative value (hyperpolarizing). Using the concept of $V = IR$ in the following form,

$$\text{resistance} \; \leftrightarrow \; R = \frac{\Delta V}{\Delta I} = \; \leftrightarrow \; \frac{\text{voltage}}{\text{current}} , \tag{8.12}$$

we can see that a $\Delta V = 10\,\text{mV}$ change in response to a $\Delta I = 5\,\text{nA}$ pulse indicates the membrane resistance is

$$R = \frac{10\,\text{mV}}{5\,\text{nA}} = 2\,\text{M}\Omega \quad (\text{megaohm}) . \tag{8.13}$$

Using the reciprocal relationship between conductance and resistance,

$$\text{conductance} \; \leftrightarrow \; g = \frac{1}{R} \; \leftrightarrow \; \frac{1}{\text{resistance}} ,$$

we see that the membrane conductance is

$$g = \frac{1}{2\,\text{M}\Omega} = 0.5\,\mu\text{S} \quad (\text{microsiemen}) .$$

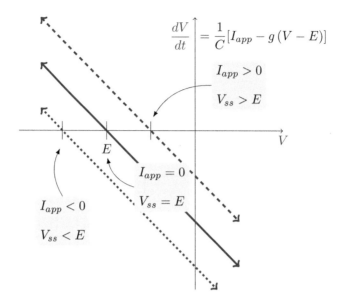

$$\frac{dV}{dt} = \frac{1}{C}[I_{app} - g(V - E)]$$

$$I_{app} > 0$$

$$V_{ss} > E$$

$I_{app} > 0$

$V_{ss} > E$

V

E

$I_{app} = 0$

$V_{ss} = E$

$I_{app} < 0$

$V_{ss} < E$

Figure 8.8   The phase diagram for the current balance equation depends on I_{app}. When $I_{app} > 0$, $V_{ss} > E$; conversely, $I_{app} < 0$, $V_{ss} < E$.

If one visually estimates the membrane time constant in Fig. 8.7 to be $\tau = C/g = 10\,\text{ms}$, then the membrane capacitance must be

$$C = g\,\tau = 0.5\,\mu\text{S} \cdot 10\,\text{ms} = 5\,\text{nF} \quad \text{(nanofarad)}.$$

Fig. 8.8 shows the phase diagram for the membrane model using three different values of applied current (cf. Fig. 8.7). In the absence of applied current ($I_{app} = 0$) the solid line intersects the voltage axis at $V_{ss} = E$. For $I_{app} > 0$ (depolarizing), $V_{ss} = E + I_{app}/g > E$ (dashed line). For $I_{app} < 0$ (hyperpolarizing), $V_{ss}E + I_{app}/g < E$ (dotted line). The plots are identical up to vertical translation, because I_{app}/C appears as an additive constant on the right side of the ODE (Eq. 8.7).

Membranes with Multiple Passive Currents

The astute reader may have noticed that the derivation of the current balance equation (Eq. 8.8) proceeded without specifying the charge carrier (potassium, sodium, etc.). The reason is that multiple ion channel types may contribute to the conductance g that determines the slope of the current-voltage relation of a passive membrane (Fig. 8.5). For a membrane permeable to both potassium and sodium, the Ohmic constitutive relation for total ionic current would have two components,

$$I_{ion} = \underbrace{g_\text{K}\,(V - E_\text{K})}_{I_\text{K}} - \underbrace{g_\text{Na}\,(V - E_\text{Na})}_{I_\text{Na}}, \tag{8.14}$$

where g_K is the potassium conductance, the reversal potential E_K is the Nernst equilibrium potential for potassium, I_K is the component of the membrane current for which potassium is the charge carrier, and similarly for g_Na, E_Na and I_Na.

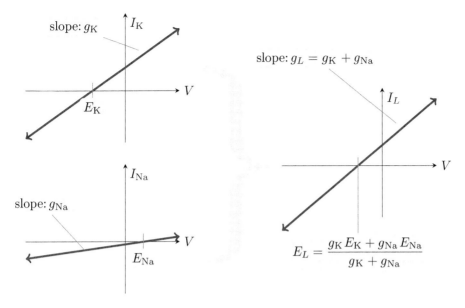

Figure 8.9   The sum of two passive currents is equivalent to a single effective current.

Fig. 8.9 (left column) shows current-voltage relations for I_K and I_{Na} with slopes indicating that $g_K > g_{Na}$, reversing at $E_K = -90\,\text{mV}$ and $E_{Na} = 60\,\text{mV}$, respectively. Fig. 8.9 (right column) illustrates that the sum of these two currents is equivalent to one effective "leakage" current

$$I_L = g_L(V - E_L)$$

where

$$g_L = g_K + g_{Na} \quad \text{and} \quad E_L = \frac{g_K E_K + g_{Na} E_{Na}}{g_K + g_{Na}} . \tag{8.15}$$

This equivalence is illustrated by Fig. 8.11 and can be confirmed by rearranging Eq. 8.14 (Problem 8.4).

8.9   Discussion

The Equivalent Circuit for the Current Balance Equation

When speaking to an electrical engineer or physicist, the ODE model of membrane voltage, the focus of this chapter, is easy to explain. Just show her a picture of the equivalent circuit for the membrane model (Fig. 8.10). The constitutive relations implied by each symbol in circuit diagrams are well known to engineers and physicists; thus, there is no ambiguity regarding the differential equation that the diagram implies (Eq. 8.1). The leads labelled ϕ_{in} and ϕ_{out} represent the intracellular and extracellular electrical potential; membrane voltage is $V = \phi_{in} - \phi_{out}$. The circuit includes three parallel pathways for current to flow across the cell plasma membrane

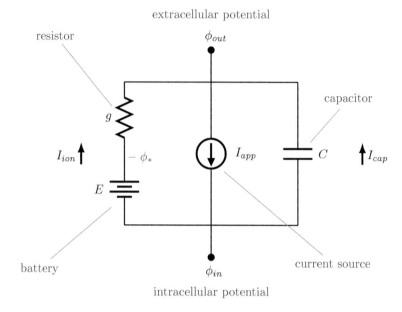

Figure 8.10 Passive sodium and potassium currents are equivalent to a single effective leakage current (Eqs. 8.8 and 8.15).

(cf. Fig. 8.3). The first of these is a current source labelled $I_{app} > 0$ with an arrow indicating that positive current will increase $V = \phi_{in} - \phi_{out}$. In parallel with the current source are two other pathways for current: a resistor and battery in series (I_{ion}) and a capacitor (I_{cap}).

Resistor and Battery

The constitutive relation for the resistor is $V_{res} = I_{res}R$ where R is resistance. Equivalently, $I_{res} = gV_{res}$ where g is the conductance. The voltage drop across the resistor is

$$V_{res} = \phi_* - \phi_{out},$$

where ϕ_* is located between the resistor and battery. The voltage drop across the battery is its electromotive force,

$$\phi_{in} - \phi_* = E.$$

The resistor and battery are in series, so the voltage drop over the resistor plus the voltage drop over the battery must equal V, that is,

$$V = \phi_{in} - \phi_{out} = \underbrace{\phi_{in} - \phi_*}_{E_K} + \underbrace{\phi_* - \phi_{out}}_{V_{res}} = E_K + V_{res}.$$

Thus, the voltage drop over the resistor is $V_{res} = V - E$ and the current flowing through the resistor is $I_{res} = g(V - E)$.

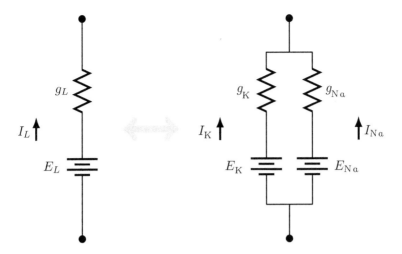

Figure 8.11   Circuit diagram showing battery, resistor, capacitor and current source. The ordinary differential equation membrane model that is the focus of this chapter can be derived from this diagram (Eq. 8.8).

Capacitor
Capacitance is the ability to store an electric charge. The capacitance of the cell plasma membrane is analogous to the mutual capacitance of a parallel-place capacitor, for which capacitance is essentially proportional to surface area and inversely proportional to the distance between the plates. Between these plates is usually a dielectric material whose properties also influence capacitance. The capacitance of cell membranes is also very nearly proportional to surface area.

  Capacitance is a measure of the voltage between plates of a capacitor – or between the inner and outer leaflet of the cell plasma membrane – that results when the plates (leaflets) have a charge of $+Q$ and $-Q$. Alternatively, one can think of capacitance as the amount of charge that is separated by a given voltage. In either case, capacitance with SI unit of farad (F) is given by the ratio of charge with SI unit of coulomb (C) to voltage with SI unit of volt (V),

$$\text{capacitance} \;\Leftrightarrow\; C = \frac{Q}{V} \;\Leftrightarrow\; \frac{\text{charge}}{\text{voltage}}. \tag{8.16}$$

Rearrange to obtain $CV = Q$, assume constant capacitance, and differentiate with respect to time,

$$\text{capacitance} \cdot \frac{\text{voltage}}{\text{time}} \;\Leftrightarrow\; C\frac{dV}{dt} = \frac{dQ}{dt} \;\Leftrightarrow\; \frac{\text{charge}}{\text{time}} = \text{current}. \tag{8.17}$$

Noting that charge per unit time is current,

$$\text{current} \;\Leftrightarrow\; I_{cap} = \frac{dQ}{dt} \;\Leftrightarrow\; \text{charge/time},$$

we obtain the constitutive relation for the capacitative membrane current given by Eq. 8.2.

Problem 8.1   Table 8.1 shows the SI units for current, capacitance, voltage and time are, respectively, amps (A), farads (F), volts (V) and seconds (s). Show that these units reconcile in Eqs. 8.16 and 8.17.

Units for the Current Balance Equation

Below is the standard form for the current balance equation of a passive membrane,

$$C\frac{dV}{dt} = I_{app} - g(V - E).$$

To perform calculations using this equation, we require units for capacitance, voltage, time, conductance and current that reconcile. There are two standard choices.

Choice 1

The first standard choice of units is: millivolt, millisecond, nanofarad, nanoamp, and microsiemen. To check that these units reconcile, write

$$nF \cdot \frac{mV}{ms} \overset{?}{=} nA \overset{?}{=} \mu S \cdot mV.$$

Next, interpret the prefixes milli ($m = 10^{-3}$), micro ($\mu = 10^{-6}$) and nano ($n = 10^{-9}$),

$$10^{-9} F \cdot \frac{10^{-3} V}{10^{-3} s} \overset{?}{=} 10^{-9} A \overset{?}{=} 10^{-6} S \cdot 10^{-3} V$$

$$10^{-9} F \cdot \frac{V}{s} \overset{?}{=} 10^{-9} A \overset{?}{=} 10^{-9} S \cdot V.$$

Multiply by 10^9 and use Table 8.1 to finish checking the equalities, as follows:

$$F \cdot \frac{V}{s} \overset{?}{=} A \overset{?}{=} S \cdot V$$

$$\frac{C}{V} \cdot \frac{V}{s} \overset{yes}{=} A \overset{yes}{=} \frac{A}{V} \cdot V. \checkmark$$

The last line uses $F = C/V$ (capacitance = charge/voltage), $A = C/s$ (current = charge/time) ,and $S = A/V$ (conductance = current/voltage).

Choice 2

The second standard choice of units is: millivolt, millisecond, microfarad per square centimeter, microamp per square centimeter, and millisiemen per square centimeter.

Problem 8.2   Show that these units reconcile,

$$\mu F/cm^2 \cdot \frac{mV}{ms} = \mu A/cm^2 = mS/cm^2 \cdot mV.$$

Observe that choice 1 involves three *extensive* quantities (capacitance, current and conductance) proportional to surface area of the cell membrane. In choice 2, these quantities are scaled by membrane surface area to yield three intensive quantities (capacitance/area, current/area, conductance/area). Choice 2 is motivated by the fact that the specific capacitance of biological membranes is relatively constant (about

$1\,\mu F/cm^2$), while the membrane surface area of neurons is variable. Choice 1 may be preferred in the context of experimental measurement of ionic membrane currents (see Chapter 12).

What is Meant by all These Terms?

The many concepts and terminologies of electrophysiology may be bewildering. Sometimes there are multiple words with exactly the same meaning, such as voltage and electrical potential. Other terms are closely related, but have subtly different meanings.

- *Voltage* and *electrical potential* are phrases with identical biophysical meaning. When the electrical context is clear, one might say *potential*. The usage of *voltage* implies an electrical context.
- *Membrane potential* = *membrane voltage* = *transmembrane potential* = *transmembrane voltage* are four phrases with identical meaning. Saying *trans*membrane emphasizes that this quantity is the difference between interior and exterior voltage (= electrical potential).
- The *resting membrane potential* is the membrane voltage (V) in a quiescent neuron or nonexcitable cell when recording with no applied current ($I_{app} = 0$).
- The *steady state membrane voltage* (V_{ss}) is found by setting $dV/dt = 0$ in a current balance equation. For Eq. 8.8, the result is $V_{ss} = E_L + I_{app}/g_L$ and $V \to V_{ss}$ at $t \to \infty$. When $I_{app} > 0$, V_{ss} is greater than the resting membrane potential (depolarized). When $I_{app} < 0$, V_{ss} is less than the resting membrane potential (hyperpolarized).
- The *Nernst equilibrium potential* for a given ion (also called the *diffusion potential*) is the voltage associated to electrochemical equilibrium of that ion, e.g., in the case of potassium: $E_K = RT/F \cdot \log([K^+]_o/[K^+]_i)$. The resting voltage of a passive membrane selectively permeable to one ion is given by the Nernst equilibrium potential for that ion.
- The *reversal potential* for an ionic current is the membrane potential that causes this ionic current to transition from negative (inward, depolarizing) to positive (outward, hyperpolarizing). The reversal potential for a highly selective current is the Nernst equilibrium potential for that ion, e.g., $I_K = g_K(V - E_K)$ reverses when $V = E_K$.

Supplemental Problems

Problem 8.3   Consider the following current balance equation

$$C\frac{dV}{dt} = I_{app} - \underbrace{g_K\,(V - E_K)}_{I_K} - \underbrace{g_{Na}\,(V - E_{Na})}_{I_{Na}}. \tag{8.18}$$

(a)   When $I_{app} < 0$ the applied current is [depolarizing / hyperpolarizing].

(b)   When $I_{Na} < 0$ the sodium current is [inward / outward].

(c)   What is the membrane time constant? Hint: Write your answer as a function of the parameters C, g_K and g_{Na}.

(d)   Assuming $I_{app} = 0$ and a potassium conductance that is ten times larger than the sodium conductance, estimate the steady state membrane potential. Assume $E_K = -90$ mV and $E_{Na} = 60$ mV.

(e)   Assuming a voltage in the range $E_K < V < E_{Na}$, are I_K and I_{Na} inward or outward? What if $V < E_K$?

Problem 8.4   Show Eq. 8.18 is equivalent to the following ODE with a single effective ionic current,

$$C\frac{dV}{dt} = I_{app} - g_L\left[V - E_L\right],$$ (8.19)

where g_L and E_L are given by Eq. 8.15.

Problem 8.5   A membrane has two passive ionic currents, one mediated by potassium, the other by sodium. Assuming $g_K = 80$ nS, $E_K = -90$ mV, $g_{Na} = 15$ nS and $E_{Na} = 50$ mV, what is the effective conductance? What is the effective reversal potential?

Problem 8.6   Suppose that a cell membrane has a resistance of 20 MΩ (megaohms) and a capacitance of 120 pF (picofarads). The initial membrane potential is -100 mV (millivolts) and the resting membrane potential is -60 mV. Hint: A farad-ohm is a second (F Ω = s).

(a)   What is the membrane conductance?

(b)   What is the membrane time constant?

(c)   What is the membrane voltage after 10 ms? Is this hyperpolarized or depolarized compared to the initial membrane voltage?

(d)   How much time must elapse before the voltage is -70 mV?

(e)   Assuming a specific capacitance of 1 μF/cm^2, estimate the membrane surface area of the neuron.

Problem 8.7   When the current balance equation for a membrane with three passive conductances,

$$C\frac{dV}{dt} = I_{app} - g_K\left(V - E_K\right) - g_{Na}\left(V - E_{Na}\right) - g_{Cl}\left(V - E_{Cl}\right),$$ (8.20)

is put in the form of Eq. 8.19, what are the effective leakage conductance (g_L) and leakage reversal potential (E_L)?

Problem 8.8   Familiarize yourself with how the sign of I_{app} influences the steady state voltage by numerically integrating Eq. 8.19 using software of your choosing. Confirm the exponential time constant for membrane voltage in your simulations is $\tau = C/g_L$. Suggested parameters: $g_L = 0.1$ mS/cm^2, $E_L = -65$ mV, $C = 1$ μF/cm^2 and $I_{app} = -5$ to 5 μA/cm^2.

Problem 8.9   One can model injection of an oscillatory current into a cell with passive membrane conductances using Eq. 8.19 with

$$I_{app} = I_0 + I_1 \cos(\omega t). \qquad (8.21)$$

In this equation, I_0 is the *direct* applied current (DC), I_1 is the *alternating* applied current (AC), $\omega = 2\pi f$ is the angular frequency, and f is the natural frequency (cycles per second).

Using parameters from Problem 8.8, numerically integrate Eqs. 8.19 and 8.21. With $I_0 = 0$ and a fixed value of I_1, explore the membrane response to oscillatory applied current at different frequencies. Create a summary plot of the amplitude of the voltage oscillations as a function of the stimulation frequency. Does the amplitude of the voltage oscillations decrease or increase as a function of stimulation frequency?

Problem 8.10   When an oscillatory applied current is written as the complex function of time, $I_{app}(t) = I_0 + I_1 e^{i\omega t}$, a complex-valued oscillatory membrane potential of the form

$$V(t) = V_0 + V_1 e^{i\omega t} \qquad (8.22)$$

satisfies Eq. 8.19. Using $e^{i\omega t} = \cos\omega t + i\sin\omega t$ show that the real part of $I_0 + I_1 e^{i\omega t}$ is equal to Eq. 8.21. The complex-valued V_1 reveals how the amplitude and phase of the oscillatory membrane response depends on stimulation frequency.

(a)   Determine V_1 as a function of the cellular (C, g_L, E_L) and stimulus (I_0, I_1, ω) parameters.
(b)   Find the amplitude of the membrane potential by calculating the complex modulus $|V_1| = \sqrt{V_1 \bar{V}_1}$ where $\bar{V}_1$ is the complex conjugate of V_1.
(c)   Calculate numerical values for $|V_1|$ from your formula using $\omega = 0.01$ and $0.1 \ \mathrm{ms}^{-1}$ (cycles per millisecond). Check that your results agree with your work in Problem 8.9.

Solutions

Problem 8.1   For Eq. 8.16, F = C/V (capacitance = charge/voltage) is found in Table 8.1. For Eq. 8.17,

$$A \overset{?}{=} F \cdot \frac{V}{time} = \frac{C}{time} = A, \ \checkmark \qquad (8.23)$$

where C/s = A (charge/time = current).

Problem 8.3   (a) Hyperpolarizing. (b) Inward. (c) The membrane time constant is $\tau = C/(g_K + g_{Na})$. This is shown by rewriting Eq. 8.18 as follows

$$\frac{dV}{dt} = \frac{1}{C} \left[I_{app} - g_K (V - E_K) - g_{Na} (V - E_{Na}) \right]$$

$$= \underbrace{\frac{I_{app} + E_L}{C}}_{J} - \underbrace{\frac{g_L}{C}}_{k = 1/\tau} V,$$

with g_L and E_L as in Eq. 8.15 and identifying $\tau = 1/k = 1/(g_L/C) = C/g_L = C/(g_K + g_{Na})$.

(d) Using $I_{app} = 0$ and $dV/dt = 0$,

$$0 = -g_K (V_{ss} - E_K) - g_{Na} (V_{ss} - E_{Na})$$

$$V_{ss} = \frac{g_K E_K + g_{Na} E_{Na}}{g_K + g_{Na}}.$$

For $g_K = 10 g_{Na}$ this becomes

$$V_{ss} = \frac{10 g_{Na} E_K + g_{Na} E_{Na}}{10 g_{Na} + g_{Na}} = \frac{10 E_K + E_{Na}}{11} = \frac{10 (-90 \text{ mV}) + 60 \text{ mV}}{11}$$

$$= \frac{-840 \text{ mV}}{11} = -76.36 \text{ mV}.$$

(e) When $E_K < V < E_{Na}$, I_K is outward (positive) and I_{Na} is inward:

$$\underset{+}{I_K} = \underset{+}{g_K} \underset{+}{(V - E_K)} \qquad \underset{-}{I_{Na}} = \underset{+}{g_{Na}} \underset{-}{(V - E_{Na})}.$$

Problem 8.4   To show Eq. 8.18 is equivalent to Eq. 8.19, perform the following algebraic steps,

$$C\frac{dV}{dt} = I_{app} - g_K (V - E_K) - g_{Na} (V - E_{Na})$$

$$= I_{app} - g_K V + g_K E_K - g_{Na} V + g_{Na} E_{Na}$$

$$= I_{app} - (g_K + g_{Na}) V + g_K E_K + g_{Na} E_{Na}.$$

Finally, factor out the sum of the potassium and sodium conductances to obtain

$$C\frac{dV}{dt} = I_{app} - (g_K + g_{Na}) \left[V - \frac{g_K E_K + g_{Na} E_{Na}}{g_K + g_{Na}} \right].$$

Problem 8.5   Following Eq. 8.15, we calculate $g_L = g_K + g_{Na} = 95 \text{ nS}$ and

$$E_L = \frac{(80 \text{ nS})(-90 \text{ mV}) + (15 \text{ nS})(50 \text{ mV})}{80 \text{ nS} + 15 \text{ nS}} \approx -68 \text{ mV}.$$

This result is consistent with Fig. 8.9 (right) that shows $E_K < E_L < E_{Na}$.

Problem 8.6   (a) The relevant formula and physical dimensions are

conductance $\Leftrightarrow g = 1/R \Leftrightarrow 1/\text{resistance}.$

Using these, the membrane conductance is found as follows

$$g = \frac{1}{R} = \frac{1}{20 \text{ M}\Omega} = \frac{1}{20 \times 10^6 \text{ } \Omega} = 5 \times 10^{-8} \text{ S} = 50 \text{ nS}.$$

(b) The relevant formula and physical dimensions are

time $\Leftrightarrow \tau = RC \Leftrightarrow$ resistance $\cdot$ capacitance.

The membrane time constant is calculated as follows:

$\tau = 20\ \text{M}\Omega \cdot 120\ \text{pF} = 20 \times 10^6\ \Omega \cdot 120 \times 10^{-12}\ \text{F} = 0.0024\ \text{s} = 2.4\ \text{ms}.$

(c) The analytical solution is Eq. 8.11 with $V_0 = -100\ \text{mV}$ and $V_{ss} = -60\ \text{mV}$. Using $t = 10\ \text{ms}$ and $\tau = 2.4\ \text{ms}$,

$$V(10\ \text{ms}) = -60\ \text{mV} + \underbrace{[-100\ \text{mV} - (-60\ \text{mV})]}_{-40\ \text{mV}} \underbrace{\exp\left(-\frac{10\ \text{ms}}{2.4\ \text{ms}}\right)}_{0.0155}$$

$$= -60\ \text{mV} - 0.62\ \text{mV} \approx -61\ \text{mV}.$$

Because 10 ms is greater than four times the membrane time constant $\tau = 2.4\ \text{ms}$, the initial deviation of $-40\ \text{mV}$ has almost completely decayed away. In fact, at $t = 10\ \text{ms}$ the fraction $(-0.62\ \text{mV})/(-40\ \text{mV}) = 0.0155$ remains (about 1.6%).
(d) Using Eq. 8.11, write $V(t) = -70\ \text{mV}$ and solve for t,

$$V(t) = -70\ \text{mV}$$

$$-60\ \text{mV} + -40\ \text{mV}\ \exp\left(-\frac{t}{2.4\ \text{ms}}\right) = -70\ \text{mV}$$

$$-40\ \text{mV}\ \exp\left(-\frac{t}{2.4\ \text{ms}}\right) = -10\ \text{mV}$$

$$\exp\left(-\frac{t}{2.4\ \text{ms}}\right) = \frac{10\ \text{mV}}{40\ \text{mV}} = \frac{1}{4}$$

$$-\frac{t}{2.4\ \text{ms}} = \overbrace{\ln\left(\frac{1}{4}\right)}^{-1.386}$$

$$t = 1.386 \cdot 2.4\ \text{ms} = 3.327\ \text{ms}.$$

(f) The relevant formula and physical dimensions are

$$\text{area} \Leftrightarrow S = \frac{C^*}{C} \Leftrightarrow \frac{\text{capacitance}}{\text{capacitance/area}}$$

where C^* is the total capacitance (an extensive quantity) and C is the specific capacitance (intensive). Using $1\ \text{cm} = 10^4\ \mu\text{m}$,

$$S = \frac{C^*}{C} = \frac{120\ \text{pF}}{1\ \mu\text{F/cm}^2} = \frac{120 \times 10^{-12}\ \text{F}}{10^{-6}\ \text{F}/(10^4\ \mu\text{m})^2} = 1.2 \times 10^4\ \mu\text{m}^2,$$

that is, about 12 000 square micrometers.

Problem 8.9 The amplitude of the voltage oscillations decreases as a function of stimulation frequency.

Problem 8.10   (a) Differentiating Eq. 8.22 we obtain,

$$\frac{dV}{dt} = V_1 e^{i\omega t} i\omega. \tag{8.24}$$

Substituting $V(t)$, dV/dt and Eq. 8.21 into Eq. 8.19 yields,

$$C\frac{dV}{dt} = I_{app} - g_L (V - E_L)$$

$$CV_1 e^{i\omega t} i\omega = I_0 + I_1 e^{i\omega t} - g_L \left(V_0 + V_1 e^{i\omega t} - E_L\right).$$

We are able to solve for V_0 and V_1 by separately equating those terms in the above expression that are constant, and those that are a function of time through the term $e^{i\omega t}$. The constant terms yield

$$0 = I_0 - g_L (V_0 - E_L)$$
$$V_0 = E_L + I_0/g_L,$$

where the value of the DC voltage V_0 is analogous to Eq. 8.10. Note that V_0 is real valued because E_L, I_0 and g_L are real valued. Next, equate the coefficients that scale the oscillatory terms,

$$CV_1 i\omega = I_1 - g_L V_1$$
$$V_1 (g_L + Ci\omega) = I_1$$

$$V_1 = \frac{I_1}{g_L + Ci\omega} = \frac{I_1/g_L}{1 + i\omega\tau}.$$

Observe that I_1/g_L is real valued, but V_1 is complex.
(b) A calculation of the complex modulus gives

$$|V_1| = \sqrt{V_1 \bar{V}_1} = \sqrt{\frac{I_1/g_L}{1 + i\omega\tau} \cdot \frac{I_1/g_L}{1 - i\omega\tau}}$$

$$= \frac{I_1}{g_L}\sqrt{\frac{1}{1 + i\omega\tau} \cdot \frac{1}{1 - i\omega\tau}} = \frac{I_1}{g_L}\sqrt{\frac{1}{1 + \omega^2\tau^2}},$$

where we have used $(1 + i\omega\tau)(1 - i\omega\tau) = 1 - i^2\omega^2\tau^2 = 1 + \omega^2\tau^2$. The complex modulus $|V_1|$ is a real-valued decreasing function of ω.
(c) Using two different values of ω, we wish to evaluate

$$|V_1| = \frac{I_1}{g_L}\sqrt{\frac{1}{1 + (\omega\tau)^2}}.$$

Using $g_L = 0.1$ mS/cm^2, $C = 1$ μF/cm^2 and $I_1 = 5$ μA/cm^2,

$$\tau = \frac{C}{g_L} = \frac{1 \ \mu\text{F/cm}^2}{0.1 \ \text{mS/cm}^2} = 10 \text{ ms} \quad \text{and} \quad \frac{I_1}{g_L} = \frac{5 \ \mu\text{A/cm}^2}{0.1 \ \text{mS/cm}^2} = 50 \text{ mV}.$$

For $\omega = 0.01$ ms^{-1}, $\omega\tau = (0.01$ ms$^{-1})(10$ ms$) = 0.1$ and

$$|V_1| = 50 \text{ mV}\sqrt{\frac{1}{1 + (0.1)^2}} = 50 \text{ mV} \cdot 0.995 = 49.75 \text{ mV}.$$

For $\omega = 0.1\ \text{ms}^{-1}$, $\omega\tau = (0.1\ \text{ms}^{-1})(10\ \text{ms}) = 1$ and

$$|V_1| = 50\ \text{mV}\sqrt{\frac{1}{1+1^2}} = 50\ \text{mV} \cdot 0.7071 = 35.36\ \text{mV}.$$

We conclude that the amplitude of the oscillation ($|V_1|$) is smaller when the stimulation frequency (ω) is larger.

Notes

1. Thanks to Benjamin Franklin, electric current is defined as positive in the direction of movement of positive charge in spite of the fact that electricity passes through metallic conductors as a flow of negatively charged electrons.
2. The direction of the arrow is significant because current is a signed quantity. The orientation indicates the direction of current flow when $I_{ion} > 0$.
3. Ionic membrane current is *resistive*, as opposed to *capacitative*.

9 GHK Theory of Membrane Permeation

The previous chapter assumed Ohmic membrane currents whose current-voltage relations are lines in the (V, I)-plane as in Fig. 8.5. The classical theory of ionic permeation of membranes predicts deviations from the simple Ohmic form $I = g(V - E)$ known as rectification.

9.1 Goldman-Hodgkin-Katz Theory – Assumptions

A good starting point for understanding the process of ionic membrane permeation is the classical theory developed by David Goldman, Alan Hodgkin and Bernard Katz. GHK theory does not provide a detailed description of ions moving through the aqueous pore of individual channels (as does contemporary theory, see Discussion). Rather, the membrane is viewed as a homogenous slab with physical properties captured by the constitutive relation for electrodiffusion (diffusion and electrophoresis combined). Ionic permeation occurs along a spatial coordinate x that is perpendicular to the plane of the membrane. Fig. 9.1 shows $x = 0$ at the interface between the cell cytoplasm and inner leaflet, while $x = L \approx 4\,\text{nm}$ is at the interface between the outer leaflet and the extracellular space. GHK theory begins with the following two assumptions.

- **Independence assumption**: physiological ions (potassium, sodium, calcium, chloride) cross the cell membrane independently without interacting with each other.
- **Constant field assumption**: the electric field within the membrane that is experienced by permeating ions is constant.

Using these assumptions and the Nernst-Planck constitutive relation for electrodiffusion, one may derive the **GHK current equation**,

$$\frac{\text{current}}{\text{length}^2} \ominus I_S = z_S F \cdot P_S \frac{z_S V}{V_\theta} \left(\frac{[S]_i - [S]_o e^{-zV/V_\theta}}{1 - e^{-zV/V_\theta}} \right), \tag{9.1}$$

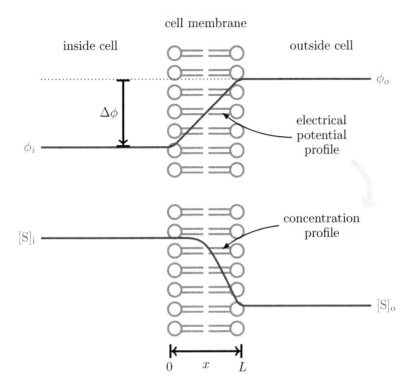

Figure 9.1 Goldman-Hodgkin-Katz current theory assumes a constant electric field, that is, the electrical potential profile within the membrane is linear. From this assumption and the values of ϕ_o, ϕ_i, $[S]_o$, $[S]_i$ the theory derives a self consistent ionic current and concentration profile.

where $V_\theta = RT/F \approx 25\,\text{mV}$ (Eq. 7.7), S is the permeating ion, and z_S, P_S, $[S]_i$ and $[S]_o$ are the ion's valence, permeability, and intracellular and extracellular concentrations. Eq. 9.1 has physical dimensions of current density (current/length^2). When multiplied by cell surface area, the GHK equation yields a bona fide current-voltage relation, that is, I_S as a function of V.

9.2 Physical Dimensions of the GHK Current Equation

Eq. 9.1 highlights the factor $z_S F$ (valence times Faraday's constant) that relates current density (current/length^2) and flux density (#/length^2),

$$\frac{\text{current}}{\text{length}^2} \ominus I = zFJ \ominus 1 \cdot \frac{\text{charge}}{\#} \cdot \frac{\#}{\text{time} \cdot \text{length}^2}, \tag{9.2}$$

where current = charge/time. The flux density J may be either positive or negative, corresponding to an outward or inward material flux. The valence z_S and current density I_S are also signed. The physical dimensions of the GHK flux density reconcile as follows,

$$\frac{\#}{\text{length}^2} \ominus J_S = P_S \frac{z_S V}{V_\theta} \left(\frac{[S]_i - [S]_o e^{-z_S V/V_\theta}}{1 - e^{-z_S V/V_\theta}} \right) \ominus \frac{\text{length}}{\text{time}} \cdot \frac{\text{voltage}}{\text{voltage}} \cdot \frac{\#}{\text{length}^3},$$

where $F \ominus$ charge/# and $P_S \ominus$ length/time (permeability). The quantity $z_S V/V_\theta = z_S VF/RT$ is dimensionless.[1] Both $[S]_i$ and $[S]_o$ are concentrations (conc = #/length^3). The exponentials are dimensionless, so the quantity in parentheses has physical dimensions of conc.

9.3   The Goldman-Hodgkin-Katz Current Equation

Fig. 9.2 plots the GHK current[2] (Eq. 9.1) using parameters for sodium. The current-voltage relation reverses at the Nernst equilibrium potential for sodium (E_{Na}), but it is not Ohmic.[3] Rather, the predicted sodium current is a concave function of voltage. Fig. 9.3 shows a convex current-voltage relation that uses parameters for potassium.

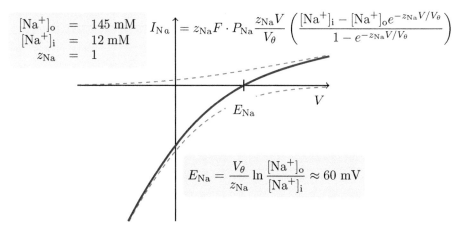

Figure 9.2   Inward rectifying (concave) GHK current-voltage relation for sodium.

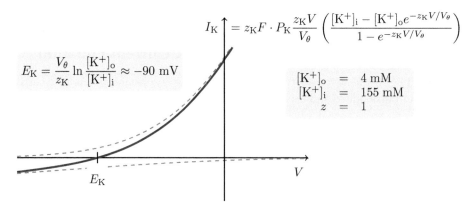

Figure 9.3   Outward rectifying (convex) GHK current-voltage relation for potassium.

The GHK currents I_{Na} and I_K exhibit **rectification**, a term referring to a convex (or concave) current-voltage relation.

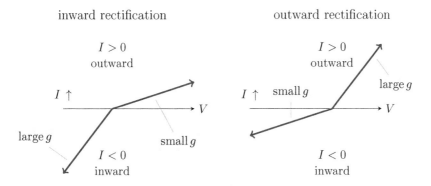

An *inward rectifying* ionic current (above) flows more easily inward. The conductance is greater when the membrane voltage is such that $I_{ion}(V) < 0$. An *outward rectifying* ionic current flows more easily outward. The conductance is greater when $I_{ion}(V) > 0$.

Problem 9.1   Figs. 9.2 and 9.3 are nonlinear current-voltage relations that reverse at the Nernst equilibrium potential for the permeating ion ($I_S = 0$ when $V = E_S$). Confirm this by setting $I_S = 0$ in Eq. 9.1 and solving for V.

9.4   Limiting Conductances Implied by GHK Theory

Fig. 9.4 shows that the GHK current-voltage relation changes with ionic concentrations inside ($[S]_i$) and outside ($[S]_o$) the cell. When the sodium concentrations are physiological, their ratio is $[Na^+]_o/[Na^+]_i \approx 16$ and I_{Na} reverses at $E_{Na} \approx 60\,mV$. However, if $[Na^+]_o$ is decreased, E_{Na} decreases (see Fig. 9.4). The inward rectification of the GHK sodium current is also attenuated by decreasing $[S]_o$. When $[S]_o/[S]_i \approx 1$, I_{Na} exhibits no rectification and $E_{Na} = 0\,mV$.

Problem 9.2   Plot Eq. 9.1 using the parameters of Fig. 9.3 but with a larger value of $[K^+]_o$. Does E_K increase or decrease? Does the outward rectification of I_K become more or less pronounced?

The GHK current equation (Eq. 9.1) predicts inward or outward rectification depending on the relative sizes of $[S]_i$ and $[S]_o$ and the sign of z_S. This may be confirmed analytically by taking the second derivative of I_S with respect to V. A simpler approach is to calculate (and then compare) the slopes of Eq. 9.1 when V is either very hyperpolarized ($V \to -\infty$) or very depolarized ($V \to \infty$), as follows.

Assume for the moment that $z_S > 0$. In that case, when V is extremely depolarized (i.e., large compared to V_θ), the exponential will be a small positive number $0 < e^{-z_S V/V_\theta} \ll 1$ and, consequently, the term in the parentheses is well approximated

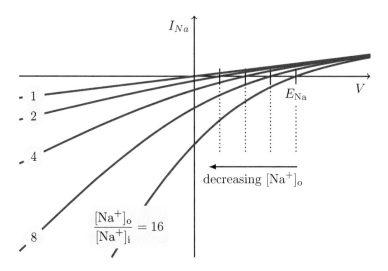

Figure 9.4   GHK theory predicts concentration dependence of the current-voltage relation.

by $[S]_i$. When V is very hyperpolarized, the term in the parentheses is well approximated by $[S]_o$. Thus, we conclude that for cations ($z_S > 0$),

$$I_S \to z_S F \cdot P_S \frac{z_S V}{V_\theta}[S]_i \text{ as } V \to \infty$$

$$I_S \to z_S F \cdot P_S \frac{z_S V}{V_\theta}[S]_o \text{ as } V \to -\infty.$$

Observe that both limiting expressions for I_S are proportional to voltage V. The proportionality constants $z_S^2 F P_S [S]_i / V_\theta$ and $z_S^2 F P_S [S]_o / V_\theta$ are the limiting slopes of the current-voltage relation. The expressions are identical except that one involves $[S]_i$ and the other involves $[S]_o$. This means you can sketch a qualitatively correct GHK current-voltage relation by drawing two lines through the origin whose slopes have ratio $[S]_o/[S]_i$. For example, in Fig. 9.5, $[Na^+]_o \approx 8 \times [Na^+]_i$ so the limiting slope of the sodium current when inward ($I_{Na} < 0$) is eight times as large as the limiting slope when outward ($I_{Na} > 0$). To see this, compare the different rises of the dashed lines for the same run. In Fig. 9.6, $[K^+]_i \approx 40 \times [K^+]_o$ and the limiting slope when $I_K > 0$ (outward) is forty times as large as the limiting slope when $I_K < 0$ (inward).

Problem 9.3   Show that $z_S^2 F P_S [S]/V_\theta \ominus$ conductance/length^2.

Problem 9.4   Do you expect the GHK current-voltage relation for calcium to be inward rectifying or outward rectifying?

In the case of negatively charged ions such as chloride, the roles of $[S]_i$ and $[S]_o$ are reversed. For permeating anions ($z_S < 0$),

$$I_S \to z_S F \cdot P_S \frac{z_S V}{V_\theta}[S]_o \text{ as } V \to \infty$$

$$I_S \to z_S F \cdot P_S \frac{z_S V}{V_\theta}[S]_i \text{ as } V \to -\infty.$$

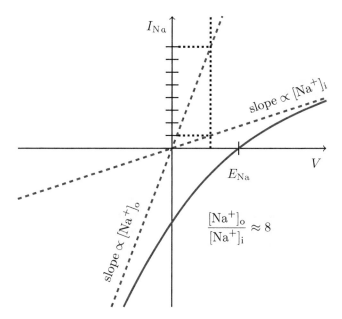

Figure 9.5   Method for sketching a GHK-type current voltage relation.

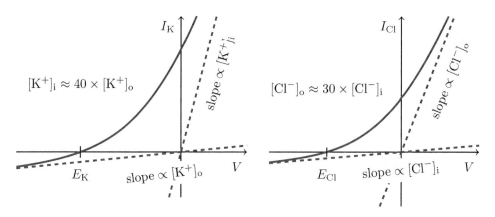

Figure 9.6   GHK potassium (cation) and chloride (anion) currents are similar.

Observe that the sign of the rectification (inward versus outward) for anions is opposite that of cations with the same intracellular and extracellular concentrations. For example, $[S]_o > [S]_i$ for both sodium and chloride, but the GHK current-voltage relation for chloride is outward rectifying, not inward rectifying, as for sodium (see Fig. 9.2).

9.5   Derivation of the GHK Current Equation

How do the independence and constant field assumptions yield the GHK current equation (Eq. 9.1)?

9.5.1   Constant Electric Field

GHK theory assumes the electrical potential varies in one spatial dimension transverse to the plane of the plasma membrane (horizontally in Fig. 9.1). Because the
electric field is the opposite of the spatial derivative of the electrical potential, the GHK
constant field assumption is

$$E = -\frac{d\phi}{dx} = \text{constant}. \tag{9.3}$$

The negative sign is necessary because the direction of the electric field is defined to
be the direction of spontaneous migration of a positively charged particle.

In Fig. 9.1, the cell interior is on the left and the exterior is on the right. For a
resting neuron, the interior and exterior electric potentials satisfy $\phi_i < \phi_o$. Thus, the
spatial derivative of the electrical potential is positive ($d\phi/dx > 0$) and the electric field
is negative ($E < 0$). The spatial derivative (slope = rise/run) of the electrical potential is

$$\frac{d\phi}{dx} = \frac{\phi_o - \phi_i}{L} = -\frac{\phi_i - \phi_o}{L} = -\frac{V}{L} > 0.$$

We conclude that in resting cells the electric field given by $E = -d\phi/dx = V/L < 0$ is
pointing inward. A positively charged ion has lower electrical potential energy near
the inner leaflet than near the outer leaflet of the cell plasma membrane.

9.5.2   The Nernst-Planck Constitutive Relation

The derivation of the GHK current equation uses a constitutive relation for electrodiffusion known as the Nernst-Planck equation:

$$J_S = -D_S \left(\frac{d[S]}{dx} + \frac{z_S}{V_\theta}[S]\frac{d\phi}{dx} \right), \tag{9.4}$$

where $V_\theta = RT/F$. In this expression, $D_S \ominus \text{length}^2/\text{time}$ is a diffusion constant, $[S]$
denotes the concentration of the permeating ion S within the membrane along the
coordinate x, and $d[S]/dx$ is the spatial derivative of this concentration profile.

The Nernst-Planck equation is the combination of two constitutive relations,

$$\frac{\#}{\text{time} \cdot \text{length}^2} \ominus J_S = J_S^{diff} + J_S^{drift}.$$

One of these is Fick's law for diffusion,

$$J_S^{diff} = -D_S\frac{d[S]}{dx} \ominus \frac{\text{length}^2}{\text{time}} \cdot \frac{\#/\text{length}^3}{\text{length}}, \tag{9.5}$$

that assumes the flux density due to diffusion is proportional to the opposite of the
spatial derivative of concentration (compare Eq. 2.32). The second constitutive relation
accounts for electrophoresis, that is, migration of the permeating S ions in response to
the electric field,

$$J_S^{drift} = -D_S\frac{z_S}{V_\theta}[S]\frac{d\phi}{dx} \ominus \frac{\text{length}^2}{\text{time}} \cdot \frac{1}{\text{voltage}} \cdot \frac{\#}{\text{length}^3} \cdot \frac{\text{voltage}}{\text{length}}. \tag{9.6}$$

Multiplying the Nernst-Planck flux density (Eq. 9.4) by $z_S F$ yields the current density,

$$I_S = -z_S F \cdot D_S \left(\frac{d[S]}{dx} + \frac{z_S F}{RT}[S]\frac{d\phi}{dx} \right) . \qquad (9.7)$$

Using the constant field assumption $E = d\phi/dx = -V/L$, and the boundary conditions

$$[S](x = 0) = [S]_i \quad \text{and} \quad [S](x = L) = [S]_o , \qquad (9.8)$$

for the inner and outer leaflets, respectively, Eq. 9.7 may be integrated to obtain the GHK current equation (Eq. 9.1).

Problem 9.5 Complete the derivation of the GHK current equation. Hint: Rearrange Eq. 9.7 to obtain

$$\frac{d[S]}{dx} = \alpha[S] - \beta , \qquad (9.9)$$

where $\alpha = z_S V/(LV_\theta)$ and $\beta = I_S/(z_S F D_S)$. Integrate and use Eq. 9.8 to obtain Eq. 9.1.

Problem 9.6 Explain why the negative signs in the drift and diffusion constitutive relations (Eqs. 9.5 and 9.6) are required for the correct sign of flux densities J_S^{diff} and J_S^{drift}.

9.6   Further Reading and Discussion

A modern perspective on the relationship between structure and function of ion channels can be found in Nonner et al. (1999). A classical text on electrodiffusion of ions in solution is Rubinstein (1990).

GHK Theory Prediction when $V = 0$

For some intracellular membranes such as the endoplasmic reticulum (ER), the transmembrane potential is essentially zero ($V = 0$). What current density does GHK theory predict in this case? This question cannot be answered simply by substitution of $V = 0$ into

$$I_S = z_S F \cdot P_S \frac{z_S V}{V_\theta} \left(\frac{[S]_i - [S]_o e^{-z_S V/V_\theta}}{1 - e^{-z_S V/V_\theta}} \right) ,$$

because both the numerator $z_S F P_S z_S V \left([S]_i - [S]_o e^{-z_S V/V_\theta} \right)/V_\theta$ and denominator $1 - e^{-z_S V/V_\theta}$ evaluate to zero when $V = 0$. However, the result can be obtained using L'Hopital's rule and the observation that

$$\lim_{x \to 0} \frac{ax}{1 - e^{-ax}} = \lim_{x \to 0} \frac{[ax]'}{[1 - e^{-ax}]'} = \lim_{x \to 0} \frac{a}{ae^{-ax}} = \lim_{x \to 0} e^{ax} = 1.$$

Using this limit it can be shown that the GHK current equation evaluated at $V = 0$ is

$$\lim_{V \to 0} I_S = z_S F \cdot P_S \left([S]_i - [S]_o \right) ,$$

and, consequently, the flux density $J_S = I_S/z_S F$ is

$$\underbrace{\frac{\#}{\text{time} \cdot \text{length}^2}}_{} \ominus J_S = P_S ([S]_i - [S]_o) \ominus \underbrace{\frac{\text{length}}{\text{time}}}_{} \cdot \underbrace{\frac{\#}{\text{length}^3}}_{} . \tag{9.10}$$

We conclude that when $V = 0$ the flux density predicted by the GHK equation is proportional to the concentration difference ($[S]_i - [S]_o$), consistent with the constitutive relation $j_{leak} = v_{leak}(c_{er} - c_{cyt})$ used in Problem 2.23.

Nernst Equilibrium Potential Redux

The Nernst-Planck constitutive relation (Eq. 9.4) can be used as the starting point in a derivation of the Nernst equilibrium potential. To simplify the notation, define a dimensionless potential $\hat{\phi} = \phi/V_\theta$ where $V_\theta = RT/F$. The Nernst-Planck flux density (Eq. 9.4) is then

$$J = -D \left(\frac{d[S]}{dx} + z[S]\frac{d\hat{\phi}}{dx} \right) .$$

Equilibrium implies zero flux density ($J = 0$),

$$0 = \frac{d[S]}{dx} + z[S]\frac{d\hat{\phi}}{dx} . \tag{9.11}$$

Multiplying by dx and separating variables,

$$\frac{d[S]}{[S]} = -z \, d\hat{\phi} ,$$

where both $[S](x)$ and $\hat{\phi}(x)$ are functions of the independent variable x. Integrating from the inner ($x = 0$) to outer ($x = L$) leaflet of the cell plasma membrane,

$$\int_{[S]_i}^{[S]_o} \frac{d[S]}{[S]} = -z \int_{\hat{\phi}_i}^{\hat{\phi}_o} d\hat{\phi}$$

$$\ln[S]_o - \ln[S]_i = -z(\hat{\phi}_o - \hat{\phi}_i)$$

where we have used $[S]_i = [S](x = 0)$, $[S]_o = [S](x = L)$, $\hat{\phi}_i = \hat{\phi}(x = 0)$ and $\hat{\phi}_o = \hat{\phi}(x = L)$. Now identifying the Nernst equilibrium potential as the potential difference $E = \phi_i - \phi_o$ we have $\hat{\phi}_o - \hat{\phi}_i = -E/V_\theta$ and thus

$$\ln \frac{[S]_o}{[S]_i} = \frac{z}{V_\theta} E \Rightarrow E = \frac{V_\theta}{z} \ln \frac{[S]_o}{[S]_i} . \tag{9.12}$$

Goldman-Hodgkin-Katz Voltage Equation

The GHK current density equation can be used to derive an equation for the equilibrium potential of a membrane permeable to multiple *monovalent* cations and anions. The **GHK voltage equation** for sodium, potassium and chloride is

$$E_{ghk} = \frac{RT}{F} \ln \left(\frac{P_{Na}[Na^+]_o + P_K[K^+]_o + P_{Cl}[Cl^-]_i}{P_{Na}[Na^+]_i + P_K[K^+]_i + P_{Cl}[Cl^-]_o} \right) . \tag{9.13}$$

Observe that the extracellular cationic (sodium and potassium) concentrations are in the numerator, while the extracellular anionic (chloride) concentration is in the denominator.

Problem 9.7 Beginning with the GHK current equation in the form

$$I_S = P_S z_S^2 F \hat{V} \left(\frac{[S]_i - [S]_o e^{-z_S \hat{V}}}{1 - e^{-z_S \hat{V}}} \right) \quad \text{where} \quad \hat{V} = V/V_\theta, \tag{9.14}$$

use the zero flux condition $0 = J_{Na} + J_K + J_{Cl}$ for the steady state voltage to derive Eq. 9.13.

Poisson-Nernst-Planck

The Goldman-Hodgkin-Katz theory of membrane permeation is relatively straight-forward because of the independence and constant field assumptions. The modern theory of membrane permeation makes neither of these assumptions and utilizes many of the mathematical principles of semiconductor physics. GHK theory accounts for the influence of the electrical potential on concentration, but not for the effect of the concentration on potential (Fig. 9.1, gray arrow). In the modern theory of ion permeation, the electrical potential solves a constitutive relation known as Poisson's equation,

$$\frac{d^2 \hat{\phi}}{dx^2} = \alpha \sum_S z_S [S], \tag{9.15}$$

where $\hat{\phi} = \phi / V_\theta$ and $\alpha = F^2 / \epsilon RT$, where ϵ is the permittivity. The Poisson equation is solved simultaneously with a collection of Nernst-Planck equations,

$$\text{for each ion S:} \quad J_S = -D_S \left(\frac{d[S]}{dx} + z_S [S] \frac{d\hat{\phi}}{dx} \right). \tag{9.16}$$

In Poisson-Nernst-Planck theory, the boundary conditions for the electric potential (Eq. 9.15) are $\phi(0) = \phi_i$ and $\phi(L) = \phi_i$ and, consequently, the transmembrane potential $V = \phi_i - \phi_o$ is known. The boundary conditions for ionic concentrations used in Eq. 9.16 are $[S](0) = [S]_i$ and $[S](L) = [S]_o$. Solutions of Eqs. 9.15 and 9.16 are numerically calculated to yield the spatial profiles of potential and concentration. The total current density given by

$$I_T(V) = \sum_S z_S F J_S(V)$$

is the predicted current-voltage relation. The Poisson-Nernst-Plank method can predict the effects of changes in the fixed charge in the aqueous pore of ion channels, the influence of intracellular and extracellular ionic concentrations on current-voltage relations, and much more.

Supplemental Problems

Problem 9.8   Show the GHK current (Eq. 9.1) reverses at the Nernst equilibrium potential for the permeating ion $V = E_S$. Hint: The reversal potential V solves $I_S(V) = 0$.

Problem 9.9   What are the physical dimensions of slope on a plot of a GHK current density as a function of voltage? Is this quantity extensive or intensive?

Problem 9.10   When the reversal potential is negative, the GHK current is outward rectifying. Conversely, when the reversal potential is positive, the GHK current is inward rectifying. Explain.

Problem 9.11   Show that if both (1) the sign of the membrane potential is changed ($V \rightarrow -V$) and (2) intracellular and extracellular concentrations are exchanged ($[S]_i \leftrightarrow [S]_o$), the GHK current changes sign ($I_S \rightarrow -I_S$).

Problem 9.12   Show that if both (1) intracellular and extracellular concentrations are reversed ($[S]_i \leftrightarrow [S]_o$) and (2) the sign of z is changed (cation $\leftrightarrow$ anion), the GHK current does *not* change.

Problem 9.13   The effective equilibrium potential for two Ohmic ionic currents is a weighted average (Chapter 8). In terms of permeabilities,

$$E_{ohm} = \frac{P_{Na}E_{Na} + P_K E_K}{P_{Na} + P_K}, \tag{9.17}$$

where $E_{Na} = (RT/z_{Na}F) \ln([Na^+]_o/[Na^+]_i)$ and similarly for E_K. The corresponding GHK voltage is given by Eq. 9.13 with $P_{Cl} = 0$,

$$E_{ghk} = \frac{RT}{F} \ln \left(\frac{P_{Na}[Na^+]_o + P_K[K^+]_o}{P_{Na}[Na^+]_i + P_K[K^+]_i} \right). \tag{9.18}$$

Eqs. 9.17 and 9.18 are similar but not equal. Using plotting software of your choosing, graph E_{ghk} and E_{ohm} as a function of the relative permeability for sodium $0 \le r_{Na} \le 1$ where

$$r_{Na} = \frac{P_{Na}}{P_{Na} + P_K}.$$

Assume the total permeability $P_{Na} + P_K$ is constant so that $r_K = 1 - r_{Na} = P_K/(P_{Na} + P_K)$. Discuss the similarities and differences that you observe. Why is E_{ohm} as a function of r_{Na} a positively sloped line? Why do E_{ghk} and E_{ohm} agree when $r_{Na} = 0$ or 1?

Problem 9.14   The GHK equation for an arbitrary number of monovalent ions can be written as follows

$$E_{ghk} = \frac{RT}{F} \ln \left(\frac{\sum_A P_A[A]_o + \sum_B P_B[B]_i}{\sum_A P_A[A]_i + \sum_B P_B[B]_o} \right),$$

where the A and B are indices for the various ions. Which set of indices corresponds to cations, and which to anions?

Problem 9.15 Eq. 9.13 assumes *monovalent* ions. The mathematically inclined student might try to derive a version of the GHK voltage equation that includes the possibility of *divalent* ions.

Solutions

Problem 9.1 For $V \neq 0$ in Eq. 9.1, $I_S = 0$ implies that

$$0 = [S]_i - [S]_o e^{-z_S V / V_\theta} .$$

Solving for potential gives $V = (V_\theta / z_S) \ln([S]_o / [S]_i) = E_S$.

Problem 9.2 Physiological values of $[K^+]_o$ and $[K^+]_i$ are such that $[K^+]_o / [K^+]_i \approx 1/40$. Higher values of $[K^+]_o$ have the effect of (1) increasing E_K from -90 to $0\,\mathrm{mV}$ and (2) attenuating outward rectification:

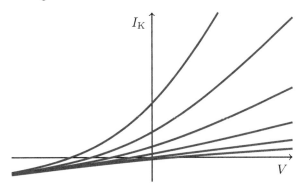

Problem 9.3 The physical dimensions of both $z_S^2 F P_S [S] / V_\theta$ are

$$\frac{\text{conductance}}{\text{length}^2} = \frac{1 \cdot \dfrac{\text{charge}}{\#} \cdot \dfrac{\text{length}}{\text{time}} \cdot \dfrac{\#}{\text{length}^3}}{\text{voltage}} . \tag{9.19}$$

Problem 9.4 Calcium is a cation ($z_S > 0$) with $[S]_o / [S]_i \gg 1$, similar to sodium, so I_{Ca} is inward rectifying. Below are GHK current-voltage relations for calcium (blue) and sodium (black) with equal permeabilities ($P_{Ca} = P_{Na}$).

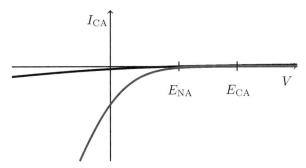

Problem 9.5 Eq. 9.9 is a linear ODE that may be solved by multiplying by the integrating factor $e^{-\alpha x}$,

$$\frac{d[S]}{dx}e^{-\alpha x} = \alpha[S]e^{-\alpha x} - \beta e^{-\alpha x}$$

and using the identity

$$\frac{d}{dx}\left([S]e^{-\alpha x}\right) = \frac{d[S]}{dx}e^{-\alpha x} + [S]e^{-\alpha x}(-\alpha)$$

to obtain

$$\frac{d}{dx}\left([S]e^{-\alpha x}\right) = -\beta e^{-\alpha x}.$$

The differential form of this expression can now be integrated from $x = 0$ to L,

$$\int_0^L d\left([S]e^{-\alpha x}\right) = \int_0^L -\beta e^{-\alpha x}dx$$

$$[S]e^{-\alpha x}\big|_0^L = \frac{\beta}{\alpha}e^{-\alpha x}\bigg|_0^L$$

$$[S]_o e^{-\alpha L} - [S]_i = \frac{\beta}{\alpha}\left(e^{-\alpha L} - 1\right),$$

where in the last step we used concentrations of S at $x = 0$ and $x = L$ (Eq. 9.8). Rewriting the result of the integration as follows,

$$\beta = \alpha\left(\frac{[S]_i - [S]_o e^{-\alpha L}}{1 - e^{-\alpha L}}\right).$$

Reverting to the original parameters, we obtain

$$\frac{I_S}{z_S F D_S} = \frac{z_S V}{L V_\theta}\left(\frac{[S]_i - [S]_o e^{-z_S V/V_\theta}}{1 - e^{-z_S V/V_\theta}}\right)$$

$$I_S = z_S F \cdot \underbrace{\frac{D_S}{L}}_{P_S} \cdot \frac{z_S V}{V_\theta}\left(\frac{[S]_i - [S]_o e^{-z_S V/V_\theta}}{1 - e^{-z_S V/V_\theta}}\right).$$

Observe that the membrane permeability (P_S) is the ratio of the diffusion coefficient D_S and the membrane thickness (L),

$$\frac{\text{length}}{\text{time}} \ominus P_S = \frac{D_S}{L} \ominus \frac{\text{length}^2/\text{time}}{\text{length}}.$$

Problem 9.6 The negative signs are required to obtain the correct sign for the net material flux. To see this, note that a positive flux density for diffusion ($J^{diff} > 0$) corresponds to a net movement of ions from lesser to greater values of the spatial variable x. The spatial variable x increases from left to right and, consequently, positive flux density corresponds to ions moving from left to right. The flux density is negative (right to left) when $d[S]/dx$ is positive. Hence, Eq. 9.5 includes a negative sign. There is a similar explanation for the negative sign in the constitutive relation for drift (Eq. 9.6). In the presence of an electrical potential increasing from left to right ($d\phi/dx > 0$), that is, a leftward oriented electric field, positively charged cations ($z_S > 0$) will move in the direction of the electric field (right to left). The net flux density due to electrophoresis is negative ($J^{drift} < 0$). Hence, Eq. 9.6 includes a negative sign.

Problem 9.7   Observe that for monovalent cations ($z_S = 1$), the GHK current (Eq. 9.14) is

$$I_S = P_S F \hat{V} \left(\frac{[S]_i - [S]_o e^{-\hat{V}}}{1 - e^{-\hat{V}}} \right).$$

The corresponding expression for monovalent anions ($z_S = -1$) is

$$I_S = P_S F \hat{V} \left(\frac{[S]_i - [S]_o e^{\hat{V}}}{1 - e^{\hat{V}}} \right) = P_S F \hat{V} \left(\frac{[S]_i e^{-\hat{V}} - [S]_o}{1 - e^{-\hat{V}}} \right).$$

Thus, the zero flux condition $0 = J_{Na} + J_K + J_{Cl}$ for the steady state voltage is

$$0 = P_{Na}([Na^+]_i - [Na^+]_o e^{-\hat{V}}) + P_K([K^+]_i - [K^+]_o e^{-\hat{V}}) + P_{Cl}([Cl^-]_i e^{-\hat{V}} - [Cl^-]_o)$$

where we have divided by $F\hat{V}/(1-e^{-\hat{V}}) \neq 0$. Rearranging this expression gives the GHK voltage equation for monovalent ions (Eq. 9.13). The GHK voltage equation is more complicated when divalent ions are included (see Problem 9.15).

Problem 9.9   This slope is a specific conductance (conductance/length^2), an intensive quantity.

Problem 9.10

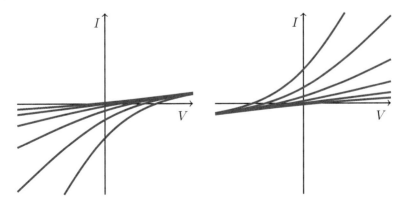

Problem 9.11   Let an asterisk represent the changed condition so that

$$V^* = -V \qquad [S]_i^* = [S]_o \qquad [S]_o^* = [S]_i.$$

Substituting $V = -V^*$, $[S]_o = [S]_i^*$ and $[S]_i = [S]_o^*$ into the GHK current density equation (Eq. 9.20), gives

$$I_S^* = z_S F \cdot P_S \frac{z_S(-V^*)}{V_\theta} \left(\frac{[S]_o^* - [S]_i^* e^{-z_S(-V^*)/V_\theta}}{1 - e^{-z_S(-V^*)/V_\theta}} \right)$$

$$\vdots$$

$$= -z_S F \cdot P_S \frac{z_S V^*}{V_\theta} \left(\frac{[S]_i^* - [S]_o^* e^{-z_S V^*/V_\theta}}{1 - e^{-z_S V^*/V_\theta}} \right) = -I_S.$$

Problem 9.12   Let an asterisk represent the changed condition so that

$$z_S^* = -z_S \qquad [S]_i^* = [S]_o \qquad [S]_o^* = [S]_i.$$

Substituting $z_S = -z_S^*$, $[S]_o = [S]_i^*$ and $[S]_i = [S]_o^*$ into the GHK current density equation (Eq. 9.20), gives

$$I_S^* = -z_S^* F \cdot P_S \frac{-z_S^* V}{V_\theta} \left(\frac{[S]_o^* - [S]_i^* e^{-(-z_S^*)V/V_\theta}}{1 - e^{-(-z_S^*)V/V_\theta}} \right)$$

$$\vdots$$

$$= z_S^* F \cdot P_S \frac{z_S^* V}{V_\theta} \left(\frac{[S]_i - [S]_o^* e^{-z_S^* V/V_\theta}}{1 - e^{-z_S^* V/V_\theta}} \right) = I_S.$$

Problem 9.14   The A are cations and the B are anions.

Notes

1. Consistent with its appearance in the exponential e^{-zV/V_θ} (Eq. 9.1).
2. Because there are no vertical axis labels, it is of no consequence that I_{Na} is a current density as opposed to a current.
3. Not a line in the (V, I_{Na})-plane.

PART III
Voltage-Gated Currents

10 Voltage-Gated Ionic Currents

The experimentally observed nonlinearities of many ionic membrane currents are not explained by the passive theory of membrane permeation (Chapter 9). Rather, their current-voltage relations are shaped by active voltage-dependent processes such as ion channel gating and permeation block.

10.1 Voltage-Dependent Gating and Permeation Block

Fig. 10.1 shows an experimentally measured current-voltage relation.[1] This relation is certainly not Ohmic, because the slope is not constant. The relation is also quite different than a passive GHK calcium current (see Problem 9.4). The ionic membrane current of Fig. 10.1 is a *high-voltage activated inward current* mediated by depolarization-activated L-type calcium channels.[2]

L-type calcium channels are integral membrane proteins with an aqueous pore that is highly selective for – you guessed it – calcium. These channels have a protein domain that serves as a voltage sensor. Protein conformations that permit permeation of calcium ions are stabilized or destabilized depending on the transmembrane potential. In this way, the voltage influences the probability that individual calcium channels are permissive (open) or nonpermissive (closed).

On a cell membrane with a large number of L-type calcium channels, the fraction of open channels (and the permeability of the membrane to calcium) is a predictable function of membrane voltage.[3] This functional dependence is responsible for the observed nonlinearity of high-voltage activated calcium currents. Similar to the N-type calcium channels encountered in Fig. 2.6, L-type calcium channels may be activated (opened) by membrane depolarization leading to calcium influx into the cell cytosol.

Fig. 10.2 shows the current-voltage relation for an *inward rectifying potassium current* (I_{Kir}).[4] The channels that mediate this current are highly selective for potassium ions. However, the observed inward rectification is the opposite of the GHK prediction that passive potassium currents are outward rectifying (Fig. 9.3). The inward

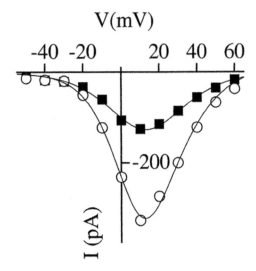

Figure 10.1 The current-voltage relation for high-voltage activated (L-type) calcium channels is nonlinear. Reproduced from McHugh, D, Sharp, EM, Scheuer, T, and Catterall, WA. 2000. Inhibition of cardiac L-type calcium channels by protein kinase C phosphorylation of two sites in the N-terminal domain. *Proceedings of the National Academy of Sciences*, **97**(22), 12334-12338. Copyright (2000) National Academy of Sciences, USA. *Question*: Filled squares show current when protein kinase C (PKC) is pharmacologically stimulated. Does this manipulation increase or decrease the magnitude of the inward calcium current?

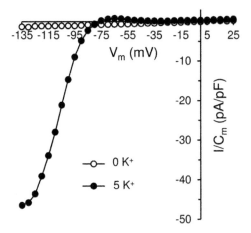

Figure 10.2 Nonlinear current-voltage relations for an inward rectifying potassium current (filled circles, $[K^+]_o = 5\,mM$). Compare the current-voltage relation for an Ohmic potassium leak (Fig. 8.5) and the GHK potassium current (Fig. 9.3). Reproduced from Masia, R, Krause, DS, and Yellen, G. The inward rectifier potassium channel Kir2.1 is expressed in mouse neutrophils from bone marrow and liver. *American Journal of Physiology: Cell Physiology*, 2015; **308**(3): C264–C276. doi:10.1152/ajpcell.00176.2014. *Question*: For the open circles there is little current. Why? What is the reversal potential for the filled and open circles?

rectification of Kir currents arises from inorganic or organic intracellular cations (e.g., magnesium or polyamines) that block Kir channels in a voltage-dependent manner.[5]

The high-voltage activated L-type calcium current (I_{Ca_V}) and inward rectifying potassium current (I_{Kir}) are two of many voltage-gated ionic membrane currents commonly expressed by neurons and myocytes. This chapter explores mathematical modeling of the nonlinear current voltage relations observed for such currents.

10.2   The L-Type Calcium Current I_{Ca_V}

The nonlinearity of the L-type calcium current-voltage relation (Fig. 10.1)[6] arises from voltage-dependent opening of plasma membrane calcium channels. At the resting voltage of a quiescent cell, most of the calcium channels responsible for the L-type current are closed and the membrane conductance mediated by these channels is negligible. Voltage-gated calcium channels open when the membrane potential is depolarized from rest (the threshold voltage is about $V_\theta = -35\,\mathrm{mV}$). Consequently, the cell membrane exhibiting a voltage-dependent conductance, denoted $g_{Ca_V}(V)$, is nearly zero for $V \approx -65\,\mathrm{mV}$, but increases in a sigmoidal fashion toward a maximum value denoted by $\bar{g}_{Ca_V}$ as V increases (see below).

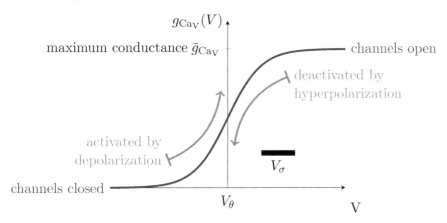

Voltage-gated ionic currents are often modeled by assuming an electrochemical driving force (physical dimensions of voltage)[7] of the form $V - E$, where E is a Nernst equilibrium potential, similar to our prior modeling of Ohmic currents (Chapter 8). Because the ion channels responsible for I_{Ca_V} are highly selective for calcium, the appropriate Nernst equilibrium potential for the L-type calcium current is E_{Ca}. The L-type calcium current is thus,

$$I_{Ca_V} = g_{Ca_V}(V)\,(V - E_{Ca}),\tag{10.1}$$

where $I_{Ca_V}(V)$ is the experimentally observed current-voltage relation (as in Fig. 10.1). Often the voltage-dependent conductance $g_{Ca_V}(V)$ is inferred from the current $I_{Ca_V}(V)$ using Eq. 10.1, as follows:

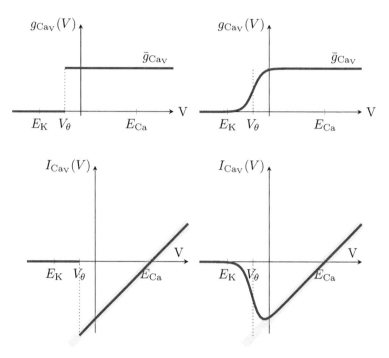

Figure 10.3   The L-type calcium ionic conductance (g_{Ca_V}, top row) and current (I_{Ca_V}, bottom row). At each voltage the product of the conductance $g_{Ca_V}(V)$ and driving force $V - E_{Ca}$ equals the current I_{Ca_V} (Eq. 10.1).

$$g_{Ca_V}(V) = \frac{I_{Ca_V}}{V - E_{Ca}}.$$

Fig. 10.3 (right) shows a theoretical current-voltage relation (bottom row, blue) given by the product of the voltage-dependent conductance $g_{Ca_V}(V)$ (top row, blue) and a driving force reversing at E_{Ca} (bottom row, gray). The conductance $g_{Ca_V}(V)$ is an increasing sigmoidal function of voltage,

$$g_{Ca_V}(V) = \bar{g}_{Ca_V} m_\infty(V), \tag{10.2}$$

where $\bar{g}_{Ca_V}$ is the maximum conductance. The voltage dependence of $g_{Ca_V}(V)$ is inherited from the **gating function** $m_\infty(V)$ that takes the form

$$m_\infty(V) = \frac{1}{2}\left[1 + \tanh\left(\frac{V - V_\theta}{V_\sigma}\right)\right]. \tag{10.3}$$

Although the hyperbolic tangent function,

$$\tanh x = \frac{e^x - e^{-x}}{e^x + e^{-x}} \tag{10.4}$$

may be unfamiliar, the important points are (i) the gating function $m_\infty(V)$ takes values between 0 and 1, (ii) the gating function is half maximum at $V = V_\theta$ (i.e., $m_\infty(V_\theta) = 1/2$), (iii) the slope of the sigmoid at half maximum is determined by the scale parameter V_σ (see below). The gating function $m_\infty(V)$ may be interpreted as the fraction of L-type calcium channels that are open at transmembrane potential V.

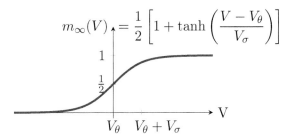

$$m_\infty(V) = \frac{1}{2}\left[1 + \tanh\left(\frac{V - V_\theta}{V_\sigma}\right)\right]$$

Observe that

$$m_\infty(V) \to 0 \quad \text{and} \quad g_{Ca_V}(V) \to 0 \quad \text{as} \quad V \to -\infty$$
$$m_\infty(V) \to 1 \quad \text{and} \quad g_{Ca_V}(V) \to \bar{g}_{Ca_V} \quad \text{as} \quad V \to \infty.$$

To understand how the voltage-dependent conductance for the L-type calcium current (Eq. 10.2) gives rise to the current-voltage relation of Fig. 10.3, it is helpful to sketch the function,

$$I_{Ca_V} = \underbrace{\bar{g}_{Ca_V} m_\infty(V)}_{g_{Ca_V}(V)} (V - E_K). \tag{10.5}$$

We begin with a caricature of $I_{Ca_V}(V)$ (left panel of Fig. 10.3)[8] for which the conductance $g_{Ca_V}(V)$ steps abruptly from 0 to $\bar{g}_{Ca_V}$ at the threshold voltage V_θ,

$$g_{Ca_V}(V) = \begin{cases} 0 & \text{for } V < V_\theta \\ \bar{g}_{Ca_V} & \text{for } V \geq V_\theta. \end{cases} \tag{10.6}$$

Because I_{Ca_V} reverses at E_{Ca} (about 115 mV), we sketch a positively sloped line that intersects the horizontal axis at E_{Ca} (gray). The slope of this line is $\bar{g}_{Ca_V}$ ⊖ conductance. Finally, we sketch the voltage-dependent calcium current I_{Ca_V}, which is zero when $V < V_\theta$ and $\bar{g}_{Ca_V}(V - E_K)$ when $V \geq V_\theta$.[9]

A more realistic plot of $I_{Ca_V}(V)$ (Fig. 10.3, right column) uses a sigmoidal function for $g_{Ca_V}(V)$ (Eq. 10.2). The scale parameter V_σ in $m_\infty(V)$ determines the curvature of I_{Ca_V}; in particular, the derivative $dm_\infty(V)/dV$ at $V = V_\theta$ is inversely proportional to V_σ.

Problem 10.1   The hyperbolic tangent function (tanh, plotted below) is a trigonometric function that can be expressed using exponentials:

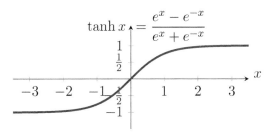

$$\tanh x = \frac{e^x - e^{-x}}{e^x + e^{-x}}$$

Using your knowledge of the limits of e^x and e^{-x} as $x \to \pm\infty$, show that (i) $\lim_{x\to\infty} \tanh x = 1$, (ii) $\lim_{x\to-\infty} \tanh x = -1$, (iii) $\tanh(0) = 0$, and (iv) $\tanh'(x) = 1 - \tanh^2(x)$.

Problem 10.2 Beginning with Eq. 10.3, show that (i) $m_\infty(V)$ is an increasing function of voltage $(dm_\infty/dV > 0)$, and (ii) the location parameter V_θ sets the voltage for which $m_\infty(V = V_\theta) = 1/2$.

Problem 10.3 V is shifted and scaled in the argument of the hyperbolic tangent that appears in the activation gating function (Eq. 10.3). Is $m'_\infty(V)$ an increasing or decreasing function of V_σ?

10.3   The Inward Rectifying Potassium Current I_{Kir}

The nonlinear current exhibited in Fig. 10.2 is referred to as an *inward rectifying* potassium current (denoted I_{Kir}) in order to highlight the contrast between this current and GHK theory prediction that potassium currents are outward rectifying (Fig. 9.3). The inward rectifying potassium current I_{Kir} is mediated by ion channels that are selective for potassium and *activated by hyperpolarization*, that is, the voltage-dependent conductance $g_{Kir}(V)$ is a decreasing sigmoidal function:

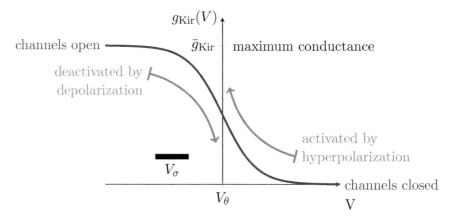

The voltage-dependence of I_{Kir} is the product of conductance and electrochemical driving force,

$$I_{Kir} = \underbrace{\bar{g}_{Kir}s_\infty(V)}_{g_{Kir}(V)}(V - E_K),$$

(10.7)

where the gating function $s_\infty(V)$ is monotone *decreasing* $(ds_\infty/dV < 0)$

$$s_\infty(V) = \frac{1}{2}\left[1 + \tanh\left(-\frac{V - V_\theta}{V_\sigma}\right)\right]$$

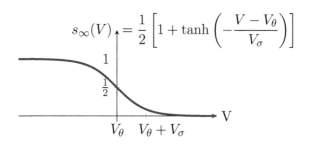

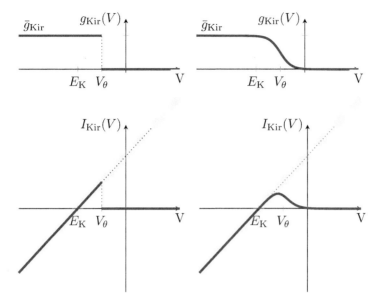

Figure 10.4 The voltage-dependent inward rectifying potassium conductance (g_{Kir}, top row) and ionic current I_{Kir} (bottom, thick solid lines) given by Eq. 10.7.

and $V_\theta \approx -80\,\text{mV}$ is about $10\,\text{mV}$ more depolarized than E_K (Fig. 10.4). Increasing voltage from rest where $s_\infty(V) \approx 1$ decreases the fraction of open channels. Thus, the inward rectifying potassium current is **hyperpolarization activated** (and deactivated by depolarization).[10]

10.4 The Hyperpolarization-Activated Cation Current I_{sag}

Fig. 10.5 shows a current-voltage relation for a hyperpolarization-activated cation current I_{sag} (also called I_h). I_{sag} is a nonspecific cation current, meaning that the ion channels responsible for I_{sag} are permeable to both sodium and potassium. Assuming $g_{Na} \approx 2g_K$ the reversal potential for I_{sag} is

$$E_{sag} = \frac{g_K E_K + g_{Na} E_{Na}}{g_K + g_{Na}} \approx 0\,\text{mV}.$$

In Fig. 10.5 the conductance $g_{sag}(V)$ is half-maximal at $V = V_\theta \approx -50\,\text{mV}$. I_{sag} is an inward rectifying current with voltage dependence similar to I_{Kir} (cf. Fig. 10.4). On the other hand, I_{sag} is a bona fide inward current, because $I_{sag} < 0$ in the physiological voltage range of -120 to $20\,\text{mV}$. This is not the case for I_{Kir}, which can be either inward or outward in this range of voltages.

10.5 The Depolarization-Activated Potassium Current I_{K_V}

Fig. 10.6 shows a depolarization-activated potassium conductance, gK_V, that yields an *outward rectifying* potassium current, I_{K_V}. The outward rectification of I_{K_V} is

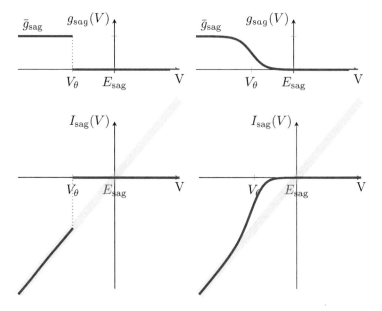

Figure 10.5   The hyperpolarization-activated nonspecific cation conductance ($g_{sag}(V)$, top row). The ionic current I_{sag} (also called I_h) is inward rectifying and reverses at $E_{sag} \approx 0\,mV$.

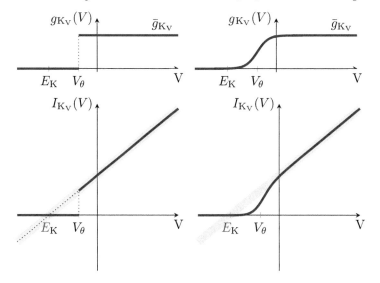

Figure 10.6   An outward rectifying voltage-gated potassium current I_{K_V} (bottom row) results from a depolarization-activated conductance (g_{K_V}, top row) and driving force that reverses at E_K.

stronger than predicted by GHK theory (Chapter 9). There is remarkable diversity of voltage-gated and calcium-regulated potassium channels, distinguished more by the details of gating than their selectivity or pharmacology (Hille, 2001, Chapter 5).[11] Depolarization-activated potassium currents that are regulated by intracellular calcium and ATP are denoted K_{Ca} and K_{ATP} currents, respectively.

10.6   Qualitative Features of Current-Voltage Relations

Fig. 10.7 (upper left) shows a current-voltage relation for a depolarization-activated sodium current (I_{Nav}).[12] Although this current is new, and the permeating ion is sodium, I_{Nav} is qualitatively similar to the voltage-gated calcium current I_{Cav} (Fig. 10.3, bottom right). This similarity occurs because the gating function for I_{Nav} is monotone increasing and, furthermore, the threshold voltage is around $V_\theta = -35\,mV$, which is hyperpolarized compared to the reversal potential ($V_\theta < E_{Na}$).

Although the voltage-gated sodium current (I_{Nav}) is outward at extremely depolarized voltages ($V > E_{Na}$), I_{Nav} is inward over the physiological range of voltages normally experienced by neurons (about -120 to $20\,mV$, light blue box in Fig. 10.7). The hyperpolarization-activated cation current I_{sag} (Fig. 10.7, bottom left) is also inward (dark blue) in the physiological range of voltages (light blue). Conversely, the depolarization-activated potassium current I_{Kv} is an outward current (gray shading). In the physiological range of voltages, the inward rectifying potassium current I_{Kir} may be either inward or outward (Fig. 10.4).

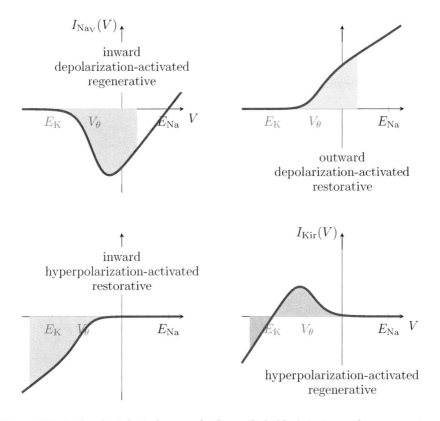

Figure 10.7   In the physiological range of voltages (light blue), ionic membrane currents can be classified as inward (blue) versus outward (gray), depolarization-activated versus hyperpolarization-activated, and regenerative versus restorative. *Question*: I_{Nav} and I_{Kir} are shown. Which current-voltage relation is typical for I_{Kv}? I_{sag}? I_{Cav}?

Both I_{Cav} and I_{Nav} are depolarization-activated inward ionic currents. These currents are **regenerative**, because depolarization from rest leads to an increase in conductance (g_{Cav} or g_{Nav}) and an increase in the magnitude of inward current (also depolarizing). That is to say, for both I_{Cav} and I_{Nav}, *depolarization recruits inward current that is further depolarizing*. The term regenerative refers to the autocatalytic (run away) aspect of voltage-dependent activation. The signature of a regenerative current is a range of voltages for which $dI_{ion}/dV < 0.$[13]

When the potential that leads to half maximum deactivation of the inward rectifying potassium conductance g_{Kir} is sufficiently depolarized ($V_\theta > E_K$), the current-voltage relation may have a negatively sloped region (see Fig. 10.4 and below right). In this range of voltages, *hyperpolarization recruits outward current that is further hyperpolarizing*. Thus, the inward rectifying potassium currents shown in Fig. 10.4 and Fig. 10.7 are regenerative. However, the plots below show that the negatively sloped region of the current-voltage relation may effectively disappear when the half-maximum activation V_θ is more hyperpolarized. On the leftmost plot below, I_{Kir} is inward, but not regenerative.

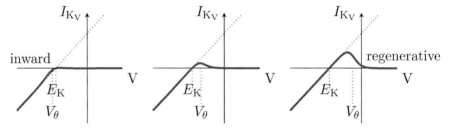

I_{Kv} and I_{sag} are **restorative** currents that do not have a negatively sloped region in their current-voltage relation (see Fig. 10.7). When these currents are activated, the evoked current resists changes in voltage that would cause further activation. When I_{Kv} is activated by depolarization, outward current is evoked; outward current is hyperpolarizing. When I_{sag} is activated by hyperpolarization, an inward current is evoked; inward current is depolarizing. The term restorative calls attention to this negative feedback.

Regenerative currents such as I_{Nav}, I_{Cav}, and I_{Kir} are, to be precise, only regenerative over that range of voltages where the current-voltage relation is negatively sloped ($dI/dV < 0$). Indeed, when the membrane potential is near the reversal potential of an ionic current, the current must be restorative. In the case of I_{Nav}, the reversal potential is $E_{Na} \approx +60\,mV$, a voltage not commonly experienced by neurons. The inward rectifier potassium current I_{Kir} reverses at $E_K \approx -85\,mV$, which is in the physiological range of voltages. I_{Kir} is restorative when $V \approx E_K$, but regenerative when $V > V_\theta$ (see Fig. 10.7, bottom right). All regenerative ionic currents have a current-voltage relation that is negatively sloped *somewhere* in the physiological range of voltages.

10.7   Further Reading and Discussion

The following two review articles on depolarization-activated calcium currents are highly recommended.

1   *A short history of voltage-gated calcium channels* by Dolphin (2006).
2   *L-type calcium channels: the low down* by Lipscombe et al. (2004).

Other reviews of voltage-gated currents that may be of interest include: Hille (2001, Chapter 4) (Ca_V), Catterall (2011) (Ca_V), and Ahern et al. (2016) (Na_V).

Expected Shapes of Current-Voltage Relations

In this chapter we have presented a variety of current-voltage relations for several ionic currents. The qualitative features of voltage-dependent currents are due to the location of the reversal potential for the ionic current (a function of the selectivity of the corresponding ion channels), the location of the half maximum for the voltage-dependent conductance (V_θ), and the scale parameter (V_σ).

It is highly recommended that students use software of their choosing to plot current-voltage relations of the form $I = g(V)(V-E)$ where the voltage-dependent conductance $g(V)$ is an increasing or decreasing sigmoidal function (Eqs. 10.2 and 10.7). Observe how the sign of $g'(V)$ changes, and whether $V_\theta < E$ or $V_\theta > E$, determine the salient aspects of the current-voltage relation: depolarization- or hyperpolarization-activated, inward or outward, inward or outward rectifying, regenerative or restorative, and so on.

Current-Voltage Relation Subtraction

The experimental method used to measure current-voltage relationships of ionic currents such as I_{Ca_V}, I_{Na_V}, I_{K_V}, I_{Kir} and I_{sag} is **voltage-clamp recording** (Chapter 12). Because most of the membranes of excitable cells usually express more than one ionic current, experiments are often performed under different conditions and the results compared in order to isolate the contribution of a current of interest.

Fig. 10.8 (left) shows whole cell ionic membrane current of brain capillary endothelial cells, an essential component of the blood-brain barrier. The Control condition (black curve) yields a current-voltage relation that is inward rectifying and reverses at 40 mV. When the extracellular medium includes 100 μM barium, the rectification is no longer observed (gray curve, Ba^{2+}). Because the inward rectifying potassium current is known to be blocked by barium, the difference of these two measurements (Fig. 10.8, right) is interpreted as I_{Kir} (Yamazaki et al., 2010).[14] This type of *subtraction method* is commonly used to determine which ionic currents are expressed by a given cell type. Often the experimental condition is a manipulation of extracellular ionic concentration(s) or pharmacological block of an ionic current.

Supplemental Problems

Problem 10.4   Plot $m_\infty(V)$ (Eq. 10.3) for different values of V_θ and V_σ.

Problem 10.5   Show that $g_{Ca_V} = \bar{g}_{Ca_V}/2$ when $V = V_\theta$ (Eq. 10.2).

Problem 10.6   Use Eq. 10.4 to show that $y = \tanh x$ solves $dy/dx = 1 - y^2$.

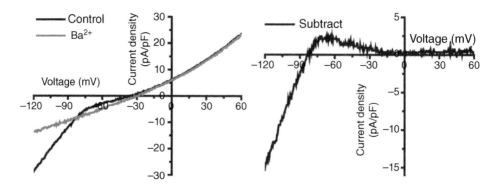

Figure 10.8   The current-voltage relations for an inward rectifying potassium current obtained by subtraction (right). Reproduced from Yamazaki, D, Kito, H, Yamamoto, S, Ohya, S, Yamamura, H, Asai, K, and Imaizumi, Y. 2010. Contribution of Kir2 potassium channels to ATP-induced cell death in brain capillary endothelial cells and reconstructed HEK293 cell model. *American Journal of Physiology: Cell Physiology*, **300**(1), C75-C86. Figure 3. *Question*: What is the reversal potential of the subtracted current? Of the control current?

Problem 10.7   Show that $m'_\infty(V) > 0$ when $V_\sigma > 0$.

Problem 10.8   Is the slope of $m_\infty(V)$ at the half maximum $(V = V_\theta)$ an increasing or decreasing function of $V_\sigma > 0$?

Problem 10.9   Show that $s_\infty(V) = 1 - m_\infty(V)$ and $s_\infty(V; V_\sigma) = m_\infty(V; -V_\sigma)$.

Problem 10.10   Sketch a qualitatively correct current-voltage relation for a voltage-gated sodium current that is activated by depolarization, that is, takes the form

$$I_{Na_V} = \bar{g}_{Na_V} m_\infty(V) (V - E_{Na})$$

where $V_\theta = -35\,\text{mV}$ and $V_\sigma = 10\,\text{mV}$.

Problem 10.11   Consider the following list of voltage-gated currents we have discussed thus far

$$I_{Ca_V}, \quad I_{Kir}, \quad I_{sag}, \quad I_{K_V}, \quad I_{Na_V}$$

and describe each one as inward, outward or neither; regenerative or restorative; inward rectifying, outward rectifying or neither.

Problem 10.12   Are the GHK currents of Chapter 9 regenerative or restorative?

Solutions

Figure 10.1   PKC stimulation decreases $|I_{Ca_V}|$ by about 50%.

Figure 10.2   Open circles: $[K^+]_o$ has been decreased and the ion channels are highly selective for potassium. The reversal potentials are about $-80\,\text{mV} \approx E_K$ (filled circles) and $0\,\text{mV}$ (open circles).

Figure 10.7 I_{Nav} upper left, I_{Kv} upper right, I_{sag} lower left.

Figure 10.8 The reversal potential of the subtracted current is about $-85\,mV \approx E_K$. The reversal potential of the control is about $-30\,mV$.

Problem 10.6 Write $d(\tanh x)/dt \overset{?}{=} 1 - (\tanh x)^2$, substitute and differentiate:

$$\left(\frac{e^x - e^{-x}}{e^x + e^{-x}}\right)' \overset{?}{=} 1 - \left(\frac{e^x - e^{-x}}{e^x + e^{-x}}\right)^2$$

$$\frac{(e^x - e^{-x})'(e^x + e^{-x}) - (e^x + e^{-x})'(e^x - e^{-x})}{(e^x + e^{-x})^2} \overset{?}{=} 1 - \frac{(e^x - e^{-x})^2}{(e^x + e^{-x})^2}$$

$$\frac{(e^x + e^{-x})(e^x + e^{-x}) - (e^x - e^{-x})(e^x - e^{-x})}{(e^x + e^{-x})^2} \overset{yes}{=} \frac{(e^x + e^{-x})^2 + (e^x - e^{-x})^2}{(e^x + e^{-x})^2} \cdot \checkmark$$

Problem 10.7

$$m_\infty = \frac{1}{2}(1 + \tanh[(V - V_\theta)/V_\sigma])$$

$$m'_\infty = \frac{1}{2}\tanh'[(V - V_\theta)/V_\sigma] = \frac{1}{2V_\sigma}\left(1 - \tanh^2[(V - V_\theta)/V_\sigma]\right)$$

$$= \frac{1}{2V_\sigma}(1 + \tanh[(V - V_\theta)/V_\sigma])(1 - \tanh[(V - V_\theta)/V_\sigma])$$

$$= \frac{m_\infty}{V_\sigma}(1 - \tanh[(V - V_\theta)/V_\sigma]) = \frac{2}{V_\sigma}m_\infty(1 - m_\infty).$$

Both m_∞ and $1 - m_\infty$ are positive; thus, $m'_\infty > 0$ when $V_\sigma > 0$. The second line uses $\tanh'(x) = 1 - \tanh^2(x)$ (Problem 10.6). The last step uses $2m_\infty = 1 + \tanh[(V - V_\theta)/V_\sigma] \Rightarrow 1 - \tanh[(V - V_\theta)/V_\sigma] = 2(1 - m_\infty)$.

Problem 10.12

I_{Cav}	inward	regenerative	neither
I_{Kir}	neither	regenerative	inward rectifying
I_{sag}	inward	restorative	outward rectifying
I_{Kv}	outward	restorative	outward rectifying
I_{Nav}	inward	regenerative	neither

Notes

1. The cardiac L-type calcium current-voltage relation shown in Fig. 10.1 was reproduced with permission from McHugh et al. (2000). Whole cell patch clamp recordings were performed using barium as the current carrier. The paper focuses on modulation of cardiac L-type calcium currents by protein kinase C. PKC activation by PMA reduced the current through heterologously expressed CaV1.2 channels (compare open circles and full squares in Fig. 10.1).
2. The distinction between high- and low-voltage activated calcium currents is emphasized in Chapter 19.

3. By assuming a large number of channels, we are able to postpone discussion of stochastic (random) aspects of ion channel gating.

4. The voltage-clamp recordings of Fig. 10.2 are whole cell currents obtained from mouse neutrophils from bone marrow and liver (phagocytic cells that play a role in innate immunity). Reverse transcription-polymerase chain reaction (RT-PCR) revealed that these neutrophils express mRNA for the Kir2 subunit Kir2.1 but not for other subunits (Kir2.2, Kir2.3, and Kir2.4). The current is blocked by barium, cesium and the Kir2-selective inhibitor ML133 (Masia et al., 2015).

5. Both Kir and IRK are common abbreviations.

6. In Masia et al. (2015) the L-type calcium channel $\alpha_1 1.2$ subunit was cloned from rabbit heart and expressed in tsA-201 cells together with the β_{1b} and $\alpha_2 \delta_1$ subunits. Barium was the charge carrier. Currents were triggered by depolarization from a holding potential of -80 mV (Chapter 12). Filled squares show decreased current in the presence of 100 µM 4α-phorbol 12-myristate 13-acetate (PMA), an activator of protein kinase C (PKC).

7. Electrochemical driving *force* is not meant as mechanical force $\ominus$ N.

8. Hille (2001) introduces current-voltage relations in this manner.

9. This type of simplified presentation is provided for many of the ionic currents discussed in this chapter.

10. We say *deactivate* rather than *inactivate*, because the words *inactivate* (and *de-inactivate*) have a technical meaning that refers to slower processes that contribute to refractoriness of excitable membranes (see Chapter 12).

11. A-type potassium currents are not mentioned here because the concept of inactivation has not yet been introduced.

12. Because there is no inactivation gating variable, this is a *persistent* sodium current (see Chapter 12).

13. Sometimes called a "negative conductance" region, though conductance is a physical quantity that is nonnegative.

14. Fig. 10.8 was obtained using a *ramp* voltage-clamp protocol (0.72 V/s) and a cell line derived from bovine brain capillary endothelial cells. RT-PCR confirmed expression of Kir2 mRNA (Yamazaki et al., 2010).

11 Regenerative Ionic Currents and Bistability

Membranes with regenerative ionic currents such as I_{Ca_V} and I_{Kir} may exhibit more than one steady state voltage. The phase diagram of the current balance equation gives insight into the phenomenon of membrane bistability in motoneurons and cerebellar Purkinje cells.

11.1 Regenerative Currents and Membrane Bistability

The dynamics of voltage in membranes that express nonlinear ionic currents can be more complex than what is observed in passive membrane models (Fig. 8.7). Consider a neuron whose membrane includes a passive leak (I_L) but also a voltage-gated Ca^{2+} current (I_{Ca_V}) mediated by high-voltage activated L-type calcium channels. The current balance equation for the model is

$$C\frac{dV}{dt} = I_{app} - \underbrace{\bar{g}_{Ca_V} m_\infty(V)(V - E_{Ca})}_{I_{Ca_V}(V)} - \underbrace{g_L(V - E_L)}_{I_L(V)} \,. \tag{11.1}$$

Steady states of the $I_{Ca_V} + I_L$ membrane model are found by setting $dV/dt = 0$ in Eq. 11.1. A steady state voltage V_{ss} solves the nonlinear algebraic expression,

$$I_{app} = \underbrace{\bar{g}_{Ca_V} m_\infty(V_{ss})(V_{ss} - E_{Ca})}_{I_{Ca_V}(V_{ss})} + \underbrace{g_L(V_{ss} - E_L)}_{I_L(V_{ss})} \,. \tag{11.2}$$

When the applied current is zero ($I_{app} = 0$), steady states of Eq. 11.1 are given by V_{ss} that solve $0 = I_{ion}(V_{ss})$ where $I_{ion} = I_{Ca_V} + I_L$ is the total ionic membrane current. Because $m_\infty(V)$ is nonlinear, analytically solving Eq. 11.2 results in an unwieldy and unenlightening expression. However, an approach based on analytical geometry leads to considerable insight.

Fig. 11.1 plots current-voltage relations for I_{Ca_V}, I_L and their sum I_{ion}. Interestingly, the total ionic current $I_{ion}(V)$ reverses *three times*. We will denote the three voltages for

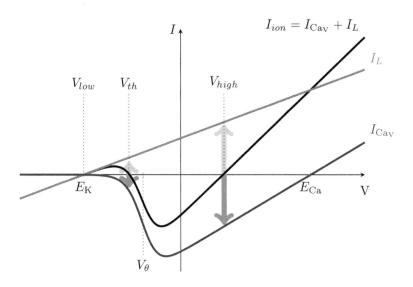

Figure 11.1 The sum of ionic currents $I_{ion} = I_{Cav} + I_L$ may have three intersections with the horizontal axis. *Question*: What are the biological meanings of the parameters V_θ and V_{th}? How would you locate these voltages on the horizontal axis?

which $I_{ion} = 0$ as V_{low}, V_{th} and V_{high} (dotted lines). When $I_{app} = 0$ these steady state voltages and the reversal potentials E_K and E_{Ca} are related as follows,

$$E_L \approx V_{low} < V_{th} < 0 < V_{high} < E_{Ca} .$$

The gray arrows in Fig. 11.1 emphasize that zero ionic membrane current, $I_{ion}(V_{ss}) = 0$, implies that I_{Cav} and I_L are equal in magnitude but opposite in sign, that is, $I_{Cav}(V_{ss}) = -I_L(V_{ss})$ for any V_{ss} solving Eq. 11.2 when $I_{app} = 0$.

Fig. 11.2 shows the phase diagram for the $I_{Cav} + I_L$ membrane model. Filled and open circles indicate the steady state voltages V_{low}, V_{th} and V_{high}. Decorating the phase line with arrows showing the sign of dV/dt reveals that V_{low} and V_{high} are stable steady states, while V_{th} is unstable. Indeed, the number and stability of steady states is easily determined from qualitative features of the phase diagram,

$$f'(V_{low}) < 0 \qquad f'(V_{th}) > 0 \qquad f'(V_{high}) < 0$$
$$\text{(stable)} \qquad \text{(unstable)} \qquad \text{(stable),}$$

where $f(V)$ is the right side of Eq. 11.1. The unstable steady state V_{th} is a **repellor** that separates the **basins of attraction** associated with the two stable steady states, both of which are **attractors** (recall Section 3.3).

Fig. 11.3 shows solutions of Eq. 11.1 with a range of initial membrane potentials (-150 to $100\,mV$). The trajectories show that the basin of attraction of the stable steady state at V_{low} is the phase line interval $(-\infty, V_{th})$ while the basin of attraction of V_{high} is (V_{th}, ∞). Note that trajectories whose initial value is close (but not equal) to the unstable steady state V_{th} increase or decrease relatively slowly (at least initially). This is consistent with the phase diagram of Fig. 11.2 that indicates $|dV/dt|$ is small when $V \approx V_{th}$.

When the initial membrane voltage is near either one of the two *stable* steady states (V_{low} or V_{high}), there is a gradual decay of the deviation from steady state,

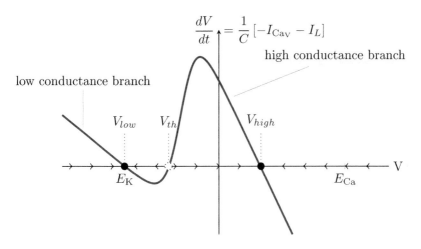

Figure 11.2   The phase diagram for a membrane that expresses I_{Ca_V} and I_L in the absence of applied current ($I_{app} = 0$). *Question*: When $V \approx V_{low}$ (or V_{high}), the dynamics are well approximated by exponential relaxation. For which steady state is the exponential time constant larger (slower relaxation)?

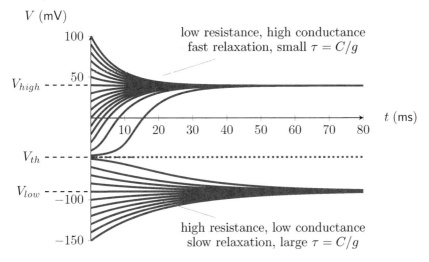

Figure 11.3   Solutions of the $I_{Ca_V} + I_L$ membrane model reveal two stable steady state membrane potentials (bistability). Relaxation of voltage is faster near the depolarized steady state (V_{high}) than near the hyperpolarized steady state (V_{low}), consistent with Fig. 11.2.

$\Delta V = V - V_{ss}$. These trajectories cannot be bona fide exponential relaxations because Eq. 11.1 is nonlinear. However, it can be shown that the relaxation of V to either stable steady state is, at least ultimately, approximately exponential (see Discussion). Furthermore, the approximate exponential time constant associated to V_{high} is smaller (faster) than the time constant associated to V_{low}. Using the passive membrane time constant formula $\tau = C/g$, we see that the conductance of the cell membrane is greater when $V \approx V_{high}$ than when $V \approx V_{low}$. Indeed, the current-voltage relation for total ionic current (Fig. 11.3) confirms this: the slope $dI_{ion}/dV \ominus$ conductance is greater at V_{high} than at V_{low}.

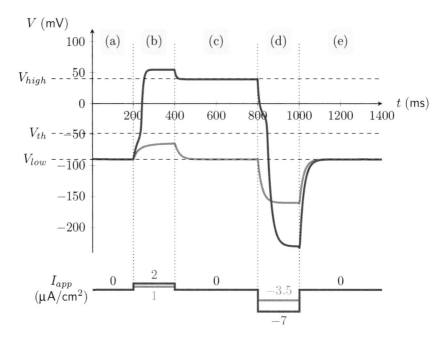

Figure 11.4 Bistability of the $I_{CaV} + I_L$ membrane model (Eq. 11.1) observed in response to pulses of applied current (I_{app}).

11.2 Response of a Bistable Membrane to Applied Current Pulses

Fig. 11.4 shows numerically integrated solutions of the $I_{CaV} + I_L$ membrane model (Eq. 11.1) when the applied current is a pulsatile function of time. $I_{app}(t)$ is zero for most of the simulation, but is depolarizing for 100 ms beginning at $t = 200$ ms, and hyperpolarizing for 100 ms beginning at $t = 800$ ms. The blue curve is the response when the first current pulse is depolarizing to such an extent ($I_{app} = 2 \mu A/cm^2$) that the membrane potential transitions from V_{low} to V_{high}. The voltage remains at V_{high} until the hyperpolarizing current pulse ($I_{app} = -7 \mu A/cm^2$) causes the membrane potential to return to V_{low}. Notice that the steady state voltage of the bistable membrane depends on the recent history of applied currents. This dynamical phenomenon is referred to as **hysteresis**. It is as though the bistable membrane is able to remember whether the most recent pulse of applied current was depolarizing or hyperpolarizing.

11.3 Membrane Currents and Fold Bifurcations

The gray curves in Fig. 11.4 repeat the sequence of depolarizing and hyperpolarizing pulses, but at half the magnitude. In this case, the depolarizing pulse is not sufficient to bring the membrane above the threshold V_{th}, so the membrane potential returns to V_{low} at the end of the first pulse. By repeating this calculation with lesser values of I_{app},

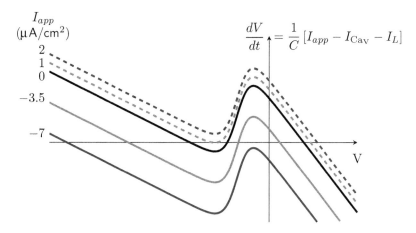

$$\frac{dV}{dt} = \frac{1}{C}[I_{app} - I_{Ca_V} - I_L]$$

Figure 11.5 Phase diagram for a membrane model that includes I_{Ca_V} and I_L (Eq. 11.1). Shown are the five values of I_{app} used in Fig. 11.4.

one may determine the minimum amount of applied current required for the membrane voltage to surpass the threshold voltage (V_{th}) and enter the basin of attraction of V_{high}. In general, the current required will depend on the pulse duration.

Fig. 11.5 shows the phase diagram of the $I_{Ca_V} + I_L$ membrane model for five values of I_{app}. For $I_{app} = 0$, the black curve indicates three steady states: two stable, one unstable (as in Fig. 11.2). When the applied current is sufficiently negative ($I_{app} = -7 \,\mu A/cm^2$), there is one hyperpolarized stable steady state (blue curve). When the applied current is sufficiently positive ($I_{app} = 2 \,\mu A/cm^2$), there is one depolarized stable steady state (blue dashed curve). The critical values of applied current lead to coalescence of a stable and unstable steady state. For example, the gray broken curve of Fig. 11.5 is nearly critical ($V_{low} \approx V_{th}$).

For current pulses of sufficiently long duration, Fig. 11.5 may be used to explain the dynamics of bistability in the $I_{Ca_V} + I_L$ membrane model. For example, the step in applied current that occurs in epoch (b) of Fig. 11.4 depolarizes the membrane from $V \approx V_{low}$ into the basin of attraction associated with V_{high} when $I_{app} = 2 \,\mu A/cm^2$ (blue curve), but not when $I_{app} = 1 \,\mu A/cm^2$ (gray curve). The phase diagrams below show how the magnitude of current applied in epoch (b) determines whether or not the transition from V_{low} to V_{high} occurs.

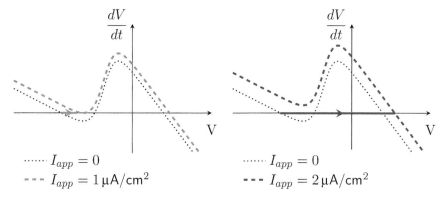

11.4 Bifurcation Diagram for the Bistable $I_{Ca_V} + I_L$ Membrane

Fig. 11.6 is a bifurcation diagram for the $I_{Ca_V} + I_L$ membrane model that summarizes how the number and stability of steady state voltages depends on I_{app}. The range of applied currents that leads to bistability is $-6.4 < I_{app} < 1.0\,\mu A/cm^2$. The critical values for applied current leading to fold bifurcations are labelled I_{app}^{**} and I_{app}^{*}.

The units and physical dimensions of Fig. 11.6 are mV (voltage, vertical axis) and $\mu A/cm^2$ (current/area, horizontal axis). Consequently, slope on this plot has physical dimensions of specific resistance (resistance/area) and units of $k\Omega/cm^2$. This allows us to classify the stable steady states (solid blue curves) as belonging to the *low resistance branch* or *high resistance branch* (top and bottom, respectively).

The negatively sloped branch of unstable steady states in the bifurcation diagram of Fig. 11.6 may be described as the *negative resistance branch* (blue dashed curve). Its existence is related to the negatively sloped region of the membrane's current-voltage relation (Fig. 11.1), due to the negatively sloped region of the current-voltage relation for I_{Ca_V} that is the signature of a regenerative current (Fig. 10.3). Of course, the membrane conductances of the L-type channels (I_{Ca_V}) and passive leak (I_L) are nonnegative quantities for any voltage. The *negative resistance* branch of steady states in the (I_{app}, V) bifurcation diagram (Fig. 11.6) and the *negative conductance* region of the current-voltage relation I_{Ca_V} (Fig. 10.3) and I_{Na_V} (Fig. 10.7) is a consequence of the voltage dependence of the conductance g_{Ca_V} and the electrochemical driving force $V - E_{Ca}$ (see Discussion).

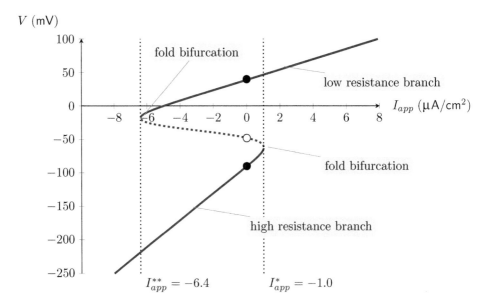

Figure 11.6   Bifurcation diagram for membrane model that includes I_{Ca_V} and I_L (Eq. 11.1). *Question:* Sketch qualitatively correct phase diagrams and phase lines for $I_{app} = I_{app}^{**}$ and I_{app}^{*}.

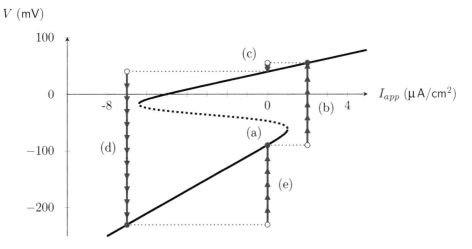

Figure 11.7 Overlaying trajectories (blue) on the bifurcation diagram (black) provides an interpretation of Fig. 11.4.

11.5   Overlaying Trajectories on the Bifurcation Diagram

Bifurcation diagrams provide useful summaries of the dynamics of membrane voltage. To illustrate, Fig. 11.7 plots a solution of the $I_{Ca_V} + I_L$ membrane model (Fig. 11.4, blue curve) on top of the (I_{app}, V) bifurcation diagram (black curve). The solution is a continuous function of voltage (vertical blue lines), but each change in applied current leads to a discontinuity in the (I_{app}, V)-plane (horizontal dotted lines).

To understand the relationship between Figs. 11.4 and 11.7, find epoch (a) on both plots ($I_{app} = 0$ and $V = V_{low}$). The depolarizing step in applied current that occurs at $t = 100$ ms causes the trajectory to shift rightward on the bifurcation diagram ($I_{app} = 0$ to $2\,\mu A/cm^2$). During epoch (b) the solution moves upward toward the only attractor available when $I_{app} = 2\,\mu A/cm^2$ (a depolarized steady state voltage). When the depolarizing current pulse ends, the trajectory instantaneously moves leftward to (c), then relaxes toward the steady state at V_{high}, and so on. Comparing Figs. 11.4 and 11.7 in this way, the response of the bistable membrane model to a sequence of applied current pulses is predictable.

11.6   Bistable Membrane Voltage Mediated by I_{Kir}

We showed that membrane potential bistability may occur in a neural membrane model (Eq. 11.1) that includes a passive leak (I_L) and an inward regenerative current (I_{Ca_V}). In this section, we show how bistability may occur in membranes with an inward rectifying potassium current (I_{Kir}). Consider the membrane model

$$C\frac{dV}{dt} = I_{app} - \underbrace{\bar{g}_{Kir}s_\infty(V)(V - E_K)}_{I_{Kir}} - \underbrace{g_L(V - E_L)}_{I_L}, \tag{11.3}$$

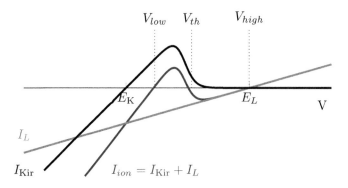

Figure 11.8   The sum of the ionic currents I_{Kir} and I_L may have three intersections with the horizontal axis $I_{ion} = I_{Kir} + I_L = 0$. Because $V_{high} \approx E_L < 0$, this type of bistability is easily distinguished from bistability mediated by I_{Cav} and I_L (compare Fig. 11.1).

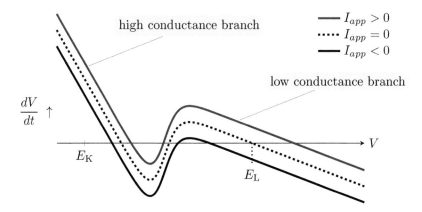

Figure 11.9   Phase diagram for the $I_{Kir} + I_L$ membrane model (Fig. 11.8).

where $g_{Kir}(V) = \bar{g}_{Kir} s_\infty(V)$ is a decreasing sigmoidal function of voltage (as in Fig. 10.4). When the reversal potential for I_{Kir} is less than the reversal potential for the leakage current ($E_K < E_L$),[1] the conductances $\bar{g}_{Kir}$ and g_L can be chosen to yield a current-voltage relation for $I_{ion} = I_{Kir} + I_L$ that reverses three times (Fig. 11.8). The slopes of the intersections with the horizontal axis indicate that the steady states at V_{low} and V_{high} are stable while V_{th} is unstable (Fig. 11.8). To see this, reverse the orientation of the vertical axis (flip up/down) and sketch arrows on the phase line (Fig. 11.9). Alternatively, one can reason that a steady state is stable if it occurs at a voltage for which the total membrane conductance is positive; for example, $dI_{ion}/dV|_{V=V_{low}} > 0$ and similarly for V_{high}. Conversely, a steady state is unstable if it occurs in a negative conductance region of the current-voltage relation where $dI_{ion}/dV < 0$.

There are qualitative similarities between the bistability mediated by I_{Kir} and I_{Cav} (compare Figs. 11.2 and 11.9). In both cases, the phase diagram has two negatively sloped branches and one positively sloped branch, separated by a local minimum and maximum. In both cases, the unstable steady state is a repellor that separates the basins of attraction of two stable steady states. Fig. 11.10 shows the salient quantitative

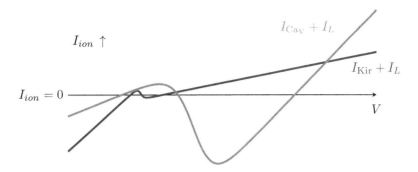

Figure 11.10   The I_{ion}-V relations for the $I_{Cav} + I_L$ and $I_{Kir} + I_L$ membrane models are qualitatively similar, but there are significant quantitative differences.

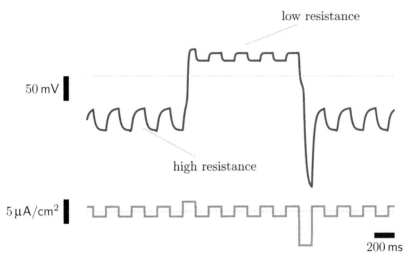

Figure 11.11   Bistability of the $I_{Cav} + I_L$ membrane model (Eq. 11.1) observed in response to pulses of applied current (I_{app}).

differences. In the $I_{Cav} + I_L$ bistable membrane, the depolarized steady state ($V_{high} \approx E_{Ca}$) is the high conductance steady state (calcium channels open). In the $I_{Kir} + I_L$ case, the depolarized steady state V_{high} is located near E_L (Kir channels closed), and the high conductance steady state is V_{low} (Kir channels open).

   Fig. 11.1 shows a simulated current-clamp recording demonstrating that the depolarized steady state is high conductance (low resistance), as in the $I_{Cav} + I_L$ membrane model.

11.7   Further Reading and Discussion

Spinal motor neurons of the brainstem and spinal cord exhibit $I_{Cav} + I_L$ bistability and **plateau potentials**, defined to be a stable membrane potential that is more depolarized than the resting membrane potential (Hounsgaard et al., 1988; Perrier et al., 2002). α-motoneurons express action potential generating currents and innervate skeletal

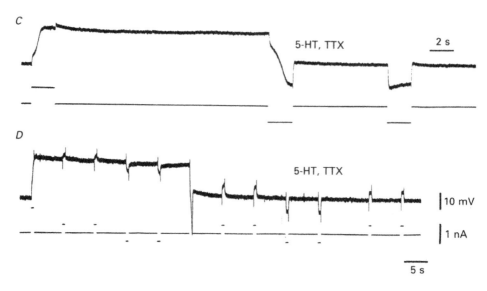

Figure 11.12   When action potentials of a spinal motoneuron are blocked by 1 μM tetrodotoxin (TTX), the neuronal membrane exhibits serotonin-induced plateau potentials (bistability). The membrane resistance is lower during the plateau than at rest. Reproduced with permission from Hounsgaard and Kiehn (1989), John Wiley and Sons.

muscle fibers; repetitive spiking of these neurons initiates muscle contraction. Plateau potentials are thought to support sustained action potential firing and to be useful in postural control (Kiehn and Eken, 1998).

Fig. 11.12 (top) shows an example in vitro intracellular recording from an α-motoneuron within a transverse section of the adult turtle spinal cord. When action potentials of a spinal motoneuron are blocked by 1 μM tetrodotoxin (TTX), the neuronal membrane exhibits plateau potentials (bistability) when stimulated by the neurotransmitter serotonin (10 μM). Fig. 11.12 (bottom) shows measurement of voltage change induced by depolarizing and hyperpolarizing current pulses of ± 0.4 nA during the plateau (the depolarized steady state V_{high}) and at rest (the hyperpolarized steady state V_{low}). Consistent with Fig. 11.11, the membrane resistance was several fold lower during the plateau (V_{high} is the high conductance steady state).[2]

The voltage dynamics of bistable neurons in vivo is complicated by the activity of presynaptic neurons. The inevitable fluctuations in the magnitude of synaptic currents can cause the postsynaptic bistable neuron to spend irregular intervals of time in the low and high voltage states, a phenomenon known as *basin hopping*.

Fig. 11.13 shows an example of basin hopping in a striatal spiny neuron and a layer V cortical pyramidal cell, recorded simultaneously. $I_{Kir} + I_L$ bistability is thought to play a role in generating these plateau potentials (Wilson, 2008). In both cells, the depolarized steady state V_{high} is only a few millivolts from the action potential threshold and occasional spikes are observed when the neuron is in the depolarized plateau state. Observe that membrane potential fluctuations around the depolarized state are of higher amplitude than those around the hyperpolarized state. This is consistent with V_{low} being the high conductance steady state with open Kir channels (Fig. 11.9).

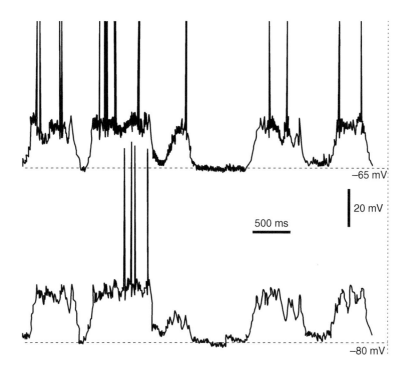

−65 mV

20 mV

500 ms

−80 mV

Figure 11.13 A simultaneously recorded layer V cortical pyramidal cell (top) and striatal spiny neuron (bottom) exhibiting plateau potentials (bistability) and basin hopping. Reproduced with permission from Charles Wilson (2008) Up and down states. Scholarpedia, 3(6):1410.

Membrane Conductance and Linearization

In the $I_{Ca_V} + I_L$ membrane model (Fig. 11.3), the stability of the steady states at V_{low} and V_{high} is obvious from flow of these trajectories. Observe that for the initial voltages near a stable steady state, the solutions are approximately exponential. For each steady state, the total ionic current reverses by crossing the horizontal axis with a particular slope that has physical dimensions of conductance or conductance/area (Fig. 11.1). On the phase diagram (Fig. 11.2), slope has physical dimensions of (voltage/time)/voltage = time^{-1}. The membrane model takes the form,

$$\frac{dV}{dt} = f(V) \tag{11.4}$$

and for each steady state V_{ss}, the function $f(V)$ crosses the horizontal axis with slope $f'(V_{ss})$. Observe that the Taylor series for $f(V)$ evaluated at each of these steady states is

$$f(V) = f(V_{ss}) + f'(V_{ss})\Delta V + f''(V_{ss})\frac{(\Delta V)^2}{2!} + f'''(V_{ss})\frac{(\Delta V)^3}{3!} + \cdots,$$

where $\Delta V = V - V_{ss}$. Remembering $f(V_{ss}) = 0$ (because V_{ss} is a steady state) and assuming that ΔV is small ($V \approx V_{ss}$), we may drop terms of order $(\Delta V)^2$ and higher to obtain

$$f(V) \approx f'(V_{ss})\Delta V = f'(V_{ss})(V - V_{ss}) .$$

Substituting this approximate expression for $f(V)$ into Eq. 11.4, we obtain a linear ODE with solution that approximates solutions of the membrane model when $V \approx V_{ss}$,

$$\frac{dV}{dt} = f'(V_{ss})(V - V_{ss}) . \tag{11.5}$$

Furthermore, because V_{ss} was assumed to be stable, we know that $f'(V_{ss}) < 0$. If we write $\tau = 1/|f'(V_{ss})| > 0$ and substitute $f'(V_{ss}) = -1/\tau < 0$ in Eq. 11.5, we obtain

$$\frac{\text{voltage}}{\text{time}} \ominus \frac{dV}{dt} = -\frac{V - V_{ss}}{\tau} \ominus \frac{\text{voltage}}{\text{time}} ,$$

where $\tau \ominus \text{time}$ is an exponential time constant.

Problem 11.1   In the $I_{Ca_V} + I_L$ membrane model (Figs. 11.1–11.3), which stable steady state (V_{low} or V_{high}) has smaller time constant τ?

The Sign of dI_{ion}/dV Determines Stability

Normally we determine the stability of steady states using the phase diagram of an ODE model. However, the stability of the steady states of a membrane model may also be determined from the current-voltage relation. Recall that the phase diagram for the current balance equation takes the form $dV/dt = f(V)$ where

$$f(V) = \frac{1}{C} \left[I_{app} - I_{ion}(V) \right] .$$

Assuming constant I_{app}, differentiating $f(V)$ with respect to voltage gives

$$f'(V) = \frac{1}{C} \left[-I'_{ion}(V) \right] . \tag{11.6}$$

Steady states occur when $f(V_{ss}) = 0$; V_{ss} is stable when $f'(V_{ss}) < 0$ and unstable when $f'(V_{ss}) > 0$. Using Eq. 11.6 we can restate this fact in biophysical terms: steady states occur when $I_{app} = I_{ion}(V_{ss})$; V_{ss} will be stable if $I'_{ion}(V_{ss}) > 0$ and unstable if $I'_{ion}(V_{ss}) < 0$. (If this seems backwards, remember that in Eq. 11.6 $I'_{ion}(V)$ and $f'(V)$ have opposite sign.) For example, in the $I_{Ca_V} + I_L$ membrane model (Fig. 11.1), the steady states at V_{low} and V_{high} are stable, because $I'_{ion}(V_{low}) > 0$ and $I'_{ion}(V_{high}) > 0$. Conversely, the steady state at V_{th} is unstable ($I'_{ion}(V_{th}) < 0$). In fact, the slope of the current-voltage relation has a contribution from each current,

$$I'_{ion}(V) = I'_{Ca_V}(V) + I'_L(V) = I'_{Ca_V}(V) + g_L .$$

Because the contribution from the leakage current to this slope is positive, $I'_{Ca_V}(V_{ss}) > 0$ is a sufficient condition for stability in the $I_{Ca_V} + I_L$ membrane model. That is, beginning at V_{low} or V_{high}, a small change in voltage results in a restorative current (depolarization $\rightarrow$ outward, hyperpolarization $\rightarrow$ inward).

Supplemental Problems

Problem 11.2   Explain why $I'_{Kir}(V_{ss}) > 0$ is a sufficient condition for stability in the $I_{Kir} + I_L$ membrane model (Figs. 11.8 and 11.9).

Problem 11.3   Answer the following questions for both the $I_{Cav} + I_L$ and $I_{Kir} + I_L$ bistable membrane models.

(a)   The leakage current is inward when V is [greater / less] than [E_L / E_{Ca}].
(b)   If I_{app} is increased the unstable steady state would move to a more [depolarized / hyperpolarized] voltage and eventually coalesce with the stable steady state associated with the [low / high] conductance branch of the current-voltage relation.
(c)   This is an example of a [transcritical / fold / pitchfork] bifurcation.

Problem 11.4   Show that for brief, strong current pulses, the voltage will almost instantly reset to $V_* = V_{low} + \Delta V$ where $\Delta V = Q/C$ and charge $\ominus$ $Q = I_{app}\,\Delta t$ $\ominus$ current · time is the integrated current.

Solutions

Figure 11.1 V_θ is the voltage at which half of the calcium channels are open, that is, $m_\infty(V_\theta) = 1/2$ (Eqs. 10.1–10.3). V_θ can be located on the horizontal axis by plotting the current that would obtain if all channels were open, $\bar{I}_{Cav} = \bar{g}_{Cav}(V - E_{Ca})$. Then identify the voltage that leads to half this maximum voltage, that is, the V_θ that satisfies $\bar{g}_{Cav}(V_\theta - E_{Ca}) = \bar{I}_{Cav}/2$.

Figure 11.2   The exponential time constant is inversely proportional to the conductance of the membrane. Lower conductance corresponds to larger values of the membrane time constant $\tau = C/g$. So relaxation to V_{low} is slower than relaxation to V_{high} (see Fig. 11.3).

Figure 11.6   The phase diagrams and phase lines for $I_{app} = I^{**}_{app}$ and I^{*}_{app} are as follows.

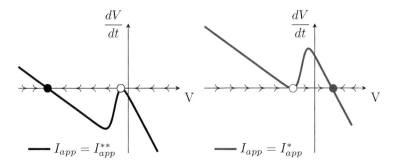

Problem 11.1   The high conductance steady state V_{high}.

Problem 11.4   To see this, rewrite the current balance equation in the following way,

$$C\frac{dV}{dt} = I_{app} - I_{ion}(V) \implies \Delta V = \Delta t \frac{I_{app} - I_{ion}(V)}{C}.$$

Here we approximate $dV/dt \approx \Delta V/\Delta t$ where ΔV is the increment in voltage caused by the current pulse and Δt is the pulse duration. Substituting $I_{app} = Q/\Delta t$ where the charge Q is fixed, we find

$$\Delta V = \frac{Q}{C} - \Delta t \frac{I_{ion}(V)}{C},$$

and taking the limit $\Delta t \to 0$ gives $\Delta V = Q/C$.

Notes

1.   This could happen, e.g., if E_L is an effective Nernst equilibrium potential given by a weighted average of E_{Na} and E_K.
2.   In Hounsgaard and Kiehn (1989) the membrane resistance was reported to be 2–4-fold lower during the plateau (4–10 mΩ) than at rest (14–18 MΩ).

12 Voltage-Clamp Recording

Current-voltage relations are obtained using the electrophysiological technique of *voltage-clamp recording*. Voltage-clamp recording reveals that ionic currents are not always instantaneous functions of membrane voltage (as assumed in previous chapters). Rather, ionic currents activate (channels open) or deactivate (channels close) on a time scale of 5–50 ms. Voltage-clamp experiments also reveal slower processes of channel closure and recovery (inactivation and de-inactivation).

12.1 Current-Clamp and Voltage-Clamp Recording

The current balance equation for a neural membrane takes the form,

$$C\frac{dV}{dt} = I_{app} - I_{ion}(V), \tag{12.1}$$

where $I_{ion}(V)$ is the total ionic membrane current. For a passive Ohmic current such as $I_{ion}(V) = I_L(V) = g_L(V - E_L)$, we are able to analytically calculate the membrane voltage $V(t)$ as a function of time (an exponential relaxation, recall Eq. 8.11). When $I_{ion}(V)$ is nonlinear, as was the case for the bistable neural membranes of Chapter 11, membrane voltage $V(t)$ may be simulated via numerical integration of the current balance equation (as in Fig. 11.4).

For both linear and nonlinear membranes, we have (up until now) thought of applied current $I_{app}(t)$ as a *stimulus*, and the resulting changes in neural membrane voltage $V(t)$ as the *response*. In Chapter 8 we learned that a positive applied current pulse will depolarize a passive membrane (Fig. 12.1, left). When the time course of applied current is chosen in advance (i.e., *clamped*), the result is a *current-clamp* recording of membrane voltage.

In **voltage-clamp recording**, the roles of applied current and membrane voltage are reversed (Fig. 12.1, right). Electronic instrumentation rapidly and precisely controls the applied current $I_{app}(t)$ so that membrane voltage $V(t)$ follows a **command voltage**, $V_{com}(t)$, that is chosen in advance to reveal the dynamics of membrane current.

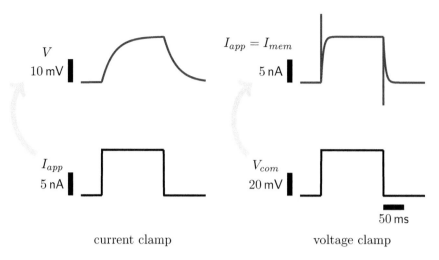

current clamp   voltage clamp

Figure 12.1   Left: In current-clamp recording, the applied current $I_{app}(t)$ is the stimulus (black) and the membrane potential $V(t)$ is the response (blue). *Question*: What is the observed membrane resistance? Right: In voltage-clamp recording, the command potential $V_{com}(t)$ is the stimulus (black) and the membrane current $I_{mem}(t)$ is the response (blue).

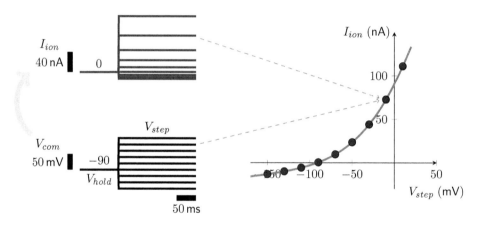

Figure 12.2   Voltage-clamp recording of a GHK-type potassium current. *Question*: Is this current inward or outward rectifying?

In voltage-clamp recording, the command voltage is the stimulus and the measured membrane current is the response (Fig. 12.1, right).

Voltage-Clamp Summary Plots and Rectification

The results of multiple voltage-clamp experiments are often presented *en masse* as an empirical current-voltage relation. Fig. 12.2 shows idealized voltage-clamp recordings of a membrane with a GHK-type potassium conductance. The **holding potential** is $V_{hold} = -90\,\mathrm{mV}$, while the **step potential** takes nine different values ranging from $V_{step} = -150$ to $10\,\mathrm{mV}$. The reversal potential for the potassium current is

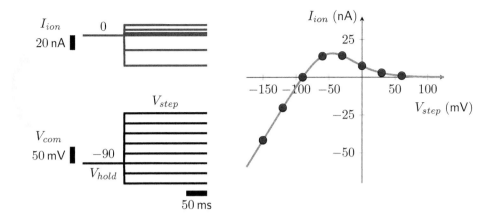

Figure 12.3   Voltage-clamp recording of an inward rectifying potassium current ($I_{ion} = I_{Kir}$). *Question:* Eight different step potentials are shown, but only six steps in ionic current are observed in the upper left panel. Why?

$E_K = -90$ mV (right panel). When the step potential is depolarized ($V_{step} > V_{hold}$), the ionic current is outward ($I_{ion} > 0$). When the step potential is hyperpolarized ($V_{step} < V_{hold}$), the ionic current is inward ($I_{ion} < 0$). For comparison, Fig. 12.3 shows *en masse* voltage-clamp recordings of a Kir current.

Capacitative Membrane Current in Voltage Clamp

Fig. 12.4 shows an idealized voltage-clamp recording of a membrane expressing a depolarization-activated I_{K_V} current (cf. Fig. 10.6). The command voltage $V_{com}(t)$ is a depolarizing pulse that increases from a holding potential of $V_{hold} = -70$ mV to a step potential of $V_{step} = -10$ mV (second panel). $V_{com}(t)$ remains at V_{step} for 300 ms before returning to V_{hold}. The blue curve (top panel) shows the changing applied current that *clamps* the membrane to the command potential.

For a deeper understanding of these changes in applied current, it is helpful to write the membrane current balance equation (Eq. 12.1) as follows,

$$I_{app}(t) = \underbrace{\overbrace{C\frac{dV}{dt}}^{I_{cap}} + I_{ion}(V)}_{I_{mem}} . \tag{12.2}$$

This clarifies that the applied current $I_{app}(t)$ that clamps the membrane voltage is a *measurement* of the membrane current $I_{mem}(t)$ elicited by $V_{com}(t)$. Eq. 12.2 reminds us that the membrane current includes capacitative (I_{cap}) and ionic (I_{ion}) components. When membrane voltage is constant ($dV/dt = 0$), the capacitative current is zero ($I_{cap} = CdV/dt$). When the command potential increases from V_{hold} to V_{step}, the membrane current includes a capacitative transient ($dV/dt > 0$, $I_{cap} > 0$) as well as the evoked ionic current I_{ion} (Fig. 12.4, gray curves). A second capacitative transient occurs when V_{com} decreases from V_{step} to V_{hold} ($dV/dt < 0$, $I_{cap} < 0$).

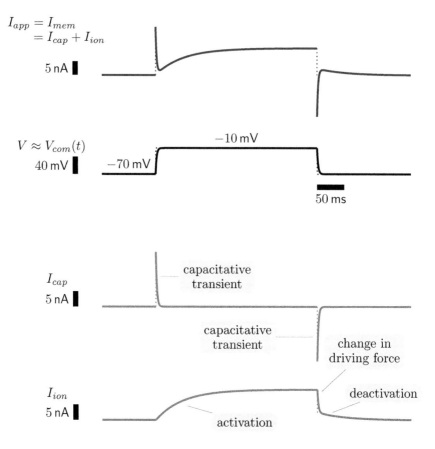

Figure 12.4 The applied current (I_{app}, blue) in the voltage-clamp technique measures the membrane current ($I_{mem} = I_{cap} + I_{ion}$, gray) in response to a depolarizing step in command voltage (V_{com}, black). The capacitative transients that occur when the voltage step begins and ends are characteristic of voltage-clamp recordings. *Question*: Is the elicited ionic current inward or outward? Why is the capacitative transient *upward* at pulse start and *downward* at pulse end?

Delayed Ionic Membrane Current

Between the capacitative transients in Fig. 12.5, the command potential is depolarized ($V_{step} = -10\,$mV) and the ionic membrane current increases from 0 to 10 nA with a time constant of about 50 ms. The ionic current is outward ($I_{ion} > 0$), consistent with a voltage-gated potassium current I_{K_V} with positive driving force ($V_{step} - E_K > 0$ so $I_{K_V} > 0$). When the command voltage returns to V_{hold}, the reduction in electrochemical driving force ($V_{hold} - E_K < V_{step} - E_K$) abruptly decreases I_{K_V}; the residual outward ionic current deactivates with $\tau \approx 50\,$ms.

Fig. 12.5 (left) shows eight voltage-clamp recordings similar to Fig. 12.4 that uses $V_{hold} = -90\,$mV and $-150 \leq V_{step} \leq 60\,$mV. The depolarizing command voltage activates ionic membrane current; for clarity, capacitative transients are not shown. Fig. 12.5 (right) shows how plotting the maximum evoked ionic membrane current

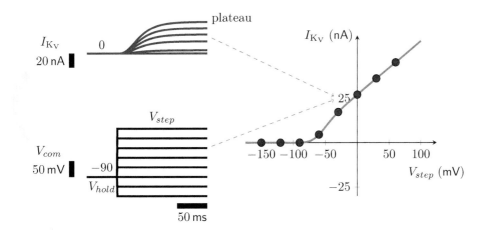

Figure 12.5   Voltage-clamp recording of a membrane expressing a depolarization-activated potassium current with delay. Notice that the plateau membrane current flows outward ($V_{step} > E_K$) more easily than inward ($V_{step} < E_K$). The time constant for activation is 10 ms. *Question*: Sketch a similar voltage-clamp experiment for a hyperpolarization-activated inward current with delay.

(the **plateau** current) as a function of V_{step} reveals the current-voltage relation for I_{K_V}. Because K_V currents are outward rectifying, those that exhibit delayed activation are referred to as *delayed rectifier* potassium currents.

Four Varieties of Persistent Ionic Currents

Because ionic membrane currents can be inward or outward, and activated by depolarization or hyperpolarization, there are four basic varieties of ionic currents and four elementary types of voltage-clamp response. Membranes may express depolarization-activated outward currents that have a delay (such as I_{K_V} in Figs. 12.4 and 12.5). There are also delayed inward currents that are activated by depolarization, and delayed inward and outward currents that are activated by hyperpolarization. Fig. 12.6 summarizes these four basic cases; more complex examples follow.

12.2   Modeling Delayed Activation of Ionic Currents

Delayed activation of ionic current is modeled using an ordinary differential equation (ODE) for gating variable dynamics. For the delayed rectifier potassium current (I_{K_V}, Fig. 12.5), the steady state fraction of ion channels activated (open) at any given voltage is a familiar increasing sigmoid (cf. Eq. 10.3),

$$w_\infty(V) = \frac{1}{2}\left[1 + \tanh\left(\frac{V - V_\theta}{V_\sigma}\right)\right]. \tag{12.3}$$

If activation and deactivation occurs with an exponential time constant τ_m, the membrane current is

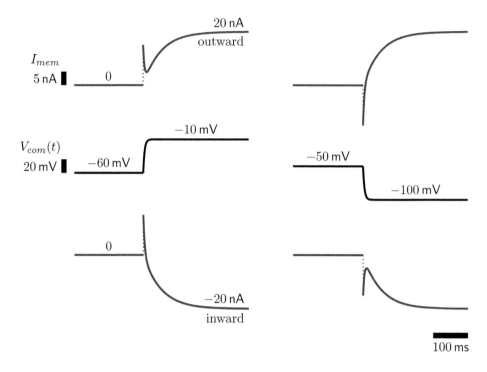

Figure 12.6   Four basic voltage-clamp measurements observed for persistent ionic currents. *Question*: Which one is a hyperpolarization-activated outward current?

$$I_{K_V} = \bar{g}_{K_V} w (V - E_K) , \tag{12.4}$$

where the fraction of open K_V channels $w(t)$ solves the first-order ODE,

$$\frac{dw}{dt} = -\frac{w - w_\infty(V)}{\tau_w} . \tag{12.5}$$

Assuming a holding potential is maintained for a long time prior to a step in V_{com} at $t = 0$, the initial condition for Eq. 12.5 is $w(0) = w_\infty(V_{hold})$. Subsequent to the command potential increase (or decrease) to V_{step}, the gating variable w will relax exponentially from $w_\infty(V_{hold})$ to the new steady state $w_\infty(V_{step})$,

$$w(t) = \left[w_\infty(V_{step}) - w_\infty(V_{hold}) \right] e^{-t/\tau_w} + w_\infty(V_{hold}) \quad \text{for} \quad t \geq 0. \tag{12.6}$$

Consequently, the potassium current evoked by step potential is,

$$I_{K_V} = \begin{cases} \bar{g}_{K_V} w_\infty(V_{hold}) (V_{hold} - E_K) & \text{for } t < 0 \\ \bar{g}_{K_V} w(t) (V_{step} - E_K) & \text{for } t \geq 0. \end{cases}$$

If V_{hold} is chosen so $w_\infty(V_{hold}) \approx 0$, the initial membrane current will be negligible and, subsequent to the voltage step, $I_{K_V}(t)$ is approximately proportional to $w(t)$.

Similar reasoning explains the deactivation phase of the delayed rectifier potassium current (I_{K_V}) observed in Fig. 12.4 (top right). The voltage step is maintained for 300 ms and the pulse ends with $w \approx w_\infty(V_{step})$. When V_{com} returns to V_{hold}, both the electrochemical driving force and potassium current decrease: $I_{K_V} = \bar{g}_{K_V} w_\infty(V_{step})(V_{hold} - E_K)$. Subsequently, I_{K_V} slowly decreases to a negligible value, following the relaxation of w from $w_\infty(V_{step})$ to $w_\infty(V_{hold}) \approx 0$.

Gating Variables with Voltage-Dependent Relaxation Times

A more flexible description of persistent ionic membrane currents may be obtained by specifying an exponential time constant for the activation gating variable that is a function of voltage. In that case, the gating variable for a persistent I_{K_V} current (Eq. 12.5) solves

$$\frac{dw}{dt} = -\frac{w - w_\infty(V)}{\tau_w(V)},$$

(12.7)

where the *voltage-dependent* time constant $\tau_w(V)$ is[1]

$$\tau_w(V) = \bar{\tau}_w / \cosh\left(\frac{V - V_\theta}{V_\sigma}\right).$$

(12.8)

In this expression, $\bar{\tau}_w \ominus$ time is a scale parameter, V_θ and V_σ are as in Eq. 12.3, and the hyperbolic cosine is $\cosh x = (e^x + e^{-x})/2$ (see Fig. 12.7).

Fig. 12.8 shows a voltage-clamp simulation of the delayed rectifier I_{K_V} modeled in this fashion. In the normalized plot below, observe that w relaxes faster for $V_{step} = +10$ (solid) than +70 mV (broken curve):

w

0.5 ▌

This is consistent with Fig. 12.7 (right) where extreme voltages (i.e., large $|V - V_\theta|/V_\sigma$) lead to small exponential time constants.

Fig. 12.9 shows simulated voltage-clamp experiments for a neural membrane expressing a hyperpolarization-activated nonspecific cation conductance I_h (also

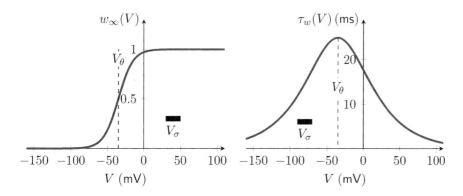

Figure 12.7 Voltage-dependent steady state and time constant for a depolarization-activated outward current I_{K_V} (Eq. 12.7). Parameters as in Morris and Lecar (1981).

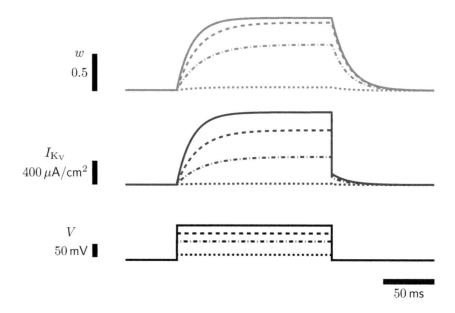

Figure 12.8 I_{K_V} with voltage-dependent time constant $\tau_w(V)$ and steady state $w_\infty(V)$ (Fig. 12.7). *Question*: Why are there no capacitative transients in the K_V current plot (blue)?

called I_{sag}). This inward persistent current is modeled with a gating variable r that has voltage-dependent steady state r_∞ and time constant τ_r,

$$I_h = \bar{g}_h r \,(V - E_h) \quad \text{where} \quad \frac{dr}{dt} = -\frac{r - r_\infty(V)}{\tau_r(V)}. \tag{12.9}$$

In a voltage-clamp study of the ionic currents expressed by thalamocortical relay neurons, investigators concluded that $\bar{g}_h = 0.03 \, \mu S$ and $E_h = -43 \, mV$ (Huguenard and McCormick, 1992). The voltage-dependent gating functions r_∞ and τ_r were determined to be

$$r_\infty(V) = \left[1 + \exp\left(\frac{V + 75 \, mV}{5.5 \, mV}\right)\right]^{-1} \tag{12.10}$$

and

$$\tau_r(V) = \frac{3900 \, ms}{\exp\left(-7.68 - 0.086 \, mV^{-1} \cdot V\right) + \exp\left(5.04 + 0.0701 \, mV^{-1} \cdot V\right)}.$$

These functions take a slightly different form than Eqs. 12.3 and 12.8, but they are qualitatively similar (compare Figs. 12.7 and 12.9).

12.3   Voltage Clamp and Transient Ionic Currents

Fig. 12.10 shows a simulated voltage-clamp recording of a membrane expressing a passive leakage current and a depolarization-activated inward regenerative ionic current.

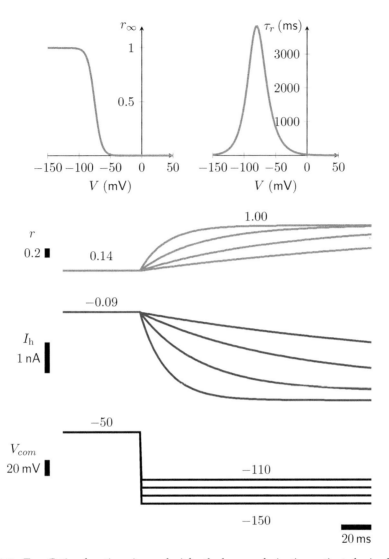

Figure 12.9 Top: Gating functions (r_∞ and τ_r) for the hyperpolarization-activated mixed cation current (I_h, also called I_{sag}). Bottom: Simulated voltage-clamp recordings. *Question*: What is the current-voltage relation? Is this current inward or outward?

The time constant for activation of this current is about 5 ms, somewhat faster than that used for the delayed rectifier I_{K_V} current (recall Fig. 12.8).

This voltage-clamp simulation illustrates a new phenomenon: ionic current **inactivation**. Because the inactivation process has a time constant of about 20 ms (slower than activation), the evoked inward current is *biphasic*. In the first (activation) phase, inward ionic current is recruited ($I_{ion} \leq 0$ and the magnitude $|I_{ion}|$ increases). Approximately 60 ms after the depolarizing voltage step begins, the **peak** inward current is achieved. The **plateau** current that persists until the voltage step ends is about 50% of the peak current in magnitude.

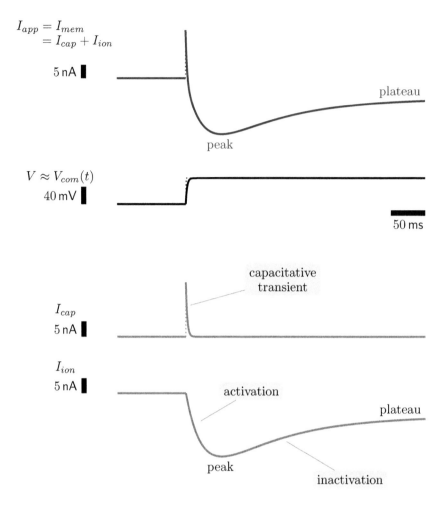

Figure 12.10   An inward current that activates with a time constant of 8 ms and inactivates with a time constant of 60 ms. The plateau current is about 50% the peak current. *Question:* Sketch the membrane current that would result from a depolarization-activated outward ionic current that has similar activation and inactivation kinetics, but a plateau/peak current ratio of 80%.

Ionic membrane currents that exhibit a phase of increasing magnitude followed by a phase of decreasing magnitude are referred to as **transient currents**. More precisely, the ionic membrane current shown in Fig. 12.10 has two components: a persistent component (about 50%) and a transient component (also about 50%).[2]

Transient ionic membrane currents may be inward or outward, and activated by depolarization or hyperpolarization, similar to persistent ionic currents. Fig. 12.11 shows these four typical cases, each with a persistent/peak current of about 20% (compare Fig. 12.6).

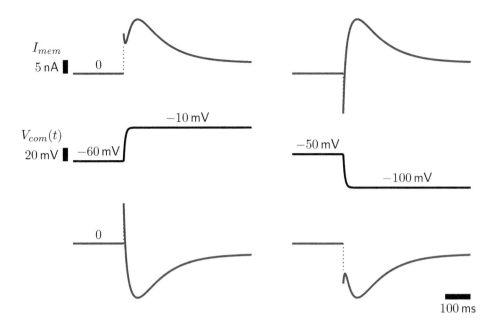

Figure 12.11   Simulated voltage-clamp recording of four varieties of transient ionic currents with a persistent/peak current of about 20%. *Question*: Sketch analogous ionic currents with persistent/peak current of 60%.

12.4   Modeling Transient Ionic Currents

Transient ionic membrane currents are modeled using voltage-dependent steady states and exponential time constants for *two* distinct gating variables, one representing the process of activation/deactivation, the other representing the process of inactivation/de-inactivation. Consider the persistent $I_{\mathrm{Ca_V}}$ current

$$I_{\mathrm{Ca_V}} = \bar{g}_{\mathrm{Ca_V}} m \, (V - E_{\mathrm{Ca}}) \,, \tag{12.11}$$

where m is an activation/deactivation gating variable with m_∞ and τ_m as in Fig. 12.12 (blue). This persistent $\mathrm{Ca_V}$ current can be made transient by including a second gating variable h with steady state h_∞ and time constant τ_h (gray),

$$I_{\mathrm{Ca_V}} = \bar{g}_{\mathrm{Ca_V}} mh \, (V - E_{\mathrm{Ca}}) \,, \tag{12.12}$$

where the gating variables m and h solve

$$\frac{dm}{dt} = -\frac{m - m_\infty(V)}{\tau_m(V)} \tag{12.13}$$

$$\frac{dh}{dt} = -\frac{h - h_\infty(V)}{\tau_h(V)} \,. \tag{12.14}$$

Fig. 12.12 shows simulated voltage-clamp recordings of this transient $\mathrm{Ca_V}$ current. Transient $I_{\mathrm{Ca_V}}$ mediated by L-type calcium channels is essential to (electrical)

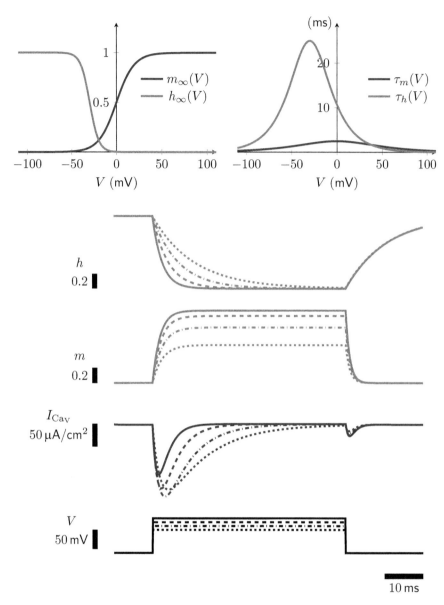

Figure 12.12 Top: Gating variables for a transient Ca_V current (Eqs. 12.11 and 12.14). Bottom: Simulated voltage-clamp recordings. *Question*: Which variable, m or h, is the inactivation gating variable? The activation gating variable? Which is faster?

excitation-contraction coupling of cardiac muscle.[3] In Chapter 13 we will discuss an analogous transient Na_V current required for neuronal action potentials (regenerative inward current is required for the upstroke). Cardiac myocytes also express I_{Na_V}, though I_{Ca_V} is the dominant inward current. In both muscle cells and neurons, the delayed rectifier I_{K_V} plays an essential role in repolarization (the action potential downstroke).

12.5   Further Reading and Discussion

Hille (2001, Chapter 3–5) and McCormick (1998) are recommended for further details on voltage-gated currents.

Current Clamp or Voltage Clamp?

When reading research papers in electrophysiology, due to interest or time constraints, you may find yourself moving directly to a figure and attempting to interpret it. Because you have skipped the relevant text, the first step in this interpretation is to determine whether the figure is reporting results from current-clamp or voltage-clamp recording. This is important because stimulus and response are reversed in voltage-clamp as compared to current-clamp recording and, unfortunately, the stimulus is not always *below* the response as is the convention in this book (see, e.g., Fig. 12.1). The presence of capacitative transients indicates voltage clamp, but these are sometimes removed prior to publishing. Stimulating currents and voltage-clamp protocols are not always drawn. For these reasons, it is usually a good idea to at least read the figure caption before attempting to decode an electrical recording.

Tail Currents

Fig. 12.13 shows a simulated voltage-clamp recording in which a persistent Ca_V current is elicited by depolarizing step potentials. This current is modeled using Eqs. 12.11 and 12.13 using the activation gating variable steady state m_∞ and time constant τ_m of Fig. 12.12 (blue curves). Inactivation is not included. Observe the downward deflection in I_{Ca_V} that occurs when the command voltage returns to V_{hold} ($\star$). This is not a capacitative transient (these are not shown), but a new phenomenon referred to as a **tail current**.

   To understand the tail current phenomenon, observe that the 80 ms depolarizing pulse is sufficiently long for the gating variable to nearly achieve its steady state $m_\infty(V_{step}) \approx 1$ (top panel). When the pulse ends, the command voltage returns to V_{hold}. This increases the magnitude of the electrochemical driving force ($|V_{hold} - E_{Ca}| > |V_{step} - E_{Ca}|$) and causes a rapid increase in inward current. This tail current ($\star$) decays as the gating variable m decays to $m_\infty(V_{hold}) \approx 0$ with exponential time constant $\tau_m(V_{hold}) \approx 1$ ms. The transient Ca_V current of Fig. 12.12 also shows a tail current.

The Form of Gating Variable Equations

The dynamics of activation and inactivation gating variables (e.g., Eqs. 12.13 and 12.14) is consistent with an ion channel transitioning between closed and open states with voltage-dependent rates,

$$\text{(closed)} \quad C \underset{\beta(V)}{\overset{\alpha(V)}{\rightleftharpoons}} O \quad \text{(open)}. \tag{12.15}$$

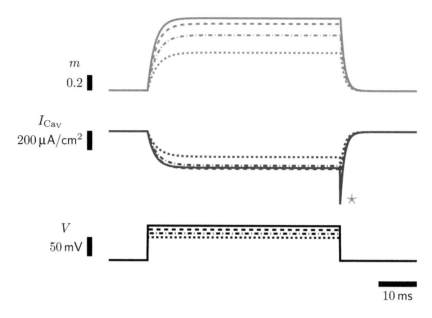

Figure 12.13 I_{Cav} exhibits a *tail current* upon depolarization (⋆). For clarity, capacitative transients are not shown. *Question:* Is the tail current inward or outward? What determines its time scale?

Denoting the fraction of closed and open channels as f_c and f_o, we write

$$\frac{df_c}{dt} = -\alpha(V)f_o + \beta(V)f_o \tag{12.16}$$

$$\frac{df_o}{dt} = \alpha(V)f_c - \beta(V)f_o, \tag{12.17}$$

ODEs that are very similar to Eqs. 4.7 and 4.8 (an isomerization reaction). Using $1 = f_c + f_o$, Eq. 12.16 can be eliminated, to obtain

$$\frac{df_o}{dt} = \alpha(V)\left[1 - f_o\right] - \beta(V)f_o. \tag{12.18}$$

This ODE may be written as

$$\frac{df_o}{dt} = -\frac{f_o - f_o^{ss}(V)}{\tau_o(V)} \tag{12.19}$$

where

$$f_o^{ss} = \frac{\alpha(V)}{\alpha(V) + \beta(V)} \quad \text{and} \quad \tau_o = \frac{1}{\alpha(V) + \beta(V)}. \tag{12.20}$$

If we assume that the rate constants are the following increasing and decreasing functions of voltage,[4]

$$\alpha(V) = k\exp\left(\frac{V - V_\theta}{V_\sigma}\right) \quad \text{and} \quad \beta(V) = k\exp\left(-\frac{V - V_\theta}{V_\sigma}\right),$$

where $k > 0$ and $V_\sigma > 0$, then the fraction of open channels at steady state (Eq. 12.20) takes the form

$$f_o^{ss} = \left[1 + \exp\left(-\frac{2(V - V_\theta)}{V_\sigma}\right)\right]^{-1} = \frac{1}{2}\left[1 + \tanh\left(\frac{V - V_\theta}{V_\sigma}\right)\right],$$

consistent with Eq. 12.3, and τ_o is given by Eq. 12.8 with $\bar{\tau}_o = 1/(2k)$.[5]

Supplemental Problems

Problem 12.1   Use Eq. 12.10 to determine the voltage for which the steady state conductance of I_h is half maximum.

Problem 12.2   Why is the tail current in Fig. 12.12 smaller in magnitude than the one in Fig. 12.13?

Problem 12.3   When $V_{step} = -110\,\text{mV}$ in Fig. 12.9, what is the time constant for the gating variable r?

Solutions

Figure 12.1   The membrane resistance is $20\,\text{mV}/10\,\text{nA} - 2\,\text{M}\Omega$.

Figure 12.2   The GHK potassium current is outward rectifying. The membrane current flows outward ($I_{ion} > 0$ when $V_{step} > E_K = -90\,\text{mV}$) more easily than inward ($I_{ion} < 0$ when $V_{step} < E_K$).

Figure 12.3   The ionic currents associated to $V_{step} = -60$ and $-30\,\text{mV}$ are both about $I_{ion} = 14\,\text{nA}$. Similarly, $I_{ion} = 3\,\text{nA}$ for $V_{step} = +60\,\text{mV}$ and, in the plot, not distinguishable from $V_{step} = -90\,\text{mV}$ ($I_{ion} = 0$).

Figure 12.4   The evoked ionic membrane current is outward ($I_{ion} > 0$). The direction of the capacitative current at pulse start is explained by considering each term in Eq. 12.2 just after V_{com} is increased to V_{step}, when $dV/dt > 0$ is large (so $I_{cap} > 0$ large), but outward I_{ion} increases more slowly ($\tau \approx 50\,\text{ms}$),

$$\Uparrow I_{app} = C \overbrace{\frac{dV}{dt}}^{\Uparrow I_{cap}} + I_{ion} \uparrow .$$

Reverse the arrows for the capacitative transient at pulse end.

Figure 12.5   Below is shown a voltage-clamp recording of a membrane expressing the hyperpolarization-activated nonspecific cation current I_h and passive leak I_{leak}. The holding potential used here is the resting potential of the membrane, which is much closer to $E_K = -90\,\text{mV}$ than $E_h = 0\,\text{mV}$ because I_h is not activated at resting membrane potentials (deactivated, ion channels closed). The summary current-voltage relation shows that the plateau membrane current exhibits inward rectification. Because the activation time constant for I_h is about 20 ms, I_h could be described as a delayed inward rectifying current.

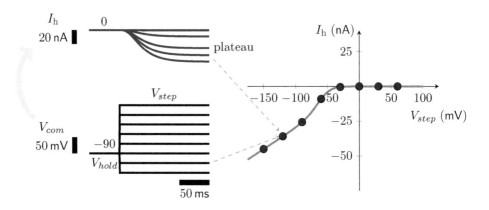

Figure 12.6   The hyperpolarization-activated outward current is top right.

Figure 12.8   The symbol I_{K_V} refers to the *ionic* membrane current due to voltage-gated potassium channels; the capacitative membrane current is not included. If total membrane current (I_{mem}) were plotted, it would be similar to Fig. 12.5.

Figure 12.9   The current is inward. The current-voltage relation is:

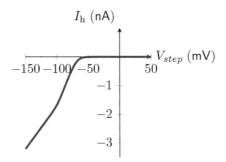

Figure 12.10   A plateau/peak current ratio of 80%:

Figure 12.11   The plateau/peak ratio below is zero (solid) and 60% (dashed).

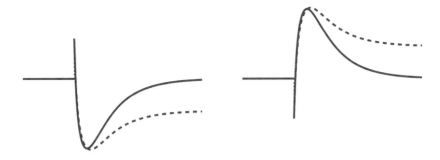

Figure 12.12   The fast variable m represents activation; the slow variable h represents inactivation.

Figure 12.12   The tail current is inward. The duration of the tail current is determined by the *deactivation* time constant of the Ca_V current, given by $\tau_m(V_{hold})$.

Problem 12.1 I_h is half maximum at $V_\theta = -75\,mV$ (note negative sign).

Problem 12.2   Fig. 12.12 is a transient Ca_V current. At the time the tail current begins, the inactivation gating variable h is nearly zero (Eq. 12.12).

Problem 12.3 r_∞ at $-110\,mV$ is about 2 s (quite long).

Notes

1. The functional form of $\tau_w(V)$ and $w_\infty(V)$ is discussed on p. 211.
2. The persistent currents I_{K_V} and I_h that activate and deactivate, but do not exhibit slow inactivation, have no transient component.
3. L-type calcium channels are sometimes referred to as dihydropyridine receptors (DHPRs), because they are blocked by dihydropyridine (DHP). In cardiac myocytes, depolarization leads to calcium influx through DHPRs. This calcium influx triggers the release of calcium from the sarcoplasmic reticulum (SR) via calcium activation of RyRs (an example of ligand-mediated gating discussed in Chapter 4). *Skeletal* myocytes have a high concentration of DHPRs, but calcium influx is minimal due to the *slow activation* (τ_m large). In skeletal muscle, DHPRs essentially function as voltage sensors, and there is evidence of physical coupling between skeletal DHPRs and RyRs. Calcium influx is not required for skeletal muscle contraction.
4. This is similar to an Arrhenius equation for the dependence of a rate constant of a chemical reaction on an activation energy, $a = a_0 e^{-E_a/RT}$.
5. The hyperbolic tangent can be written $\tanh x = (1 - e^{-2x})/(1 + e^{-2x})$ and thus

$$\frac{1}{2}(1 + \tanh x) = \frac{1}{2}\left(1 + \frac{1 - e^{-2x}}{1 + e^{-2x}}\right) = \frac{1}{2}\left(\frac{1 + e^{-2x} + 1 - e^{-2x}}{1 + e^{-2x}}\right) = \frac{1}{1 + e^{-2x}}.$$

13 Hodgkin-Huxley Model of the Action Potential

Alan Hodgkin and Andrew Huxley's Nobel Prize winning studies included seminal experimental observations of the action potential in the squid giant axon. Hodgkin and Huxley introduced the mathematical framework for modeling ionic currents of excitable membranes that has become standard.

13.1 The Squid Giant Axon

Alan Hodgkin and Andrew Huxley's studies of the dynamics of the action potential in neural membranes included the refinement of an experimental preparation: the squid giant axon.[1] The giant axon of a (normal sized) squid is a motoneuron that participates in escape reflex circuitry. The axon is many centimeters long (about half the squid's length) – originating in the stellate ganglion and innervating muscle cells in the mantle – and nearly a millimeter in diameter. The squid giant axon's diameter is a specialization that allows rapid propagation of action potentials.[2] The giant axon enabled experimental manipulation of intracellular and extracellular solutions and control of ionic concentrations on both sides of the nerve membrane. A wire could be threaded down the length of the large diameter axon leading to a *spatially clamped* electrical recording mode in which the giant axon was essentially uni-potential. This meant that wave phenomena related to the propagation of action potentials down the length of the axon would not confound electrical recordings.

Current-Clamp Recording

Hodgkin and Huxley used both current-clamp and voltage-clamp recording techniques to study the action potential of the squid giant axon. To illustrate, Fig. 13.1 shows how the excitability of the squid giant axon may be triggered by brief pulses of depolarizing applied current ($I_{pulse} > 0$). For a **subthreshold** current pulse, the membrane responds passively (increasing while $I_{pulse} > 0$ and decreasing when I_{pulse} returns to zero). On the other hand, when **superthreshold** current is applied, an

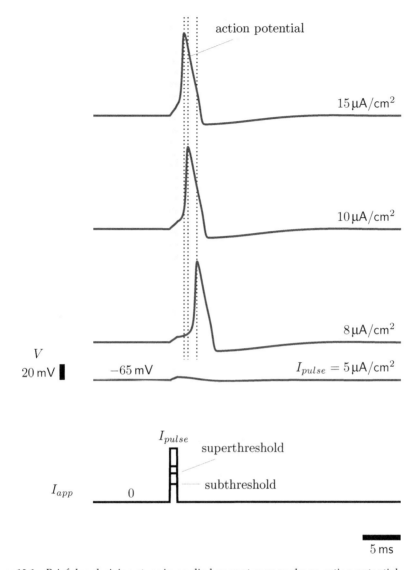

Figure 13.1   Brief depolarizing steps in applied current may evoke an action potential.

action potential is evoked. Under some experimental conditions, the action potential is essentially all-or-none. If the applied current pulse is only slightly over threshold, the latency between the pulse and the action potential increases slightly (vertical lines), but the action potential is full sized (see Fig. 13.1).

Voltage-Clamp

Fig. 13.2 shows simulated voltage-clamp recordings similar to those performed in Hodgkin and Huxley's Nobel Prize winning research. From a holding potential of $V_{hold} = -65\,\text{mV}$, hyperpolarization to a step potential of $V_{step} = -100\,\text{mV}$ elicits little

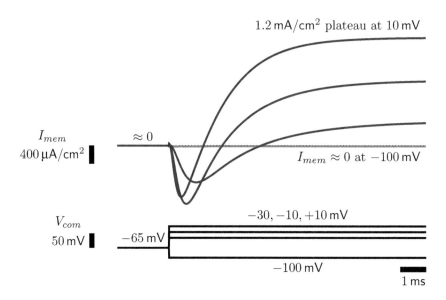

Figure 13.2   Voltage-gated membrane currents (sodium, potassium and leak) of the Hodgkin-Huxley model.

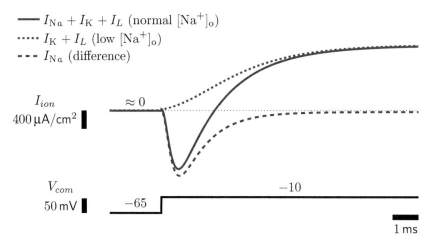

Figure 13.3   The inward/outward membrane current (solid curve) evoked in voltage-clamp mode can be apportioned between I_{Na_V} (dashed) and I_{K_V} (dotted) by subtracting the delayed outward current observed in low extracellular sodium.

ionic membrane current ($I_{mem} \approx 0$). However, depolarizing voltage steps of $V_{step} = -30, -10$ and $10\,mV$ evoke significant ionic membrane current that is inward for 1 to 2 ms with a peak of $-800\,\mu A/cm^2$. Subsequently, the current reverses and the outward current gradually increases to a plateau of $1.2\,mA/cm^2$.

Hodgkin and Huxley disentangled the contributions of I_{Na_V} and I_{K_V} to the depolarization-induced biphasic membrane current by performing voltage-clamp recordings with low extracellular sodium (Fig. 13.3). Low $[Na^+]_o$ eliminated the

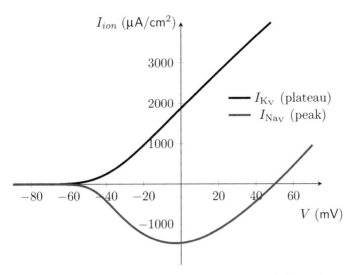

Figure 13.4 Current-voltage relations for the peak Na_V current and plateau K_V current in the Hodgkin-Huxley model.

inward phase of the evoked ionic membrane current. The persistent outward current that remained was attributed to the delayed rectifier I_{K_V}, and the difference was interpreted as I_{Na_V}.[3] This dissection of ionic membrane currents allowed Hodgkin and Huxley to estimate current-voltage relations for both the transient sodium current I_{Na} and the persistent potassium current I_{K_V} (Fig. 13.4).

13.2   The Hodgkin-Huxley Model

The celebrated **Hodgkin-Huxley model** of the squid giant axon action potential is the following system of four ODEs:

$$C\frac{dV}{dt} = I_{app} - \overbrace{g_{Na}m^3h\,(V - E_{Na})}^{I_{Na_V}} - \overbrace{g_K n^4\,(V - E_K)}^{I_{K_V}} - \overbrace{g_L\,(V - E_L)}^{I_L} \tag{13.1a}$$

$$\frac{dm}{dt} = -\frac{m - m_\infty\,(V)}{\tau_m\,(V)} \tag{13.1b}$$

$$\frac{dh}{dt} = -\frac{h - h_\infty\,(V)}{\tau_h\,(V)} \tag{13.1c}$$

$$\frac{dn}{dt} = -\frac{n - n_\infty\,(V)}{\tau_n\,(V)}. \tag{13.1d}$$

The current balance equation (Eq. 13.1a) includes applied current (I_{app}), a passive leak (I_L), and the two voltage-gated Na_V and K_V currents discussed above. I_{Na_V} is inward and transient (Fig. 12.10). I_{K_V} is outward and persistent (Figs. 12.4 and 12.5).

   The Hodgkin-Huxley ODE model was constrained by voltage-clamp recordings similar to those shown in Fig. 13.3. Assuming $I_{K_V} = g_K(V)\,(V - E_K)$, the empirical current-voltage relation for the plateau I_{K_V} current (Fig. 13.4, black) allowed back-calculation of the voltage-dependent potassium conductance,

$$g_K(V) = \frac{I_{K_V}(V)}{(V - E_K)}.$$

Hodgkin and Huxley had reason to believe that potassium channels had multiple voltage sensors responsible for channel opening and closing (i.e., voltage-dependent gating). Assuming four independent activation gates, the potassium current was modeled as

$$I_{K_V} = \bar{g}_K n^4 (V - E_K), \tag{13.2}$$

where n satisfied Eq. 13.1d (compare Eqs. 12.15–12.19).[4] The steady state gating function $n_\infty(V)$ was fit to the empirical plateau current-voltage relation (Eq. 13.4, black),[5] and the time constant $\tau_n(V)$ was chosen to match the observed delayed activation of I_{K_V}. Fig. 13.5 shows the response of the activation gate n and the I_{K_V} current when a voltage-clamped membrane is depolarized from rest.

In the Hodgkin-Huxley model, the transient I_{Na_V} current is

$$I_{Na_V} = g_{Na} m^3 h (V - E_{Na}),$$

where m and h are activation/deactivation and inactivation/de-inactivation gates solving Eqs. 13.1b and 13.1c. Hodgkin and Huxley devised elaborate voltage-clamp protocols that allowed back-calculation of the voltage-dependent steady states, $m_\infty(V)$, $h_\infty(V)$, and time constants, $\tau_m(V)$, $\tau_h(V)$. Fig. 13.6 shows the dynamics of these gating variables and the I_{Na_V} current when a voltage-clamped membrane is depolarized from rest. The gating functions are plotted in Fig. 13.7.

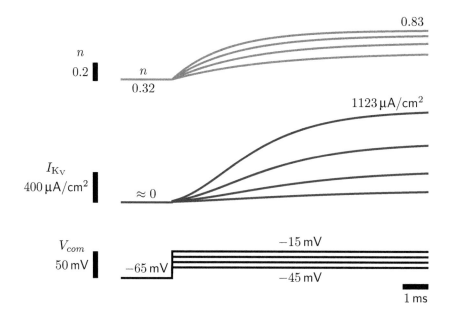

Figure 13.5 Voltage-gated potassium current of the Hodgkin-Huxley model. Parameters as in Hansel et al. (1993).

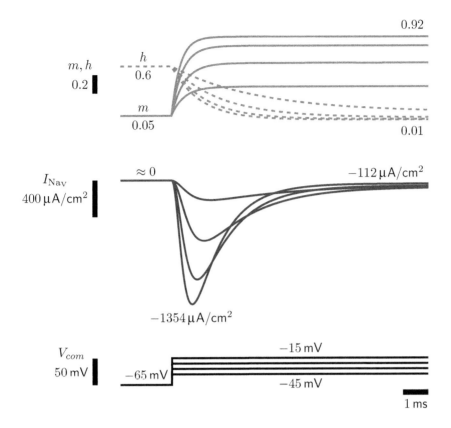

Figure 13.6 Voltage-gated sodium current of the Hodgkin-Huxley model. The initial sodium current is $-1.22\,\mu A/cm^2$.

13.3 Excitability in the Hodgkin-Huxley Model

The Hodgkin-Huxley model is a good starting point to understand the dynamics of neuronal membrane excitability. The first 2 ms of Fig. 13.8 shows the quiescent steady state of the Hodgkin-Huxley model simulated in current-clamp mode ($V = -65\,\text{mV}$ when $I_{app} = 0$). The resting voltage is a steady state of the current balance equation leading to balance of the sodium, potassium and leakage currents ($I_{Nav} + I_{Kv} + I_L = I_{app} = 0$).[6]

In Fig. 13.8, an action potential is evoked by an instantaneous increase in voltage, as would occur during an extremely brief but strong application of depolarizing current ($I_{app} > 0$). This depolarization starts a process of activation of both I_{Nav} and I_K (increasing m and n), and the slower process of sodium current inactivation (h decreasing). The smaller time constant for the sodium current activation gating variable is one of the factors that causes I_{Nav} to increase faster than I_{Kv}, which leads to a negative value of the total ionic current $I_{Nav} + I_{Kv} + I_{leak}$ and depolarization of the membrane. The maximum of 37 mV occurs at 3.3 ms after initiation of the action potential. Following the action potential the membrane voltage is lower than rest for

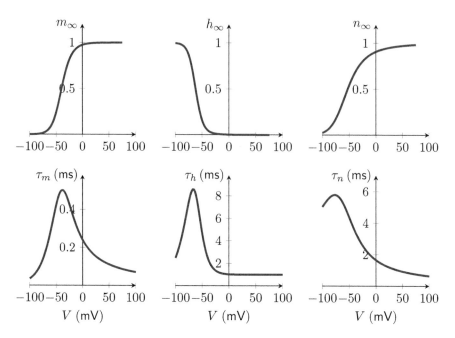

Figure 13.7   Gating variables for the Hodgkin-Huxley model.

several milliseconds. This phase of the action potential is referred to as the after-hyperpolarization (AHP).

Fig. 13.9 shows current-clamp simulations of the Hodgkin-Huxley model where action potentials are evoked by depolarizing current steps of amplitude $10\,\mu A/cm^2$ and duration 1 ms. When the two pulses are separated by 15 ms, both induce full sized action potentials. However, when the time between pulses (i.e., the inter-pulse interval) is reduced to 10 ms, the second pulse does not evoke an action potential. The reason is that the inactivation gating variable for the sodium current h is near zero at the end of the first action potential. It has recovered to $h = 0.6$ (40% inactivation) in 15 ms, but only to $h = 0.48$ (52% inactivation) in 10 ms. More inactivation (smaller value of h) means less inward sodium current is evoked by the second pulse, and this inward ionic current must be sufficiently strong for the regenerative nature of I_{Nav} to initiate the second action potential.

Anodal Break Excitation

The dynamics of inactivation in the Hodgkin-Huxley model lead to an interesting and unexpected phenomenon. As shown in Fig. 13.10, hyperpolarizing current steps may lead to responses that are nearly passive (as expected, bottom two traces). However, applied current pulses that are sufficiently hyperpolarizing are followed by an action potential upon release of the membrane from hyperpolarization. The explanation for this phenomenon is that the inactivation gating variable steady state $h_\infty(V)$ increases at voltages that are hyperpolarized compared to rest ($V < -65\,mV$, see Fig. 13.7). In Fig. 13.10, the deepest hyperpolarization causes h to relax from

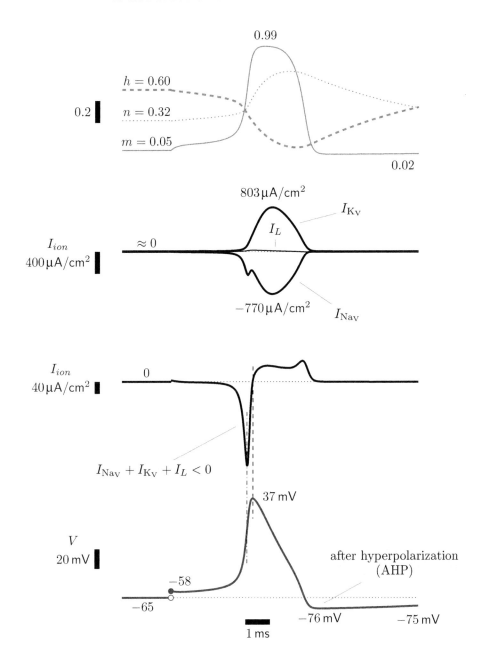

Figure 13.8 Hodgkin-Huxley model action potential in current-clamp mode: voltage (blue), currents (black), and gating variables (gray). *Question*: The full width at half maximum is about 2 ms. What is the characteristic time for changes in voltage? What is the initial fraction of activated potassium and sodium channels? What is the initial fraction of inactivated sodium channels?

0.60 to 0.75 (40% to 25% inactivation). This *recovery from inactivation* causes the sodium current (which is inward as V recovers to resting values) to be greater than it otherwise would be. If sufficiently de-inactivated, the recovered inward I_{Nav} current can lead

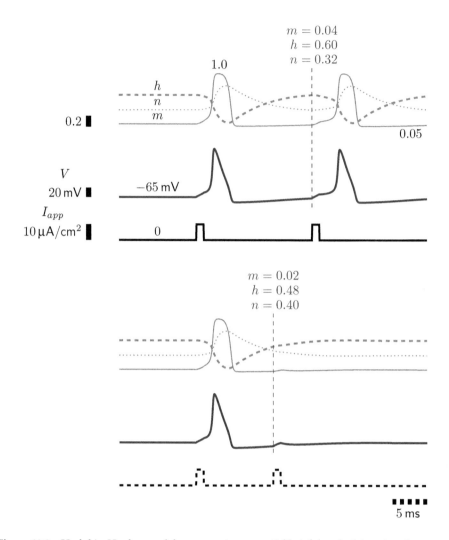

Figure 13.9 Hodgkin-Huxley model response to sequential brief depolarizing steps in applied current may evoke an action potential.

to **anodal break excitation**, an old-fashioned phrase meaning "excitation that occurs after hyperpolarization."[7]

13.4   Repetitive Spiking (Oscillations)

Fig. 13.11 shows membrane voltage, action potential generating currents, and gating variables, during a current-clamp simulation of the Hodgkin-Huxley model response during prolonged application of depolarizing applied current ($I_{app} = 12\,\mu A/cm^2$). The membrane model responds with a repetitive sequence of action potentials, a phenomenon that is observed in many excitable cell types. Because repetitive action potentials are periodic, such spike trains may be referred to as *neural oscillations*.

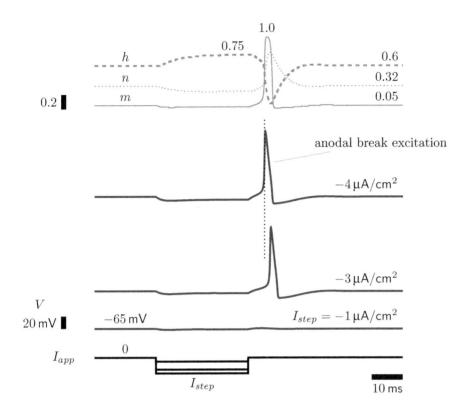

Figure 13.10   Hodgkin-Huxley model response to a hyperpolarizing step in applied current may exhibit anodal break excitation.

Fig. 13.12 shows the dynamics of the Hodgkin-Huxley model to applied current steps that are 300 ms in duration and 4 to 180 μA/cm^2 in amplitude. The response depends very much on the amplitude of the applied current. For very small positive values of applied current, subthreshold depolarizations occur (not shown). A slightly greater value of applied current ($I_{app} = 4\,\mu$A/cm^2) triggers a single action potential. When I_{app} is sufficiently depolarizing, the membrane model responds with repetitive action potentials. $I_{app} = 8\,\mu$A/cm^2 evokes thirteen spikes during the 200 ms pulse (about 65 Hz); the spiking ceases when the pulse ends. $I_{app} = 12\,\mu$A/cm^2 evokes about 75 spikes per second. When the depolarizing current step is quite large ($I_{app} = 180\,\mu$A/cm^2), there is a complicated transient (one spike and some ringing) ending at a depolarized steady state voltage of about -42 mV, a phenomenon referred to as **depolarization block**. Repetitive action potentials cease because I_{Nav} is not able to recover from depolarization when the minimum voltage during the oscillations is too depolarized.

13.5   Further Reading and Discussion

There are many excellent brief histories of the Hodgkin-Huxley theory of the action potential including Häusser (2000), Bean (2007) and Schwiening (2012). Nelson and

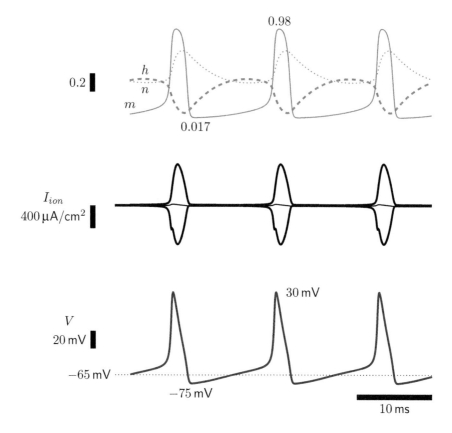

Figure 13.11   Hodgkin-Huxley model action potential currents and gating variables during repetitive oscillations with $I_{app} = 12\,\mu A/cm^2$. *Question*: What is the frequency of this periodic response?

Rinzel's chapter in *The Book of GENESIS* (Bower and Beeman, 1995, Chapter 4) is an approachable mathematical perspective. Intrepid students may wish to read the first two chapters of the graduate level text *Ionic Channels of Excitable Membranes* (Hille, 2001) that focuses on the classical biophysics of the squid giant axon.

Voltage-Clamp Measurement of $\tau_h(V)$

The voltage-dependent time constant τ_h for inactivation and de-inactivation of I_{Nav} is generally larger than the activation/deactivation time constants τ_m (sodium) or τ_n (potassium). Recovery of the inactivation gating variable h (de-inactivation) is the explanation for the phenomenon of anodal break excitation (Fig. 13.10), and for the refractory period of an excitable membrane (Fig. 13.9). The voltage dependence of $\tau_h(V)$ may be probed using a *two-pulse* voltage-clamp protocol. First, depolarizing V_{step} evokes an inward current that reverses after a few milliseconds (as in Fig. 13.3). For a second, identical V_{step} that is 10 ms or more later, the evoked membrane current is nearly identical to the first. However, for a second V_{step} a few milliseconds subsequent

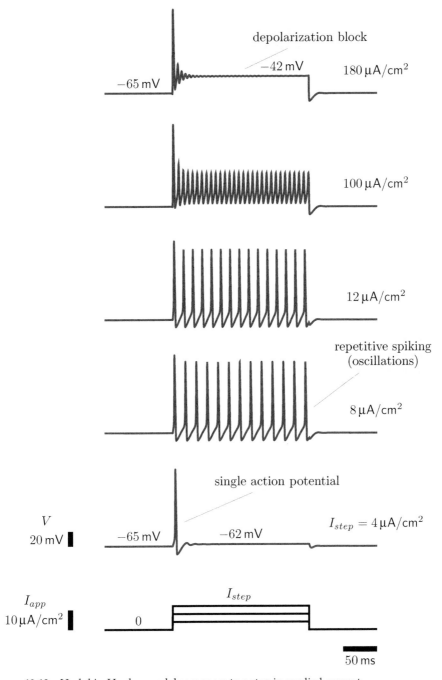

Figure 13.12 Hodgkin-Huxley model responses to a step in applied current.

to the first, the inward phase of the evoked membrane current is attenuated. $\tau_h(V_{step})$ is revealed by a *recovery plot* of attenuation versus inter-pulse interval, and $\tau(V)$ is found by varying V_{step} (see Hille 2001, Fig. 2.15).

13.5.1   How to do the Hodgkin-Huxley Macarena

The Hodgkin-Huxley macarena[8] is a rendition of the dynamics of the I_{Na} gating variables m and h during a depolarization-induced action potential.

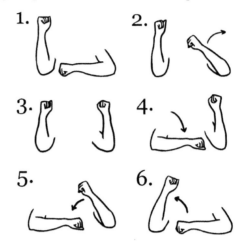

(1) At rest the sodium current activation gate m is closed and the inactivation gate h is open. (2) Depolarization causes the activation gate m to begin to open. (3) The activation gate m is fully open. Inward sodium current causes depolarization of the membrane. (4) The inactivation gate h closes in response to depolarized membrane voltage. (5) After the membrane potential returns to values near rest due to the outward potassium current, the m gate of the sodium current closes (deactivation). (6) Recovery of the h gate of the sodium current occurs on a slower time scale (de-inactivation). The Hodgkin-Huxley macarena is a caricature. How is the macarena different than the dynamics of m and h observed in Fig. 13.8 (top panel)? Modify the Hodgkin-Huxley macarena to include n as well as m and h.

Options for Simulating the Hodgkin-Huxley Equations

If you are interested in simulating the Hodgkin-Huxley model of the action potential, there are several options. (1) An online search will yield several interactive web pages. (2) The software package XPPAUT written by Bard Ermentrout at University of Pittsburgh includes a .ode file for the Hodgkin-Huxley model entitled hh.ode. Students may download this XPP and perform numerical integration using hh.ode as the input file. (3) Below is the MATLAB script that was used to create Fig. 13.1.

```
function hhsim
total = 40;
v0=-65; m0=0.053; h0=0.596; n0=0.317;
i0=0; i1=8; ton=10; toff=11;
[t x] = ode45(@odefun,[0:0.01:total],[v0 m0 h0 n0],[],i0,i1,ton,toff);
iapp = i0+i1*(t>ton).*(toff>t);
subplot(3,1,1); plot(t,x(:,1)); ylabel('V (mV)')
subplot(3,1,2); plot(t,x(:,2:end)); ylabel('m, h, n')
```

```
subplot (3,1,3); plot(t,iapp); ylabel('iapp'); xlabel('time (ms)')
return

function dxdt = odefun(t,x,i0,i1,ton,toff)
v = x(1); m = x(2); h = x(3); n = x(4);
c=1; gnabar=120; gkbar=36; gl=0.3; ena=50; ek=-77; el=-54.4;
alpham = 0.1*(v+40)./(1-exp(-(v+40)/10));
betam = 4*exp(-(v+65)/18);
minf = 1./(1+betam./alpham);
taum = 1./(alpham+betam);
alphah = 0.07*exp(-(v+65)/20);
betah = 1./(1+exp(-(v+35)/10));
hinf = 1./(1+betah./alphah);
tauh = 1./(alphah+betah);
alphan = 0.01*(v+55)./(1-exp(-(v+55)/10));
betan = 0.125*exp(-(v+65)/80);
ninf = 1./(1+betan/alphan);
taun = 1./(alphan+betan);
iapp = i0+i1*(t>ton).*(toff>t);
dvdt = (iapp-gnabar*m.^3.*h.*(v-ena) ...
               -gkbar*n.^4.*(v-ek) ...
               -gl.*(v-el))/c;
dmdt = -(m-minf)./taum;
dhdt = -(h-hinf)./tauh;
dndt = -(n-ninf)./taun;
dxdt = [ dvdt;  dmdt ; dhdt ; dndt ];
return
```

The parameters follow Hansel et al. (1993).

Supplemental Problems

Problem 13.1   Sketch the equivalent circuit for the Hodgkin-Huxley model.

Problem 13.2   The gating variable q that solves

$$\frac{dq}{dt} = -\frac{q - q_\infty(V)}{\tau_q(V)} \quad \text{also solves} \quad \frac{dq}{dt} = \alpha(V)(1-q) - \beta(V)q.$$

Express $\alpha(V)$ and $\beta(V)$ in terms of $q_\infty(V)$ and $\tau_q(V)$.

Problem 13.3   Using a Hodgkin-Huxley simulator of your choosing (see p. 228), perform the following numerical experiments.

(a)   Determine the threshold I_{app} leading to excitation to 3 significant figures. Hint: The binary search technique works well.

(b)   Plot the action potential frequency (spikes per second) as a function of I_{app}. Hint: Perform multiple simulations using different values of I_{app}, and count spikes.

(c)   Show that an action potential can be triggered by deactivating I_{K_V} just slightly. Explain why.

Solutions

Figure 13.8 The characteristic time is about 0.1 ms. At -65 mV, the activation gating variable for the potassium current is $n = 0.32$, that is, I_{K_V} is about 32% activated (68% deactivated). The inactivation gating variable for the sodium current is $h = 0.60$, that is, I_{Na_V} is 40% inactivated (60% de-inactivated). If this is confusing, recall that $h = 0$ is full inactivation, while $h = 1$ means no inactivation (full deactivation). At -65 mV, the sodium current is 5% activated at rest (95% deactivated, $m = 0.05$).

Figure 13.11 There are about 3 oscillations per 40 ms, which is an action potential frequency of 75 Hz.

Problem 13.1

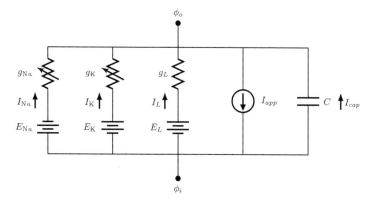

Problem 13.2 $\alpha (V) = q_\infty (V) / \tau_q (V)$ and $\beta (V) = [1 - q_\infty (V)] / \tau_q (V)$.

Notes

1. This experimental preparation was first developed by JZ Young in England.
2. The rate of propagation of action potentials along an axon depends on many things including the axon diameter and whether or not the axon is myelinated (Stone, 1983, Chapter 1).
3. I_L is negligible compared to the sodium and potassium currents ($|I_L| \ll |I_{Na_V}|$ and similarly for I_{K_V}).
4. The factor n^4 is consistent with four identical and independent voltage gates, all of which are required to be permissive for the channel to be open.
5. The relationship of the plateau current $I_{K_V}^\infty$ and the steady state gating function n_∞ is found by inverting Eq. 13.2,

$$n_\infty(V_{step}) = \left(\frac{I_{K_V}^\infty (V_{step})}{\bar{g}_K \left(V_{step} - E_K \right)} \right)^{1/4}.$$

6. In $\mu A/cm^2$, $I_{Nav} = -1.22$, $I_{Kv} = 4.40$, and $I_L = -3.18$. These resting ionic currents are small in magnitude compared to the peak sodium and potassium currents that occur midway through the action potential ($I_{Nav} \approx -770$, $I_{Kv} \approx 803$).

7. This terminology was used prior to the development of microelectrodes and intracellular recording. Cathode (+) and anode (−) refer to simulation via *extracellular* electrodes. Cathodal stimulation decreases ϕ_o (depolarizing, because $V = \phi_i - \phi_o$). Anodal extracellular stimulation increases ϕ_o and is hyperpolarizing. It was observed that neurons could be excited by turning on a cathodal stimulus or, alternatively, turning off an anodal stimulus.

8. I learned the Hodgkin-Huxley macarena from Jim Keener. Olivia Walch drew the illustration.

PART IV
Excitability and Phase Planes

14 The Morris-Lecar Model

Membrane excitability is explored in the context of the Morris-Lecar model of barnacle muscle fiber. The ionic currents included in the Morris-Lecar model are I_{Ca_V} (inward, regenerative), the delayed rectifier I_{K_V}, and a passive leak. Unlike I_{Na_V} in the Hodgkin-Huxley model, the Ca_V current in the Morris-Lecar model does not inactivate. When I_{Ca_V} activation is rapid, membrane excitability and oscillations are analyzed in a graphical fashion.

14.1 The Morris-Lecar Model

The Morris-Lecar model of barnacle muscle fiber was introduced in a 1981 *Biophysical Journal* article by Cathy Morris and Harold Lecar. Fig. 14.1 shows the equivalent circuit. The model includes a leakage current and two voltage-gated currents: an inward regenerative Ca_V current and an outward restorative K_V current, both of which are activated by depolarization,

$$C\frac{dV}{dt} = I_{app} - \underbrace{\bar{g}_{Ca}m\,(V - E_{Ca})}_{I_{Ca_V}} - \underbrace{\bar{g}_K w\,(V - E_K)}_{I_{K_V}} - \underbrace{g_L\,(V - E_L)}_{I_L}$$

$$\frac{dm}{dt} = -[m - m_\infty(V)]/\tau_m(V) \qquad\qquad (14.1)$$

$$\frac{dw}{dt} = -[w - w_\infty(V)]/\tau_w(V).$$

The gating variable m represents activation of I_{Ca_V}, and w represents activation of I_{K_V}. As in the Hodgkin-Huxley model (Eq. 13.1), these gating variables take values between 0 (all channels closed) and 1 (all channels open). The parameters $\bar{g}_{Ca}$ and $\bar{g}_K$ are the maximum conductances of I_{Ca_V} and I_{K_V}. Fig. 14.2 shows the voltage dependence of the gating functions $m_\infty(V)$, $\tau_m(V)$, $w_\infty(V)$ and $\tau_w(V)$ that appear in Eq. 14.1. Observe that the activation function for I_{Ca_V} is slightly steeper than the activation

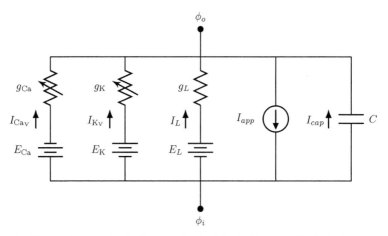

Figure 14.1 The equivalent circuit diagram for the Morris-Lecar model includes a capacitative current (I_{cap}), an applied current (I_{app}), and three ionic currents (I_{Ca_V}, I_{K_V}, I_L). *Question:* Which of these ionic currents are passive? Which are voltage gated?

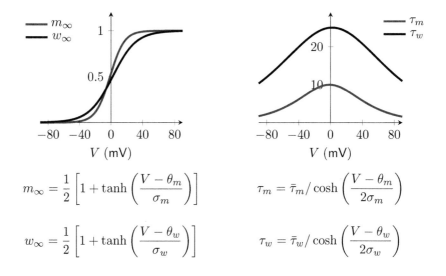

$$m_\infty = \frac{1}{2}\left[1 + \tanh\left(\frac{V - \theta_m}{\sigma_m}\right)\right]$$

$$\tau_m = \bar{\tau}_m / \cosh\left(\frac{V - \theta_m}{2\sigma_m}\right)$$

$$w_\infty = \frac{1}{2}\left[1 + \tanh\left(\frac{V - \theta_w}{\sigma_w}\right)\right]$$

$$\tau_w = \bar{\tau}_w / \cosh\left(\frac{V - \theta_w}{2\sigma_w}\right)$$

Figure 14.2 Gating functions for the Morris-Lecar model. Parameters: $\theta_m = -1.2$ mV, $\sigma_m = 18$ mV, $\bar{\tau}_m = 10$ ms, $\theta_w = 2$ mV, $\sigma_w = 30$ mV, $\bar{\tau}_w = 25$ ms. *Question:* Which gating function, m_∞ or w_∞, has the more depolarized half maximum? At $V = -40$ mV, which gating variable is faster, m or w?

function for I_{K_V} (18 mV $= \sigma_m < \sigma_w = 30$ mV) and the maximum time constant for I_{Ca_V} is smaller than the maximum time constant for I_{K_V} (10 ms $= \bar{\tau}_m < \bar{\tau}_w = 25$ ms).

The Morris-Lecar model is of interest to neuroscientists because the electrophysiology of muscle cells parallels that of neurons. Indeed, the Morris-Lecar model of barnacle muscle fiber exhibits many of the phenomena associated with membrane excitability that were observed in squid giant axon, including action potentials (Fig. 14.3) and repetitive spiking (Fig. 14.4).

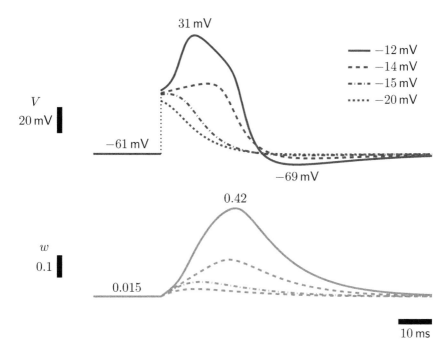

Figure 14.3 Different initial voltages in the Morris-Lecar model may lead to action potentials of different amplitude (or no action potential at all).

14.2 The Reduced Morris-Lecar Model

The I_{Ca_V} current in the Morris-Lecar model does not inactivate and, consequently, this current requires only one gating variable (m). In contrast, Hodgkin and Huxley's representation of I_{Na_V} includes both an activation gating variable (m) and an inactivation gating variable (h). Thus, the original Morris-Lecar model (Eq. 14.1) involves three ODEs, while the Hodgkin-Huxley model involves four (V, m, h and n in Eq. 13.1). Mathematicians who studied the Morris-Lecar model pointed out that a *reduced* version with only two ODEs behaved similarly, and is easier to analyze.

This reduced Morris-Lecar model eliminates the ODE for the Ca_V current activation gating variable, and uses the algebraic expression for the steady state $m_\infty(V)$ in the current balance equation,

$$C\frac{dV}{dt} = I_{app} - I_{Ca_V}(V) - I_{K_V}(w, V) - I_L(V) \tag{14.2}$$

$$\frac{dw}{dt} = -[w - w_\infty(V)]/\tau_w(V) . \tag{14.3}$$

This reduced model is obtained from the original by decreasing the time constant $\bar{\tau}_m$, that is, making the faster of the two activation gating variables even faster.[1] When $\bar{\tau}_m$ is small compared to the membrane time constant, the activation/deactivation of I_{Ca_V} rapidly equilibrates with membrane voltage ($m \approx m_\infty(V)$, rendering the ODE for m superfluous).

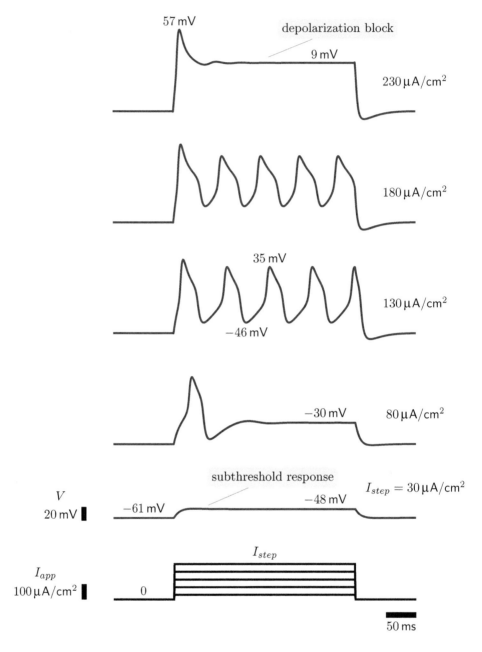

Figure 14.4   In response to depolarizing applied current, the Morris-Lecar model exhibits membrane oscillations (repetitive action potentials) similar to the Hodgkin-Huxley model (cf. Fig. 13.12).

The reduced Morris-Lecar model may be studied via numerical integration (as in Figs. 14.3 and 14.4). However, the advantage (and true significance) of the two-variable Morris-Lecar model is that the dynamics of membrane excitability and oscillations can be explored using graphical techniques that focus on the geometry of solutions in the (V, w)-plane.

14.3   The Morris-Lecar Phase Plane

The dynamics of membrane excitability and oscillations in the Morris-Lecar model (Figs. 14.3 and 14.4) may be understood through **phase plane analysis**. The **phase plane** for a two-variable ODE system is analogous to the phase diagram for a scalar ODE system (such as Eq. 11.2). Because there are two dependent variables in the reduced Morris-Lecar model, the state space is not a *line* (i.e., the phase line of membrane voltages V), but rather the (V, w) *plane*. Each point in the plane is a possible system state, and the (V, w)-plane in its entirety is the state space of the dynamical system.[2] To every location in the (V, w)-plane, there are two associated rates of change (dV/dt and dw/dt) given by the governing ODEs (Eqs. 14.2 and 14.3).

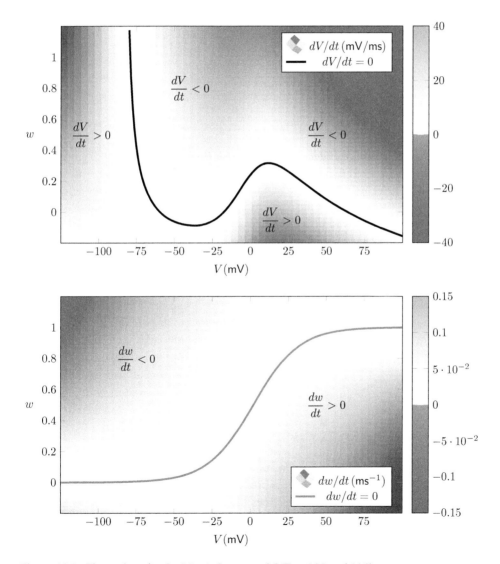

Figure 14.5   Phase plane for the Morris-Lecar model (Eqs. 14.2 and 14.3).

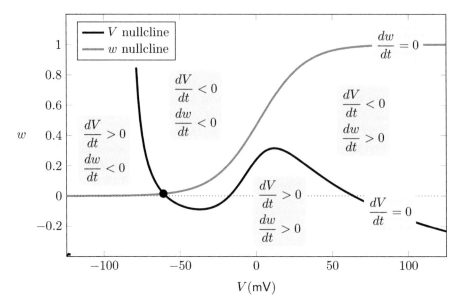

Figure 14.6 V and w nullclines of the Morris-Lecar phase plane. Parameters as in Fig. 14.3.

Using the same parameters as Fig. 14.3, Fig. 14.5 shows the regions in the plane where the rates of change dV/dt and dw/dt are positive (blue) and negative (gray). The solid curves in Fig. 14.5 are **nullclines** that indicate locations in the phase plane where the rate of change of an independent variable is zero. The V nullcline is the set of points (V, w) that lead to $dV/dt = 0$ in Eq. 14.2 (top panel, black curve). The w nullcline is the set of points (V, w) that lead to $dw/dt = 0$ in Eq. 14.3 (bottom panel, gray curve).

The phase plane shown in Fig. 14.6 emphasizes the relationship between the V and w nullclines. The nullclines delimit regions of the (V, w)-plane where dV/dt and dw/dt are positive versus negative. The rate of change of w is negative ($dw/dt < 0$) *above* the w nullcline where $w > w_\infty$, and positive ($dw/dt > 0$) *below* ($w < w_\infty$). Similarly, $dV/dt < 0$ to the right of the V nullcline, and $dV/dt > 0$ to the left.

The shape and position of the Morris-Lecar nullclines depend on parameter values used in Eqs. 14.2 and 14.3. For example, the w nullcline is an increasing sigmoidal function in the (V, w)-plane, because setting $dw/dt = 0$ in Eq. 14.3 gives[3]

$$0 = -\frac{w - w_\infty(V)}{\tau_w(V)} \Rightarrow w = w_\infty(V).$$

The w nullcline is essentially a plot of the gating variable steady state $w_\infty(V)$ with location and scale parameters θ_w and σ_w (Fig. 14.2). The shape of the V nullcline depends on the parameters of the current balance equation (Eq. 14.2), which includes the applied current (Fig. 14.5 uses $I_{app} = 0$).

Fig. 14.6 shows the V and w nullclines intersect at the point $(V_{ss}, w_{ss}) = (-61\,\text{mV}, 0.01)$. Because this point is on both nullclines, it is a steady state (fixed point) of the Morris-Lecar model ($dV/dt = 0$ and $dw/dt = 0$). The steady state is stable, because small deviations $\Delta V = V - V_{ss}$ and $\Delta w = w - w_{ss}$ decay (not shown). In physiological

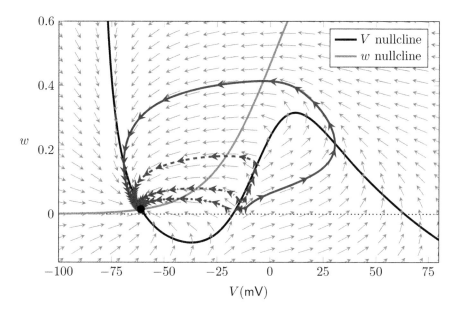

Figure 14.7   The phase plane (nullclines and direction field) for the Morris-Lecar model (Eqs. 14.2 and 14.3). Four solutions indicate the dynamic phenomenon of membrane excitability (compare Fig. 14.3).

terms, the resting membrane potential of the Morris-Lecar membrane is −61 mV. At this voltage, most potassium channels are closed ($w \approx 0.01$).

The gray arrows of Fig. 14.7 show the **direction field** of the Morris-Lecar model. The arrows indicate the relative rate of change of V and w at each location. For example, an arrow pointing north west at 135 degrees indicates that the rates of change have the same magnitude (absolute value), but dV/dt is negative and dw/dt is positive. Arrows that cross the V nullcline do so vertically (upward or downward), because on the V nullcline the rate of change of V is zero. Similarly, arrows that cross the w nullcline do so horizontally (either leftward or rightward).

14.4   Phase Plane Analysis of Membrane Excitability

Fig. 14.3 shows action potentials exhibited by the Morris-Lecar model of barnacle muscle fiber when the membrane voltage is instantaneously increased from rest ($V = -61$ mV) to four different initial voltages from −20 to −12 mV. The blue curves in Fig. 14.7 are these same four solutions plotted in the phase plane (where they are called **trajectories** or **orbits**). Each trajectory begins with a different initial voltage $V(0)$, but in all cases $w(0) = 0.01$ (most potassium channels closed). Blue arrowheads separated by 3 ms placed along each trajectory (Fig. 14.7) indicate how quickly V and w are changing.

Both Figs. 14.3 and 14.7 show the Morris-Lecar model exhibiting action potentials of graded amplitude. The initial voltage $V(0)$ that is most depolarized (−12 mV) results in the action potential with largest excursion in voltage (solid blue curve).

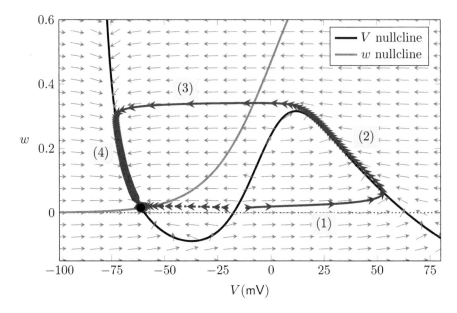

Figure 14.8   Phase plane analysis of membrane excitability in the Morris-Lecar model with separated time scales ($\bar{\tau}_w = 250$ ms here, 25 ms in Fig. 14.7).

A slightly lower initial voltage of -14 mV yields an action potential with a less depolarized peak voltage (dashed blue curve). All four trajectories in Figs. 14.3 and 14.7 have a positive initial rate of change for the gating variable w, representing slow activation of potassium channels. In three of four cases the initial rate of change of voltage is positive. Compare Figs. 14.3 and 14.7 and consider the advantages and disadvantages of these two different perspectives on excitability exhibited by the Morris-Lecar model.

Fast/Slow Analysis of Excitability

The dynamics of membrane excitability in the Morris-Lecar model are illuminated by considering a parameter regime for which the characteristic times for changes in V and w are separated. This can be achieved by increasing the parameter $\bar{\tau}_w$ from 25 to 250 ms, thereby slowing the dynamics of the potassium channels. Fig. 14.8 shows that this parameter change does not relocate the V and w nulllines in the phase plane. However, it does cause many of the direction field arrows to point horizontally (leftward or rightward), indicating that $|dw/dt| \ll |dV/dt|$ for most of the (V, w)-plane.

The phase plane of Fig. 14.8 shows trajectories of the "fast V/slow w" Morris-Lecar model for subthreshold and superthreshold initial voltages (compare Fig. 14.9). When the initial voltage is $V(0) = -20$ mV (dashed curves), the voltage relaxes back to rest in a monotone fashion. When the initial voltage is -12 mV (solid curves), the behavior of the trajectory in the phase plane can be partitioned into four distinct epochs.

(1) The first epoch begins with the superthreshold voltage rapidly increasing, consistent with the rightward pointing arrows that indicate $dV/dt > 0$. Depolarization is

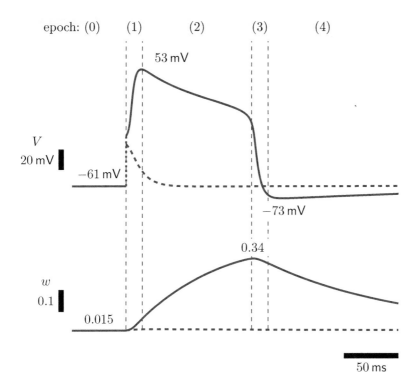

epoch: (0)   (1)   (2)   (3)   (4)

53 mV

V
20 mV

−61 mV

−73 mV

0.34

w
0.1

0.015

50 ms

Figure 14.9   Excitability in the Morris-Lecar model with separated time scales (cf. Fig. 14.8).

due to rapid activation of I_{Ca_V}; the gating variable for this current, $m_\infty(V)$, increases from 0.23 at the initial voltage of −12 mV to nearly 1 at the peak voltage of 53 mV. Epoch (1) ends when the trajectory nears the voltage nullcline, turns upward, and crosses the voltage nullcline vertically (upward).

(2) In the second epoch, the voltage trajectory moves upward (w increasing) and leftward (V decreasing), following the depolarized branch of the voltage nullcline. Increasing w results in increasing outward I_{K_V} current and slow hyperpolarization of membrane voltage ($dV/dt < 0$). Epoch (2) ends when the knee of the depolarized voltage nullcline is reached.

(3) The third epoch begins with rapidly decreasing voltage, consistent with the leftward pointing arrows ($dV/dt < 0$). This hyperpolarization is due to rapid deactivation of I_{Ca_V} (calcium channels closing) from $m_\infty(V) = 0.91$ when $V = 20$ mV to nearly zero at −73 mV. Epoch (3) ends when the decreasing voltage nears the hyperpolarized branch of the voltage nullcline, $|dV/dt|$ becomes comparable to $|dw/dt|$, and the voltage nullcline is crossed vertically downward.

(4) The voltage trajectory moves downward (w decreasing) and rightward (V increasing) along the hyperpolarized branch of the voltage nullcline. During epoch (4), decreasing w corresponds to deactivation of the potassium current I_{K_V}, which results in decreasing outward current and slow depolarization of membrane voltage ($dV/dt > 0$). Epoch (4) ends with V approaching the resting membrane potential from below.

Fast/Slow Analysis of Response to an Applied Current Pulse

Figs. 14.7–14.9 illustrate the dynamics of membrane excitability in the Morris-Lecar model by instantaneously increasing the membrane voltage. This is an appropriate way to model a very brief and strong pulse of applied current. Fast/slow analysis can also be used to understand the dynamics of membrane excitability induced by longer duration, lower amplitude, current pulses.

Using separated time scales for V and w, Fig. 14.10 (top panel) shows a phase plane, nullclines and trajectories for the membrane response to a $50\,\mu A/cm^2$ pulse of applied current (time course of V and w shown in the bottom panel). The action potential is initiated by depolarizing applied current that *lifts* the voltage nullcline sufficiently high for the trajectory to escape (skirt under) the hyperpolarized knee of the voltage nullcline. The slowing down of voltage dynamics in the middle of epoch (2) is explained by how near (in the phase plane) the trajectory passes to the knee. After escaping the knee, activation of I_{Ca_V} causes the voltage to approach, cross, and follow the voltage nullcline that is relevant when $I_{app} = 50\,\mu A/cm^2$ (black dashed curve). About 20 ms after the peak voltage of 56 mV is achieved, the depolarizing pulse of applied current ends. The voltage quickly hyperpolarizes in response; in the phase plane, this is a leftward turn toward the voltage nullcline that is relevant when $I_{app} = 0$ (black solid curve). The transition from epoch (3) to (4) occurs for the same reasons as in Fig. 14.8. This is followed by an after-hyperpolarization to -73 mV and a slow depolarization returning the voltage to rest (-61 mV).

14.5   Phase Plane Analysis of Membrane Oscillations

Fig. 14.11 shows membrane potential oscillations (repetitive action potentials) in the (V, w)-plane. Because $I_{app} = 130\,\mu A/cm^2$, the voltage nullcline has moved upward (solid curve) compared to the $I_{app} = 0$ location (dotted curves). The intersection of the V and w nullclines now occurs between the knees of the voltage nullcline, in contrast to Fig. 14.7 where the intersection occurs on the hyperpolarized branch of the V nullcline.

The steady state is unstable, as suggested by the direction field arrows pointing away from the open circle. The trajectory (blue curve) corresponds to what would be observed if the applied current remained at $I_{app} = 300\,\mu A/cm^2$ indefinitely (as opposed to a 130 ms step as in Fig. 14.4). Because the solution is periodic, the trajectory in the phase plane is a closed loop (**limit cycle**). The term limit cycle indicates that the periodic solution is an attractor, that is, nearby solutions converge asymptotically to the periodic steady state (not shown).

Fast/Slow Analysis of Oscillations

Fig. 14.12 shows how the direction field and limit cycle change when a separation of V and w time scales is introduced (as in Section 14.4). When the dynamics of membrane voltage is fast compared to potassium channel activation/deactivation, the limit cycle oscillation consists of four epochs, similar to those described on p. 242.

(1)   Voltage rapidly increases in response to activation of I_{Ca_V}.

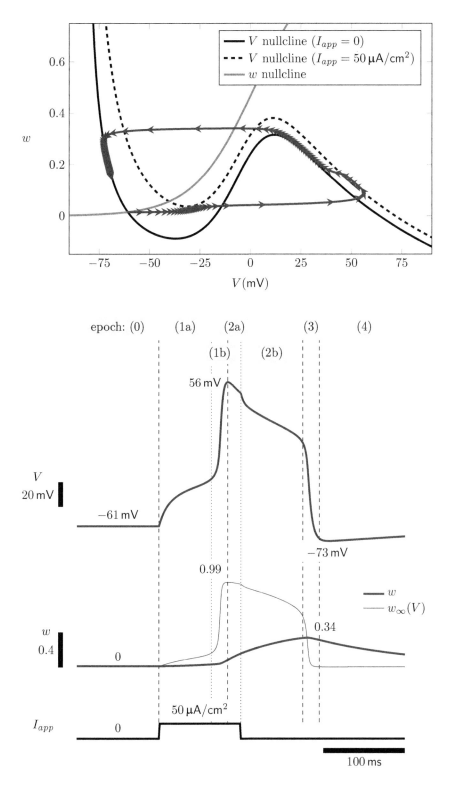

Figure 14.10 Excitability in the Morris-Lecar model with separated time scales and applied current pulse.

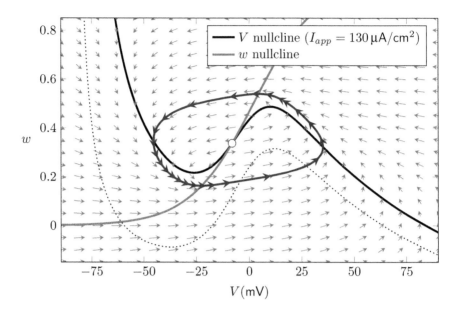

Figure 14.11   Phase plane for the Morris-Lecar model using $I_{app} = 130\,\mu A/cm^2$ (cf. Fig. 14.4). The V nullcline for this depolarizing applied current (solid curve) is positioned above the V nullcline for $I_{app} = 0$ (dotted curve, see Fig. 14.7).

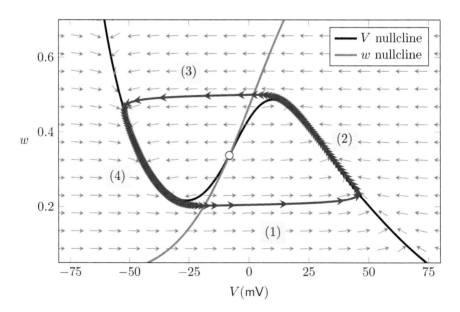

Figure 14.12   Phase plane for the Morris-Lecar model with separation of time scales and $I_{app} = 130\,\mu A/cm^2$.

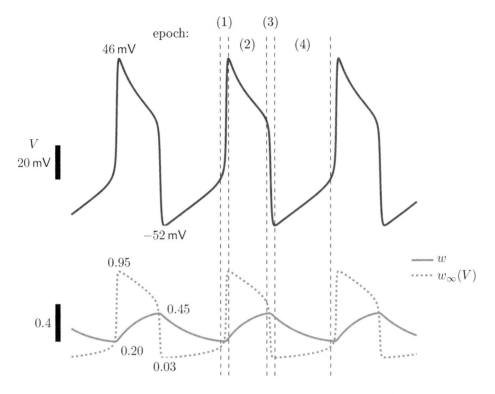

Figure 14.13   Oscillations in the Morris-Lecar model with separated time scales ($I_{app} = 130\ \mu A/cm^2$).

(2) I_{K_V} slowly activates in response to depolarized voltage ($w < w_\infty(V)$ so $dw/dt > 0$). Voltage slowly decreases with increasing I_{K_V} (outward current is hyperpolarizing).

(3)   Voltage rapidly decreases in response to deactivation of I_{Ca_V}.

(4) I_{K_V} slowly deactivates in response to hyperpolarized voltage ($w > w_\infty(V)$ so $dw/dt < 0$). Voltage slowly increases in response to deactivation I_{K_V} and decreasing outward current.

The reader is encouraged to carefully compare Figs. 14.12–14.13.

Fast/Slow Analysis Summary Diagram

In the "fast V/slow w" limit, we may summarize the oscillatory responses of the Morris-Lecar model as follows. For each value of applied current, the V and w nulllines intersect once (see Fig. 14.14). When the intersection occurs outside the knees of the voltage nullcline (i.e., on the hyperpolarized or depolarized branch), the steady state is stable. When this intersection occurs between the knees of the voltage nullcline, the steady state (fixed point) is unstable. In this case there is a limit cycle oscillation in which the voltage quickly transitions between the hyperpolarized and depolarized

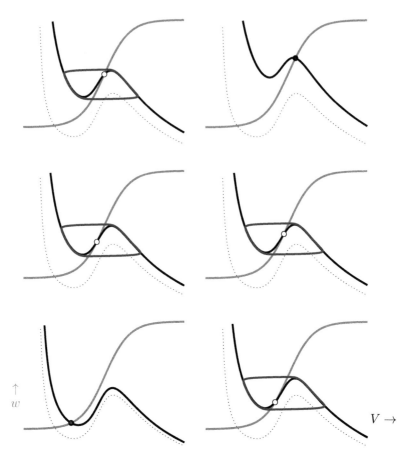

Figure 14.14   Oscillations in the Morris-Lecar model are observed for intermediate values of applied current. Shown are nullclines (solid) and trajectories (blue) for $I_{app} = 50, 90, 120,$ $160, 200$ and $250\,\mu A/cm^2$. Voltage nullcline for $I_{app} = 0$ is shown dotted. *Question*: Which values of I_{app} lead to oscillations? Which value of I_{app} leads to depolarization block?

branches of the voltage nullcline. The slow dynamics of the gating variable w lead to slow changes in the voltage during the hyperpolarized and depolarized epochs of the oscillation, and determine the period of the membrane potential oscillation.

14.6   Further Reading and Discussion

The equations and parameters used in this chapter follow Rinzel and Ermentrout (1989). Other presentations of this material can be found in Terman (2005), Friedman et al. (2005), Izhikevich (2007), Ermentrout and Terman (2010) and Sterratt et al. (2011). The Morris-Lecar model is featured as an example two-dimensional dynamical system in the online tutorial for XPPAUT written by Bard Ermentrout.

Why Does Increased I_{app} Raise the V Nullcline?

Many of the phase planes presented in this chapter have involved more than one value of applied current. In all cases, the larger value of I_{app} has the effect of raising the voltage nullcline (see, e.g., Fig. 14.9). To understand why, observe that *upward* in these phase planes is the direction of *increasing* w and greater values of I_{K_V} for positive driving force $(V > E_K)$, because $dI_{K_V}/dw = \bar{g}_{K_V}(V - E_K)$ and $\bar{g}_{K_V} > 0$. Recall that the voltage nullcline is defined as the points (V, w) in the plane for which $dV/dt = 0$, that is,

$$I_{app} = I_{Ca_V}(V) + I_{K_V}(V, w) + I_L(V). \tag{14.4}$$

For fixed V, increased I_{app} is balanced by increased I_{K_V},

$$\uparrow I_{app} = I_{Ca_V} + \uparrow I_{K_V}(w) + I_L,$$

and this corresponds to increased w. To see this, differentiate Eq. 14.4 with respect to w, as follows

$$\frac{dI_{app}}{dw} = \bar{g}_{K_V}(V - E_K).$$

Taking the reciprocal gives

$$\frac{dw}{dI_{app}} = \frac{1}{\bar{g}_{K_V}(V - E_K)}. \tag{14.5}$$

Thus, the w on the voltage nullcline for fixed V is an increasing function of I_{app} when $V > E_K = -84\,\text{mV}$. Observe that this rate of change is a decreasing hyperbolic function of V.

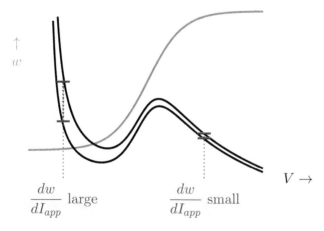

The voltage nullclines above use $I_{app} = 0$ and $50\,\mu\text{A}/\text{cm}^2$ (black curves). Observe that the difference between them (blue lines) decreases with depolarization (as in Eq. 14.5).

Supplemental Problems

Problem 14.1 Consider the following nullclines and trajectories for the fast/slow version of the Morris-Lecar model:

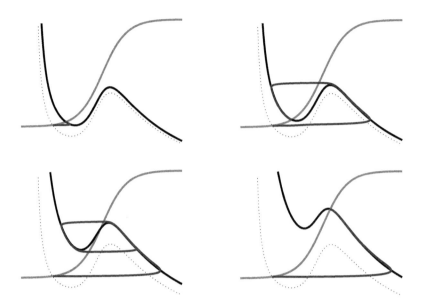

The values of I_{app} used are 40, 60, 150 and 250 μA/cm^2 (solid black curves are V nullclines). Trajectories (blue curves) start at the resting membrane potential (dotted curve is V nullcline for $I_{app} = 0$). Which of the four panels corresponds to depolarization block? A subthreshold response? Repetitive spiking? A single evoked action potential? Assign the proper value of I_{app} to each panel.

Problem 14.2 The plots below summarize oscillatory responses of the fast/slow version of the Morris-Lecar model (including those in Fig. 14.14). For each value of I_{app} (horizontal axis), the solid and dotted curves show the stable or unstable steady state voltage. The thick solid lines show the minimum and maximum voltages obtained during the limit cycle oscillation. Sketch a similar diagram that shows how applied current changes the dynamics of I_{K_V} activation/deactivation.

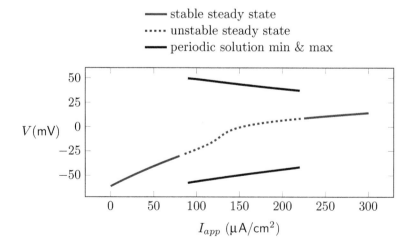

Solutions

Figure 14.1   The variable resistor symbols (g_{Ca_V} and g_{K_V}) associated with I_{Ca_V} and I_{K_V} indicate that these two currents are voltage gated. Conversely, I_L is a passive leakage current.

Figure 14.2   The half maximums of the gating functions m_∞ and w_∞ occur at $V = \theta_m$ and θ_w, respectively. Because $\theta_w > \theta_m$, we conclude that the half maximum of w_∞ is more depolarized than m_∞. Because $\tau_m < \tau_w$, we conclude that the gating variable m is faster than w at $V = -40\,\text{mV}$.

Figure 14.14   Oscillations occur for $I_{app} = 90, 120, 160$ and $200\,\mu\text{A/cm}^2$. Depolarization block occurs for $250\,\mu\text{A/cm}^2$.

Problem 14.1   Top left: Subthreshold applied current step of $40\,\mu\text{A/cm}^2$. Top right: A single evoked action potential ($I_{app} = 60\,\mu\text{A/cm}^2$). Bottom left: Repetitive spiking (limit cycle oscillations) using $I_{app} = 150\,\mu\text{A/cm}^2$. Bottom right: Depolarization block ($I_{app} = 250\,\mu\text{A/cm}^2$).

Problem 14.2   Diagram summarizing dynamics of the gating variable w:

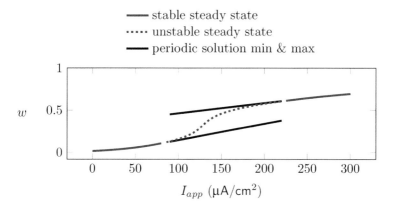

Notes

1. Eq. 14.1 → 14.2 as $\bar{\tau}_m \to 0$.
2. Values for the gating variable outside the range $0 \le w \le 1$ are nonphysical, and some voltages are physiologically unrealistic.
3. We have used $\tau_w(V) > 0$ (see Fig. 14.2).

15 Phase Plane Analysis

Phase plane analysis is a tool for understanding the dynamical behavior of two-dimensional autonomous ODE systems. We further explore the concepts of nullcline and direction field, and introduce linear stability analysis.

15.1 The Phase Plane for Two-Dimensional Autonomous ODEs

A two-dimensional autonomous ODE system takes the form,

$$\frac{dx}{dt} = f(x, y) \tag{15.1a}$$

$$\frac{dy}{dt} = g(x, y) \tag{15.1b}$$

where f and g are functions of the two independent variables x and y. We will usually not attempt to derive analytical solutions to nonlinear systems of this type. Instead, we will focus on the graphical technique of **phase plane analysis** first encountered in Chapter 14. On the other hand, we will have a lot to say about analytical solutions to *linear homogenous* systems that can be written as follows,[1]

$$\frac{dx}{dt} = ax + by \tag{15.2a}$$

$$\frac{dy}{dt} = cx + dy, \tag{15.2b}$$

where a, b, c and d are constants (a special case of Eq. 15.1). For concreteness, consider the following linear two-dimensional ODE initial value problem,

$$\frac{dx}{dt} = -\frac{1}{2}x - 3y \qquad x(0) = 1 \tag{15.3a}$$

$$\frac{dy}{dt} = 3x - \frac{1}{2}y \qquad y(0) = 0. \tag{15.3b}$$

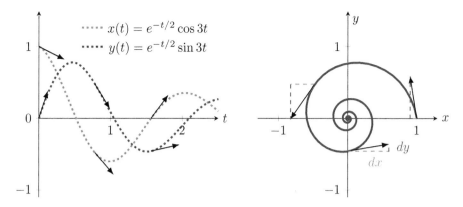

Figure 15.1   The solution of a two-dimensional autonomous ODE system such as Eq. 15.3 (left) is a trajectory $(x(t), y(t))$ in the Cartesian plane that is parameterized by time.

Methods discussed in Chapter 16 give the solution,

$$x(t) = e^{-t/2} \cos 3t \tag{15.4a}$$

$$y(t) = e^{-t/2} \sin 3t. \tag{15.4b}$$

These decaying oscillatory functions of time have a notable phase relationship; namely, when one is zero, the other is at a local minimum or maximum (Fig. 15.1, left). When $x(t)$ and $y(t)$ are plotted in the (x, y) phase plane, the trajectory is a spiral that winds inward as t increases (Fig. 15.1, right).

Problem 15.1   Show that Eq. 15.3 is solved by Eq. 15.4.

The arrows on Fig. 15.1 (left) are tangent to the solution curves $x(t)$ and $y(t)$, that is, their slope is the rate of change of the dependent variables (dx/dt and dy/dt). To confirm the slopes of the first pair of arrows located at $t = 0$, observe that the negatively sloped arrow is anchored at $x(0) = 1$ and the positively sloped arrow at $y(0) = 0$. Substituting these values of x and y into Eq. 15.3 gives values for dx/dt and dy/dt consistent with the figure,

$$\frac{dx}{dt} = -\frac{1}{2} \cdot 1 - 3 \cdot 0 = -\frac{1}{2} \tag{15.5a}$$

$$\frac{dy}{dt} = 3 \cdot 1 - \frac{1}{2} \cdot 0 = 3. \tag{15.5b}$$

The arrows on Fig. 15.1 (right) are tangent to the solution curve in the (x, y) phase plane.[2] Their slope is given by the ratio of the rates of change of $y(t)$ and $x(t)$,

$$\frac{dy/dt}{dx/dt} = \frac{dy}{dx} = \frac{\text{rise}}{\text{run}}.$$

For example, consider the arrow anchored at the initial value $(x(0), y(0)) = (1, 0)$. Using Eq. 15.5, the slope of this arrow is

$$\frac{dy/dt}{dx/dt} = \frac{3}{-1/2} = -6.$$

Although this calculation is correct, dividing the derivatives in this manner obscures the fact that x is decreasing and y is increasing.[3] For this reason, rates of change in the phase plane are usually reported as a 2×1 column **vector** such as

$$\begin{bmatrix} dx/dt \\ dy/dt \end{bmatrix} = \begin{bmatrix} -0.5 \\ 3 \end{bmatrix}.$$

Problem 15.2 On Fig. 15.1 (right), confirm the orientation of the arrow anchored at $(x(t), y(t)) = (-0.4755, 0.4755)$ where $t = \pi/4$.

Phase Plane Fact 1: Trajectories do not Cross

The first and most fundamental phase plane fact is the following.

- Solutions of two-dimensional autonomous ODE systems do not cross when plotted in the phase plane.

For example, the spiral shown in Fig. 15.1 (right) begins at $(1, 0)$ and winds into the origin *without ever coming into contact with itself*. This can be shown by converting the solution (Eq. 15.4) to polar coordinates via $x = r \sin \theta$ and $y = r \sin \theta$. The result is $\theta = 3t$ and $r = e^{-t/2}$. The expression for θ implies that the solution moves counterclockwise around the origin. The expression for r indicates that the distance from the origin is a monotone decreasing function and, for this reason, we conclude that the solution never crosses itself.[4]

As a counterexample, consider the two polynomial functions of time shown in Fig. 15.2 (left).[5] When these functions are plotted against one another in the (x, y)-plane, the parameterized curve crosses itself at the origin (Fig. 15.2, right). The first time the curve meets the origin, $t = t_a$, dy/dt is positive, and dx/dt is negative. The second time the curve meets the origin, $t = t_b$, dy/dt is positive, and dx/dt is *positive*. We conclude that the trajectory $(x(t), y(t))$ is not the solution of a two-dimensional

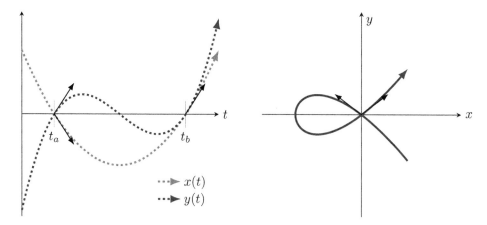

Figure 15.2 The functions of time $x(t)$ and $y(t)$ shown on the left cannot be solutions of an autonomous ODE system such as Eq. 15.1, because the solution crosses itself in the (x, y) phase plane.

autonomous ODE system (Eq. 15.1), because it is not possible for a given state (x, y) to yield different dynamical outcomes at different times.

15.2   Direction Fields of Two-Dimensional Autonomous ODEs

Using Eq. 15.1, the rates of change of x and y may be calculated at any location in the (x, y)-plane. As discussed above, these values give a vector tangent to a trajectory passing through that location. Using a computer, this slope can be calculated for many values of x and y and the results summarized using arrows of fixed length. Such a plot is referred to as a **direction field** for the ODE system.

Fig. 15.3 (left) shows the direction field for Eq. 15.3, a linear ODE system whose solutions are inwardly winding spirals. The small gray arrows of Fig. 15.3 are analogous to the three black arrows on Fig. 15.1 (right). That is, for any point (x, y) where an arrow is anchored, the arrow points in the direction implied by the governing ODE system.[6] Fig. 15.3 (right) includes the solution to Eq. 15.4 and three others with initial conditions $(0, 1)$, $(-1, 0)$ and $(0, -1)$. All these solutions asymptotically approach the origin $(0, 0)$ and – as we argued in the previous section – the trajectories do not cross in the phase plane.

Fig. 15.4 shows example direction fields for two-dimensional autonomous *nonlinear* ODE systems (Eq. 15.1). The black arrows are located at integer values of x and y, making it easy to confirm their direction. For example, in Fig. 15.4 (top) the arrow located at $(x, y) = (2, 2)$ points rightward, consistent with $dx/dt = y^2 - x + 2 = 2^2 - 2 + 2 = 4$ and $dy/dt = x^3 - y^3 = 2^3 - 2^3 = 0$. Direction fields indicate the *flow* of solutions in the phase plane. One may sketch representative trajectories by following the direction field arrows.

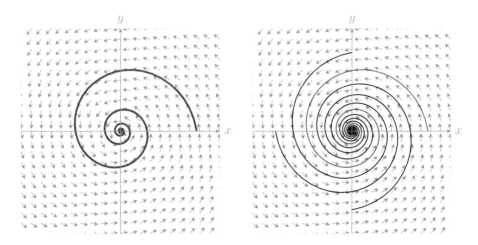

Figure 15.3   Direction fields for Eq. 15.3 with example solutions plotted as curves in the (x, y) phase plane.

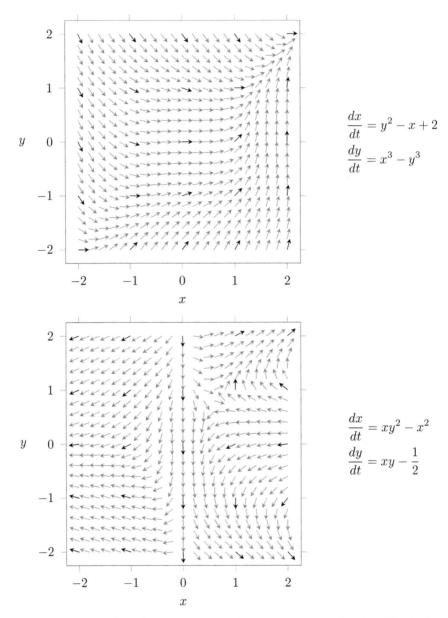

$$\frac{dx}{dt} = y^2 - x + 2$$

$$\frac{dy}{dt} = x^3 - y^3$$

$$\frac{dx}{dt} = xy^2 - x^2$$

$$\frac{dy}{dt} = xy - \frac{1}{2}$$

Figure 15.4 Direction fields for two-dimensional ODE systems. *Question:* Top: What is the slope of the arrow anchored at $(-2, 2)$? Bottom: Why do all the arrows point downward when $x = 0$?

15.3 Nullclines for Two-Dimensional Autonomous ODEs

Trajectories for two-dimensional autonomous ODE systems (Eq. 15.1) are constrained by special curves in the phase plane referred to as **nullclines**. The x nullcline is the set of points in the (x, y)-plane for which the rate of change dx/dt is zero. For typographical efficiency, we will often use Newton's dot notation and write $\dot{x} = 0$ rather than

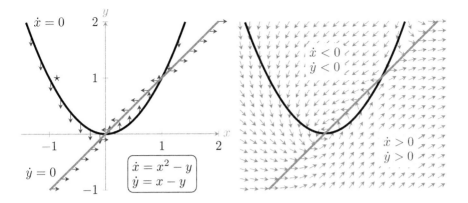

Figure 15.5   Phase plane analysis of Eq. 15.7. *Question*: The x and y nullclines partition the plane into five regions. What is the sign of $\dot{x}$ and $\dot{y}$ in the three regions that are not labelled?

$dx/dt = 0$ to indicate the x nullcline. Similarly, the y nullcline is the set of points in the (x, y)-plane for which $dy/dt = \dot{y} = 0$. Saying the same thing in another way, the x and y nullclines are the locations in the phase plane where the functions defining an ODE system $-f(x, y)$ and $g(x, y)$ in Eq. 15.1 – evaluate to zero, that is,

$$x \text{ nullcline} = \{(x, y) : 0 = f(x, y)\} \tag{15.6a}$$
$$y \text{ nullcline} = \{(x, y) : 0 = g(x, y)\}. \tag{15.6b}$$

For example, Fig. 15.5 shows nullclines for the following ODE system,

$$\dot{x} = x^2 - y \tag{15.7a}$$
$$\dot{y} = x - y. \tag{15.7b}$$

The x nullcline is the set of points (x, y) for which $0 = x^2 - y$, that is, the parabola $y = x^2$ (black curve), while the y nullcline is the line $y = x$ (gray). Plotting the nullclines reveals two intersections: the points $(0, 0)$ and $(1, 1)$. Because $\dot{x} = 0$ on the x nullcline and $\dot{y} = 0$ on the y nullcline, the points at any intersection are on both nullclines. We conclude that $(x_{ss}, y_{ss}) = (0, 0)$ and $(x_{ss}, y_{ss}) = (1, 1)$ are steady states of Eq. 15.7.

Phase Plane Fact 2: Trajectories may only Cross Nullclines Vertically or Horizontally

Because the x nullcline is the set of points in the (x, y)-plane for which the rate of change dx/dt is zero (Eq. 15.6), it is evident that the direction of flow associated with any point on the x nullcline – that is *not* on the y nullcline – must be vertical. The associated direction field arrow must point either up or down, depending on the sign of $\dot{y}$.

- Trajectories that cross the x nullcline do so vertically ($\dot{x} = 0$).

Conversely, the direction of flow associated with any point on the y nullcline – that is *not* on the x nullcline – must be horizontal. The associated direction field arrow must point either left or right, depending on the sign of $\dot{x}$.

- Trajectories that cross the y nullcline do so horizontally ($\dot{y} = 0$).

For example, in Fig. 15.5 the point $(-1, 1)$ is on the x nullcline ($\dot{x} = x^2 - y = (-1)^2 - 1 = 0$) and the sign of $\dot{y}$ is negative ($\dot{y} = x - y = -1 - 1 = -2$); thus, solutions leave $(-1, 1)$ downward (blue arrow labelled with ⋆). In fact, this is true for any point on the x nullcline for which $x < 0$. To see this, substitute the condition for being on the nullcline (i.e., $y = x^2$) into $\dot{y} = g(x, y) = x - y$ to obtain $\dot{y} = x - x^2$, which is negative when $x < 0$. Using arguments of this kind, it is usually straightforward to establish whether the direction field points up, down, left or right for any point on either nullcline (Fig. 15.5, blue arrows). Because $f(x, y)$ and $g(x, y)$ are continuous functions, the signs of $\dot{x}$ and $\dot{y}$ cannot change within a region of the phase plane delimited by nullclines. This fact allows one to determine (analytically) whether x is increasing or decreasing in each region of the plane, and similarly for y.

15.4   How to Sketch a Phase Plane

To perform phase plane analysis for an ODE system such as Eq. 15.1, draw the axes of the (x, y)-plane and proceed as follows.

(1x)   Draw the x nullcline. Begin by setting $f(x, y) = 0$ and determine the relationship(s) between x and y that solve this equation. Then sketch the points in the plane that satisfy these relationships. Label the x nullcline with $\dot{x} = 0$. You may wish to draw vertical hash marks on the x nullclines to remind yourself that trajectories may only cross it vertically.

(1y)   Draw the y nullcline in a similar manner using $g(x, y) = 0$. Label the y nullcline $\dot{y} = 0$ and draw horizontal hash marks on the y nullcline.

(2)   Find and label points where the x and y nullclines intersect. These intersections are steady states, that is, points (x_{ss}, y_{ss}) that simultaneously solve $0 = f(x_{ss}, y_{ss})$ and $0 = g(x_{ss}, y_{ss})$.

(3x)   Focusing on the x nullcline where $\dot{x} = 0$, determine where on this nullcline that the flow is upward ($\dot{y} > 0$) versus downward ($\dot{y} < 0$). Often this can be done by substituting the equation for the x nullcline into $g(x, y)$ and evaluating the sign of $\dot{y}$. Indicate your result using arrows anchored on the x nullcline.

(3y)   Focusing on the y nullcline where $\dot{y} = 0$, determine where on this nullcline that the flow is leftward ($\dot{x} < 0$) versus rightward ($\dot{x} > 0$). Often this can be done by substituting the equation for the y nullcline into $f(x, y)$ and evaluating the sign of $\dot{x}$. Indicate your result using arrows anchored on the y nullcline.

(4x)   Label the regions of the (x, y)-plane where $\dot{x} < 0$ and where $\dot{x} > 0$. Because $f(x, y)$ is continuous, these regions will be separated by x nullclines.

(4y)   Label the regions of the (x, y)-plane where $\dot{y} < 0$ and where $\dot{y} > 0$. Because $g(x, y)$ is continuous, these regions will be separated by y nullclines.

Example 1

Let us practice phase plane analysis for the following linear ODE system,

$$\dot{x} = -2x + 3y \tag{15.8a}$$
$$\dot{y} = 4x - y, \tag{15.8b}$$

by drawing the axes of the (x, y)-plane and following the steps discussed above.

(1x)   To identify the x nullcline set $\dot{x} = 0$ in Eq. 15.8a,

$$x\text{-nullcline}: \quad 0 = -2x + 3y \implies 3y = 2x \implies y = \frac{2}{3}x. \tag{15.9}$$

The x nullcline is a line through the origin $(0,0)$ with slope 2/3 (see black line below).

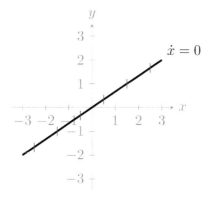

The vertical hash marks on the x nullcline indicate that trajectories that cross the x nullcline do so vertically ($\dot{x} = 0$).

(1y)   To identify the y nullcline, set $\dot{y} = 0$ in Eq. 15.8b,

$$y\text{-nullcline}: \quad 0 = 4x - y \implies y = 4x. \tag{15.10}$$

The y nullcline is a line through the origin $(0,0)$ with slope 4 (see gray line below).

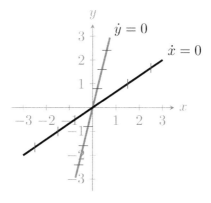

Trajectories that cross the y nullcline must move horizontally ($\dot{y} = 0$).

(2)   Steady states of Eq. 15.8 are intersections of the x and y nullclines. There is only one intersection; it is located at the origin $(0,0)$.

(3x)   By substituting the equation for the x nullcline (Eq. 15.9) into the equation for $\dot{y}$ (Eq. 15.8b), we conclude that $\dot{y} = 4x - y = 4x - \frac{2}{3}x = \frac{10}{3}x$. That is, on the x nullcline, dy/dt and x have the same sign. This is indicated by adding arrowheads to the hashmarks on the x nullcline.

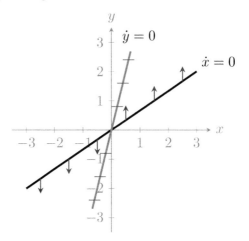

(3y)   Similarly, substituting the equation for the y nullcline (Eq. 15.10) into the equation for $\dot{x}$ (Eq. 15.8a) shows that, on the y nullcline, $\dot{x}$ has the same sign as y ($\dot{x} = -2x + 3y = -2 \cdot 4y + 3y = 5y$). This is indicated by adding arrow heads to the hashmarks on the y nullcline.

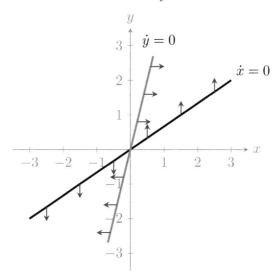

(4x&4y)   The nullclines divide the plane into four regions. The signs of $\dot{x}$ and $\dot{y}$ cannot change from positive to negative without evaluating to zero and this only happens on nullclines. Using this fact, it is straightforward to determine whether x and y are increasing or decreasing in each of the four regions (see Fig. 15.6).

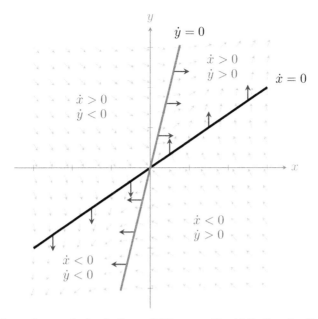

$\dot{y} = 0$

$\dot{x} > 0$
$\dot{y} > 0$

$\dot{x} = 0$

$\dot{x} > 0$
$\dot{y} < 0$

$\dot{x} < 0$
$\dot{y} > 0$

$\dot{x} < 0$
$\dot{y} < 0$

Figure 15.6   Phase plane analysis of a linear ODE system (Eq. 15.8). *Question*: Based on the orientation of the direction field arrows, is the steady state at $(0,0)$ stable or unstable?

Example 2

Phase plane analysis of the following nonlinear ODE system,

$$\dot{x} = xy - 1 \qquad\qquad (15.11a)$$
$$\dot{y} = x^2 - y, \qquad\qquad (15.11b)$$

proceeds in a similar manner.

(1)   The relations between x and y that define the nullclines are:

$$x\text{-nullcline}: 0 = xy - 1 \Rightarrow xy = 1 \qquad\qquad (15.12a)$$
$$y\text{-nullcline}: 0 = x^2 - y \Rightarrow y = x^2. \qquad\qquad (15.12b)$$

Fig. 15.7 shows the x nullcline (the black hyperbola) and the y nullcline (a gray parabola).

(2)   There is one steady state located at the intersection of the parabola and the hyperbola $(x, y) = (1, 1)$, as can be shown as follows,

$$y = x^2 \text{ and } xy = 1 \Rightarrow \underbrace{\frac{1}{x}}_{y} = x^2 \Rightarrow x^3 = 1 \Rightarrow x = 1,$$

which using either $y = x^2$ or $y = 1/x$ implies $y = 1$.

(3x)   On the x nullcline $xy = 1$ and the rate of change of y is

$$\dot{y} = x^2 - y = x^2 - \frac{1}{x} = \frac{x^3 - 1}{x} \Rightarrow \begin{cases} \dot{y} < 0 & (0 < x < 1) \\[2mm] \dot{y} > 0 & (x < 0 \text{ or } x > 1). \end{cases}$$

(3y)   On the y nullcline $y = x^2$ and the rate of change of x is

$$\dot{x} = xy - 1 = x \cdot x^2 - 1 = x^3 - 1 \Rightarrow \begin{cases} \dot{x} < 0 & (x < 1) \\[2mm] \dot{x} > 0 & (x > 1). \end{cases}$$

(4)   Fig. 15.7 shows that the x and y nullclines divide the phase plane into five regions. In each region we can determine the sign of dx/dt and dy/dt. For example, the region that includes the fourth quadrant (where $x > 0$ and $y < 0$) is below the parabola ($y < x^2$). Using Eq. 15.11b we see that

$$y < x^2 \Rightarrow dy/dt = x^2 - y > 0,$$

that is, the rate of change of y is positive (upward flow). Using Eq. 15.11a we see that

$$x > 0 \text{ and } y < 0 \Rightarrow xy < 0 \Rightarrow dx/dt = xy - 1 < 0,$$

that is, the rate of change of x is negative (leftward flow). The numerically calculated direction field confirms the sign of dx/dt and dy/dt in the fourth quadrant.

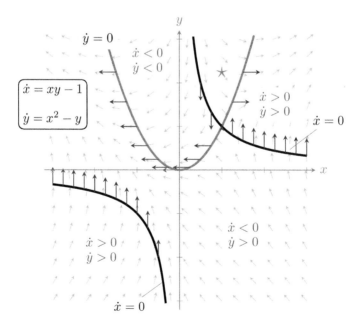

Figure 15.7   Phase plane analysis of Eq. 15.11. The x nullcline has two branches. *Question:* What is the sign of $\dot{x}$ and $\dot{y}$ in the region labelled ⋆?

15.5   Phase Planes and Steady States

The previous sections have discussed the multiple steps involved in phase plane analysis and provided several specific examples. We have noted that steady states of two-dimensional autonomous ODE systems are those points in the phase plane that are simultaneously on both the x and y nullclines.

Consider the point $(1, 1)$ in Fig. 15.7. This point is a steady state because it is on both the x nullcline ($\dot{x} = 0$) and the y nullcline ($\dot{y} = 0$). Is this steady state stable or unstable? The direction field nearby $(1, 1)$ suggests that this fixed point is unstable, because the x component of the flow is negative (leftward) when $x < 1$ and positive (rightward) when $x > 1$. While it is often possible to glean the stability of a fixed point by considering the flow of the local (nearby) direction field, a systematic analytical approach is needed.

In scalar autonomous ODEs, $dy/dt = f(y)$, fixed points solve $0 = f(y_{ss})$, and $f'(y_{ss}) < 0$ is a necessary and sufficient condition for stability. For two-dimensional autonomous ODE systems

$$\frac{dx}{dt} = f(x, y) \tag{15.13a}$$

$$\frac{dy}{dt} = g(x, y), \tag{15.13b}$$

steady states solve

$$0 = f(x_{ss}, y_{ss}) \tag{15.14a}$$

$$0 = g(x_{ss}, y_{ss}). \tag{15.14b}$$

For any such system, one may calculate four partial derivatives $\partial f/\partial x$, $\partial f/\partial y$, $\partial g/\partial x$, and $\partial g/\partial y$. Each is potentially a function of x and y. When collected into a 2×2 matrix with rows associated to f and g and columns associated to x and y, as follows,

$$J(x, y) = \begin{bmatrix} \dfrac{\partial f}{\partial x} & \dfrac{\partial f}{\partial y} \\ \dfrac{\partial g}{\partial x} & \dfrac{\partial g}{\partial y} \end{bmatrix}, \tag{15.15}$$

the result is referred to as the **Jacobian** of the ODE system. To determine the stability of a steady state (x_{ss}, y_{ss}), evaluate the Jacobian at that point to obtain a 2×2 matrix of constants,

$$J_{ss} = J(x_{ss}, y_{ss}) = \begin{bmatrix} f_x & f_y \\ g_x & g_y \end{bmatrix}. \tag{15.16}$$

Next, calculate the following two numbers.

(a)   The *trace* of the Jacobian (the sum of the diagonal entries),

$$\mathrm{tr}J_{ss} = f_x + g_y.$$

(b)   The *determinant* of the Jacobian, given by

$$\det J_{ss} = f_x g_y - f_y g_x.$$

A necessary and sufficient condition for stability of (x_{ss}, y_{ss}) is that the trace of the Jacobian be negative *and* the determinant of the Jacobian be positive, that is,

$$(x_{ss}, y_{ss}) \text{ is } \begin{cases} \text{stable} & \text{if } \text{tr} J_{ss} < 0 \text{ and } \det J_{ss} > 0 \\ \text{unstable} & \text{otherwise.} \end{cases} \tag{15.17}$$

The examples below show this criterion for stability of steady states is straightforward to apply.

Example Stability Test

To analyze the stability of the fixed point $(1, 1)$ in Fig. 15.7 begin with the ODE system Eq. 15.11,

$$\dot{x} = f(x, y) = xy - 1$$
$$\dot{y} = g(x, y) = x^2 - y,$$

and identify the functions $f(x, y)$ and $g(x, y)$. Next, calculate the required four partial derivatives: $\partial f/\partial x = y$, $\partial f/\partial y = x$, $\partial g/\partial x = 2x$ and $\partial f/\partial y = -1$. Arranging these in a 2×2 matrix, the Jacobian is

$$J(x, y) = \begin{bmatrix} y & x \\ 2x & -1 \end{bmatrix}.$$

Evaluating the Jacobian at the fixed point $(1, 1)$ gives the constant matrix

$$J_{ss} = J(1, 1) = \begin{bmatrix} 1 & 1 \\ 2 & -1 \end{bmatrix}.$$

Because the trace and determinant of the Jacobian are

$$\text{tr} J_{ss} = 1 + (-1) = 0 \quad \text{and} \quad \det J_{ss} = 1(-1) - 1(2) = -3,$$

the criteria for stability are not met (the trace is not negative, the determinant is not positive). We conclude that the steady state at $(1, 1)$ is unstable.

Another Example Stability Test

As another example, let us return to the ODE system Eq. 15.7,

$$\dot{x} = g(x, y) = x^2 - y$$
$$\dot{y} = f(x, y) = x - y.$$

Fig. 15.5 shows two fixed points, $(0, 0)$ and $(1, 1)$, located at the intersections of the x and y nullclines. Using $\partial f/\partial x = 2x$, $\partial f/\partial y = -1$, $\partial g/\partial x = 1$ and $\partial f/\partial y = -1$, we calculate the Jacobian associated to this ODE system,

$$J(x, y) = \begin{bmatrix} 2x & -1 \\ 1 & -1 \end{bmatrix}.$$

For the steady state at $(0,0)$ we find

$$J_{ss} = J(0,0) = \begin{bmatrix} 0 & -1 \\ 1 & -1 \end{bmatrix}.$$

This constant matrix has negative trace, $\mathrm{tr} J_{ss} = 0 + (-1) = -1$, and positive determinant, $\det J_{ss} = 0(-1) - (-1)1 = 1$. We conclude that the steady state at $(0,0)$ is stable. The steady state at $(1,1)$ is unstable, because

$$J_{ss} = J(1,1) = \begin{bmatrix} 2 & -1 \\ 1 & -1 \end{bmatrix},$$

$\mathrm{tr} J_{ss} = 2 + (-1) = 1$, and $\det J_{ss} = 2(-1) - (-1)1 = -1$.

15.6   Discussion

We have introduced stability analysis for two-dimensional autonomous ODE systems (Eq. 15.1). The criterion for stability of steady states (Eqs. 15.15–15.17) will become less mysterious as we discuss **linearization** of scalar ODEs and ODE systems (see below), and explore analytical solutions of *linear* ODE systems (Chapter 16).

Two-Dimensional Linear ODE Systems

Eq. 15.2 is *linear* because solutions of this ODE system can be superposed. This may be demonstrated by rewriting it in matrix-vector form,

$$\frac{dx}{dt} = Ax, \qquad (15.18)$$

where

$$A = \begin{bmatrix} a & b \\ c & d \end{bmatrix} \quad \text{and} \quad x = \begin{bmatrix} x \\ y \end{bmatrix}.$$

If two functions of time $x_1(t)$ and $x_2(t)$, both of which are 2×1 vectors, solve Eq. 15.18, the sum $x(t) = x_1 + x_2$ does as well. To see this, write $dx_1/dt = Ax_1$ and $dx_2/dt = Ax_2$, and sum the equations to obtain,

$$\frac{dx_1}{dt} + \frac{dx_2}{dt} = Ax_1 + Ax_2$$

$$\frac{d}{dt} \underbrace{[x_1 + x_2]}_{x} = A \underbrace{[x_1 + x_2]}_{x}.$$

Conversely, a two-dimensional system of ODEs given by $dx/dt = Ax + b$ where $b \neq 0$ is not linear (see Problem 15.9).

Number of Steady States

Steady state solutions of two-dimensional linear ODE systems (Eq. 15.18) are ordered pairs (x_{ss}, y_{ss}) that lead to $dx/dt = 0$. Such an ordered pair satisfies the matrix equation $Ax = 0$, which is a compact representation of two equations for two unknowns,

$$\begin{bmatrix} a & b \\ c & d \end{bmatrix} \begin{bmatrix} x_{ss} \\ y_{ss} \end{bmatrix} = \begin{bmatrix} 0 \\ 0 \end{bmatrix} \quad \Leftrightarrow \quad \begin{aligned} ax_{ss} + by_{ss} = 0 \\ cx_{ss} + dy_{ss} = 0. \end{aligned} \tag{15.19}$$

The number of solutions depends on the **determinant** of the 2×2 matrix of constants, given by

$$\det \underbrace{\begin{bmatrix} a & b \\ c & d \end{bmatrix}}_{A} = ad - bc. \tag{15.20}$$

If the determinant is nonzero ($\det A \neq 0$), the only solution to Eq. 15.19 is the trivial solution $(x_{ss}, y_{ss}) = (0, 0)$. If $\det A = 0$, there are an infinite number of steady states.

In Eq. 15.8, for example, the determinant of the constant coefficient matrix is $-2(-1) - 3(4) = -10 \neq 0$, so the only steady state is $(x_{ss}, y_{ss}) = (0, 0)$. This is consistent with Fig. 15.6 that shows one intersection of the x and y nullclines located at the origin.

Linearization

The stability analysis introduced in Section 15.5 works because the Jacobian of an ODE system (evaluated at a steady state) yields the constant coefficient matrix of *an associated linear system that well approximates the dynamics of the nonlinear system near the steady state of interest.*

Scalar ODEs and Linearization

Consider a scalar ODE system $\dot{y} = f(y)$ with steady state y_{ss} satisfying $0 = f(y_{ss})$. Let us write $y(t) = y_{ss} + \delta y$ where y_{ss} is constant and $\delta y(t)$ is the deviation from steady state and derive the ODE that governs the dynamics of this time-dependent deviation. Differentiating with respect to time we see that $y = y_{ss} + \delta y \Rightarrow \dot{y} = \dot{\delta y}$ (y_{ss} is constant so $\dot{y}_{ss} = 0$) and thus $\dot{\delta y} = f(y_{ss} + \delta y)$. Next, expand the right side in a Taylor series,

$$f(y_{ss} + \delta y) = f(y_{ss}) + \delta y f'(y_{ss}) + \frac{1}{2}(\delta y)^2 f''(y_{ss}) + \cdots,$$

use $f(y_{ss}) = 0$ and drop terms of order $(\delta y)^2$ and higher to obtain

$$\dot{\delta y} \approx f'(y_{ss}) \delta y. \tag{15.21}$$

This ODE governs the dynamics of the deviation of a solution from a nearby steady state ($\delta y = y - y_{ss}$). The steady state y_{ss} is stable if small perturbations δy decay away, which will occur provided $f'(y_{ss}) < 0$.

When $f(y)$ is a polynomial, one can perform this linearization process in a concrete manner. Consider the ODE $\dot{y} = y^3 - y$, with steady states $y_{ss} = 0$ and 1. If we write $\delta y = y - y_{ss}$, then $y = y_{ss} + \delta y$, $\dot{y} = \dot{\delta y}$, and substitution gives,

$$\dot{\delta y} = (y_{ss} + \delta y)^3 - (y_{ss} + \delta y)$$
$$= y_{ss}^3 + 3y_{ss}^2\delta y + 3y_{ss}\delta y^2 + \delta y^3 - y_{ss} - \delta y.$$

Now using $0 = y_{ss}^3 - y_{ss}$ and dropping terms involving $(\delta y)^2$ and $(\delta y)^3$ we obtain,

$$\dot{\delta y} = 3y_{ss}^2\delta y - \delta y = (3y_{ss}^2 - 1)\,\delta y. \qquad (15.22)$$

This is equivalent to the more abstract calculation (Eq. 15.21), because the factor in parentheses is $[y^3 - y]' = 3y^2 - 1$ evaluated at steady state.

Linearization of Two-Dimensional ODE Systems

Linearization proceeds in an analogous fashion for a two-dimensional autonomous ODE system (Eq. 15.13). Of course, the calculation is more complex because there are two dependent variables and two ODEs. Any steady state (x_{ss}, y_{ss}) satisfies Eq. 15.14 and the deviations from the steady state are $\delta x = x - x_{ss}$ and $\delta y = y - y_{ss}$. The first terms of the relevant Taylor series are

$$f(x_{ss} + \delta x, y_{ss} + \delta y) = f(x_{ss}, y_{ss}) + \delta x \overbrace{\left.\frac{\partial f}{\partial x}\right|_{ss}}^{f_x} + \delta y \overbrace{\left.\frac{\partial f}{\partial y}\right|_{ss}}^{f_y} + \cdots$$

$$g(x_{ss} + \delta x, y_{ss} + \delta y) = g(x_{ss}, y_{ss}) + \delta x \underbrace{\left.\frac{\partial g}{\partial x}\right|_{ss}}_{g_x} + \delta y \underbrace{\left.\frac{\partial g}{\partial y}\right|_{ss}}_{g_y} + \cdots$$

Dropping terms of second order and higher, the deviations solve the following ODE system,

$$\begin{aligned}\dot{\delta x} &= f_x\delta x + f_y\delta y \\ \dot{\delta y} &= g_x\delta x + g_y\delta y\end{aligned} \quad\Leftrightarrow\quad \frac{d}{dt}\begin{bmatrix}\delta x\\\delta y\end{bmatrix} = \underbrace{\begin{bmatrix}f_x & f_y\\g_x & g_y\end{bmatrix}}_{J(x_{ss},y_{ss})}\begin{bmatrix}\delta x\\\delta y\end{bmatrix}, \qquad (15.23)$$

where the constants f_x, f_y, g_x and g_y are the elements of the Jacobian matrix evaluated at the steady state of interest.

Consider the two-dimensional nonlinear system Eq. 15.11 that has steady state $(x_{ss}, y_{ss}) = (1, 1)$. The Jacobian of this system is

$$J(x, y) = \begin{bmatrix}\dfrac{\partial f}{\partial x} & \dfrac{\partial f}{\partial y}\\[2mm]\dfrac{\partial g}{\partial x} & \dfrac{\partial g}{\partial y}\end{bmatrix} = \begin{bmatrix}y & x\\2x & -1\end{bmatrix}.$$

Evaluating $J(x, y)$ at the steady state $(1, 1)$ yields

$$J(1, 1) = \begin{bmatrix}1 & 1\\2 & -1\end{bmatrix}.$$

The linearized system associated with the steady state $(1, 1)$ is

$$\frac{d}{dt}\begin{bmatrix}\delta x\\\delta y\end{bmatrix} = \begin{bmatrix}1 & 1\\2 & -1\end{bmatrix}\begin{bmatrix}\delta x\\\delta y\end{bmatrix} \quad\Leftrightarrow\quad \begin{aligned}\dot{\delta x} &= \delta x + \delta y\\ \dot{\delta y} &= 2\delta x - \delta y.\end{aligned} \qquad (15.24)$$

The alternative approach to linearizing Eq. 15.11 at the steady state $(1,1)$ begins by substituting $x = x_{ss} + \delta x$, $y = y_{ss} + \delta y$,

$$\dot{x} = (x_{ss} + \delta x)(y_{ss} + \delta y) - 1 = x_{ss}y_{ss} + x_{ss}\delta y + \delta x y_{ss} + \delta x \delta y - 1$$

$$\dot{y} = (x_{ss} + \delta x)^2 - (y_{ss} + \delta y) = x_{ss}^2 + 2x_{ss}\delta x + \delta x^2 - y_{ss} - \delta y .$$

Using $0 = x_{ss}y_{ss} - 1$ and $0 = x_{ss}^2 - y_{ss}$ to eliminate several terms, and dropping higher order terms involving $(\delta x)^2$ and $\delta x \delta y$, this becomes

$$\begin{aligned}\dot{x} &= y_{ss}\delta x + x_{ss}\delta y \\ \dot{y} &= 2x_{ss}\delta x - \delta y\end{aligned} \qquad \Leftrightarrow \qquad \frac{d}{dt}\begin{bmatrix} \delta x \\ \delta y \end{bmatrix} = \underbrace{\begin{bmatrix} y_{ss} & x_{ss} \\ 2x_{ss} & -1 \end{bmatrix}}_{J(x_{ss}, y_{ss})}\begin{bmatrix} \delta x \\ \delta y \end{bmatrix} .$$

Substituting the steady state $(x_{ss}, y_{ss}) = (1,1)$ gives Eq. 15.24.

Supplemental Problems

Problem 15.3   Using Eq. 15.2 it is evident that the slope of the second black arrow of Fig. 15.1 (right) is

$$\frac{dy/dt}{dx/dt} \approx \frac{-1.6713}{-1.1938} \approx 1.4 .$$

Using the fact that (1) the functions $x(t)$ and $y(t)$ plotted in Fig. 15.1 (left) solve Eq. 15.3 and (2) $y = -x$ when $t = \pi/4$, show without approximation (or use of a calculator) that the slope is exactly $7/5$.

Problem 15.4   The origin is the only fixed point of the two-dimensional linear system $\dot{x} = Ax$ when $\det A = ad - bc \neq 0$. Show that when a and d are nonzero, the x and y nullclines are two distinct lines that intersect at the origin.

Problem 15.5   Consider the nullclines and arrows in the following phase planes. Visually determine whether the steady state at the origin is stable or unstable.

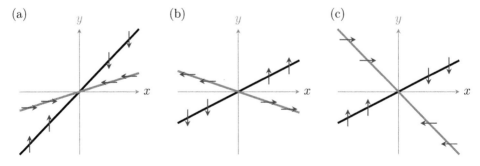

(a)   (b)   (c)

Problem 15.6   Following the steps in Section 15.4, perform phase plane analysis of the following two-dimensional autonomous ODE systems.

(a) $\dot{x} = y - xy$ (d) $\dot{x} = x(x - y)$ (g) $\dot{x} = x^2 - y$

       $\dot{y} = xy$ $\dot{y} = y(2x - y)$ $\dot{y} = x - y$

(b) $\dot{x} = x - xy$ (e) $\dot{x} = y$ (h) $\dot{x} = x^2 + y^2 - 4$

       $\dot{y} = xy - y^2$ $\dot{y} = x(1 + y) - 1$ $\dot{y} = x^2 - y^2$

(c) $\dot{x} = x - x^3$ (f) $\dot{x} = x(2 - x - y)$ (i) $\dot{x} = \alpha x - \beta xy$

       $\dot{y} = -y$ $\dot{y} = x - y$ $\dot{y} = \delta xy - \gamma y$

Problem 15.7   What is the Jacobian of Eq. 15.2?

Problem 15.8   What are the solutions to Eq. 15.19 when a, b, c and d are nonzero and the determinant given by Eq. 15.20 is zero?

Problem 15.9   Show that an inhomogenous ODE system of the form

$$\dot{x} = Ax + b \tag{15.25}$$

is not linear when the 2×1 constant vector b is nonzero ($b \neq 0$).

Solutions

Figure 15.4   Top: For the arrow located at $(x, y) = (-2, 2)$, $dx/dt = y^2 - x + 2 = 2^2 - (-2) + 2 = 8$ and $dy/dt = x^3 - y^3 = (-2)^3 - 2^3 = -16$. Consequently, the arrow is pointing south east with slope $-16/8 = -2$. Bottom: When $x = 0$, $dx/dt = xy^2 - x^2 = 0$ and $dy/dt = xy - \frac{1}{2} = \frac{1}{2}$. So the direction field is vertical downward on the line $x = 0$.

Figure 15.5

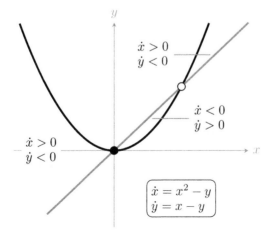

Figure 15.6   Inspection of Fig. 15.6 shows that it is *not* the case that "solutions that start near the steady state remain near the steady state." Thus, the steady state at

the origin is unstable. For example, the initial value $(1,1)$ produces a solution for which dx/dt and dy/dt are both positive, and as the trajectory moves rightward and upward it will not cross the x nullcline. In fact, for this solution beginning at $(1,1)$ both x and y increase without bound ($x \to \infty$ and $y \to \infty$ as $t \to \infty$).

Figure 15.7 The unlabelled region of the phase plane is above the upper branch of the x nullcline (hyperbola) where $xy > 1$ so $\dot{x} = xy - 1 > 0$. This region is also above the y nullcline (parabola) where $y > x^2$ so $\dot{y} = x^2 - y < 0$.

Problem 15.1 The initial conditions of Eq. 15.3 are satisfied, because when $t = 0$, $e^{-t/2} = e^0 = 1$, $\cos 3t = \cos 0 = 1$ and $\sin 3t = \sin 0 = 0$; thus, $x(0) = 1 \cdot 1 = 1$ and $y(0) = 1 \cdot 0 = 0$. To show that Eq. 15.4 satisfies the ODE system, calculate the derivatives $x'(t)$ and $y'(t)$, for example,

$$x'(t) = [e^{-t/2} \cos 3t]'$$
$$= [e^{-t/2}]' \cos 3t + e^{-t/2}[\cos 3t]'$$
$$= -\frac{1}{2}e^{-t/2} \cos 3t - 3e^{-t/2} \sin 3t,$$

substitute x, y, x' and y' into Eq. 15.3, and confirm the equality:

$$\overbrace{-\frac{1}{2}e^{-t/2} \cos 3t - 3e^{-t/2} \sin 3t}^{x'(t)} \overset{?}{=} -\frac{1}{2} \overbrace{[e^{-t/2} \cos 3t]}^{x(t)} - 3 \overbrace{[e^{-t/2} \sin 3t]}^{y(t)}$$

$$\underbrace{-\frac{1}{2}e^{-t/2} \sin 3t + 3e^{-t/2} \cos 3t}_{y'(t)} \overset{?}{=} 3 \underbrace{[e^{-t/2} \cos 3t]}_{x(t)} - \frac{1}{2} \underbrace{[e^{-t/2} \sin 3t]}_{y(t)}.$$

Problem 15.2 The south west pointing arrow in Fig. 15.1 (left) is anchored at $t = \pi/4$, that is, $3t = 3\pi/4$ radians $= 135$ degrees. According to Eq. 15.4, the values of x and y at this time are

$$\begin{bmatrix} x \\ y \end{bmatrix} = \begin{bmatrix} e^{-\pi/8} \cos(3\pi/4) \\ e^{-\pi/8} \sin(3\pi/4) \end{bmatrix} \approx \begin{bmatrix} -0.4775 \\ 0.4775 \end{bmatrix}.$$

Substituting these values of x and y into Eq. 15.3 gives

$$\begin{bmatrix} dx/dt \\ dy/dt \end{bmatrix} = \begin{bmatrix} -\frac{1}{2}(-0.4775) - 3 \cdot 0.4775 \\ 3(-0.4775) - \frac{1}{2} \cdot 0.4775 \end{bmatrix} \approx \begin{bmatrix} -1.1938 \\ -1.6713 \end{bmatrix}.$$

Observe that $x' < 0$ and $y' < 0$, consistent with the south west orientation of the tangent arrow.

Problem 15.3 Using Eq. 15.3 the slope of a direction field arrow is

$$\frac{dy/dt}{dx/dt} = \frac{-x/2 - 3y}{3x - y/2}.$$

When $t = \pi/4$, $y = -x$ and the above ratio evaluates to

$$\frac{-x/2 + 3x}{3x + x/2} = \frac{5/2}{7/2} = 5/7.$$

Problem 15.4   The nullclines of Eq. 15.2 are found by setting $dx/dt = 0$ and $dy/dt = 0$. Assuming a and d are nonzero, divide Eq. 15.2a by a and divide Eq. 15.2b by d to obtain,[7]

$$x \text{ nullcline:} \quad 0 = \frac{a}{b}x + y \implies x = -\frac{b}{a}y$$

$$y \text{ nullcline:} \quad 0 = \frac{c}{d}x + y \implies y = -\frac{c}{d}x.$$

If we substitute the equation for the y nullcline into the ODE for x (and vice versa), we obtain,

$$\text{on } y \text{ nullcline:} \quad \frac{dx}{dt} = ax + b\left(-\frac{c}{d}x\right) = \left(a - \frac{bc}{d}\right)x$$

$$\text{on } x \text{ nullcline:} \quad \frac{dy}{dt} = c\left(-\frac{b}{a}y\right) + dy = \left(d - \frac{bc}{a}\right)y.$$

Factoring out $1/d$ and $1/a$ gives

$$\text{on } y \text{ nullcline:} \quad \frac{dx}{dt} = \frac{1}{d}(ad - bc)\,x$$

$$\text{on } x \text{ nullcline:} \quad \frac{dy}{dt} = \frac{1}{a}(ad - bc)\,y$$

where $\det A \neq 0 \implies ad \neq bc$. The above equations imply that the direction of flow on both nullclines reverses at the origin (as in Fig. 15.6).[8]

Problem 15.5   (a) stable; (b) unstable; (c) stable.

Problem 15.6

(a)

$$\dot{x} = y - xy$$
$$\dot{y} = xy$$

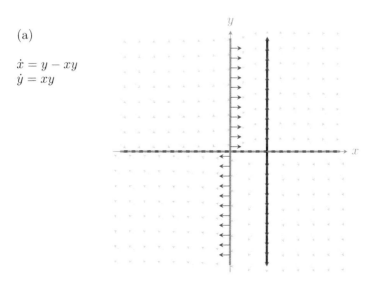

(b)

$$\dot{x} = x - xy$$
$$\dot{y} = xy - y^2$$

(c)

$$\dot{x} = x - xy$$
$$\dot{y} = xy - y^2$$

(h)

$$\dot{x} = x^2 + y^2 - 4$$
$$\dot{y} = x^2 - y^2$$

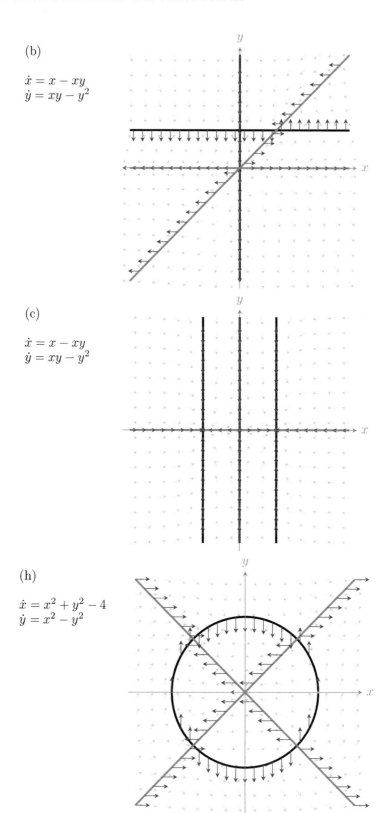

Problem 15.7   Eq. 15.1 is equivalent to Eq. 15.2 when $f(x,y) = ax + by$ and $g(x,y) = cx + dy$. Calculating the four required partial derivatives (Eq. 15.15), we find $\partial f/\partial x = a$, $\partial f/\partial y = b$, $\partial g/\partial x = c$ and $\partial f/\partial y = d$. Thus, the Jacobian associated to Eq. 15.2 is the 2×2 matrix of constants,

$$J = \begin{bmatrix} a & b \\ c & d \end{bmatrix}.$$

Problem 15.8   Because b and d are both nonzero, we can divide $0 = ax_{ss} + by_{ss}$ by b and divide $0 = cx_{ss} + dy_{ss}$ by d,

$$0 = \frac{a}{b}x_{ss} + y_{ss} \Rightarrow y_{ss} = -\frac{a}{b}x_{ss} \qquad (15.26a)$$

$$0 = \frac{c}{d}x_{ss} + y_{ss} \Rightarrow y_{ss} = -\frac{c}{d}x_{ss}. \qquad (15.26b)$$

The determinant is zero, $ad = bc \Rightarrow a/b = c/d$, so the second equation in Eq. 15.26 is equivalent to the first and gives no new information. Thus, there are an infinite number of solutions given by the single-parameter family of ordered pairs $(x_{ss}, -\frac{a}{b}x_{ss})$ for all x_{ss}. The steady states are a line in the (x,y)-plane. In fact, both the x and y nullclines are given by this line.

Problem 15.9   If x_1 and x_2 solve Eq. 15.25 for $b \neq 0$, the sum $x = x_1 + x_2$ does not. To see this, sum $\dot{x}_1 = Ax_1 + b$ and $\dot{x}_2 = Ax_2 + b$ to obtain $\dot{x}_1 + \dot{x}_2 = Ax_1 + Ax_2 + 2b \Rightarrow \dot{x} = Ax + 2b \Rightarrow 0 = b$, which contradicts our assumption that $b \neq 0$.

Notes

1. Such systems are also written in the following way,

$$\frac{dx_1}{dt} = a_{11}x_1 + a_{12}x_2$$

$$\frac{dx_2}{dt} = a_{21}x_1 + a_{22}x_2,$$

where a_{11}, a_{12}, a_{21} and a_{22} are constants. This has the advantage that the notation is easily extended to three-dimensional systems (and higher).
2. In general, dx/dt and dy/dt have different physical dimensions than the dependent variables x and y. Consequently, the length of the arrows in the phase plane depends on a choice of a scale parameter with units of time. In Fig. 15.1, we have chosen this parameter so that the arrows have length 0.5. A normalized (unit length) vector pointing in the same direction as $\begin{bmatrix} x \\ y \end{bmatrix}$ is given by $\frac{1}{\ell}\begin{bmatrix} x \\ y \end{bmatrix}$ where $\ell = \sqrt{x^2 + y^2}$.
3. The -6 slope of the tangent line is silent on whether the arrow ought to point in the direction of decreasing x (uphill) or increasing x (downhill).
4. It also does not make contact with itself tangentially.
5. The functions shown are $x(t) = t^2 - 1$ and $y(t) = t(t^2 - 1)$ on the interval $t \in [-1.5, 1.5]$. Because the trajectory in the (x,y)-plane crosses itself, we conclude that there is no two-dimensional ODE system solved by these polynomials.

6. The process of constructing a direction field is rather tedious when done by hand, but easily automated using a computer. For clarity, all the arrows have the same length and do not overlap.

7. In our first example of phase plane analysis (Eqs. 15.9 and 15.10), the equations for the nullclines took this form.

8. One salient feature of Fig. 15.5 (a nonlinear system) is that the direction of flow on a nullcline reverses at steady states (located at intersections with the other nullcline). This will occur provided the determinant of the Jacobian evaluated at the intersection of the nullclines is nonzero.

16 Linear Stability Analysis

The stability analysis introduced in Chapter 15 works because the Jacobian of an ODE system (evaluated at a steady state) yields the constant coefficient matrix of an associated linear system. This linear system well approximates the dynamics of the nonlinear system near the steady state of interest. Solutions of two-dimensional linear ODE systems may be categorized into five qualitatively different types.

16.1 Solutions for Two-Dimensional Linear Systems

Two-dimensional (2d) linear systems take the form $\dot{x} = Ax$,

$$\underbrace{\frac{d}{dt}\begin{bmatrix} x \\ y \end{bmatrix}}_{\dot{x}} = \underbrace{\begin{bmatrix} a & b \\ c & d \end{bmatrix}}_{A}\underbrace{\begin{bmatrix} x \\ y \end{bmatrix}}_{x} \quad \Leftrightarrow \quad \begin{aligned} \dot{x} &= ax + bx \\ \dot{y} &= cx + dx, \end{aligned} \tag{16.1}$$

where a, b, c and d are constants. We assume that $\det A = ad - bc \neq 0$, thereby ensuring that Eq. 16.1 has two distinct nullclines that intersect at the origin, which is the unique steady state of the linear system, because linearization of 2d autonomous ODE systems yields equations of this form (see p. 266).

Because Eq. 16.1 is linear, it is natural to seek solutions of the form

$$x(t) = ve^{\lambda t} \quad \Leftrightarrow \quad \begin{bmatrix} x(t) \\ y(t) \end{bmatrix} = \begin{bmatrix} v_x \\ v_y \end{bmatrix}e^{\lambda t} \tag{16.2}$$

where λ is a scalar constant and v is a constant 2×1 column vector. It can be shown that Eq. 16.2 solves Eq. 16.1 provided that

$$Av = \lambda v \quad \Leftrightarrow \quad \begin{bmatrix} a & b \\ c & d \end{bmatrix}\begin{bmatrix} v_x \\ v_y \end{bmatrix} = \lambda \begin{bmatrix} v_x \\ v_y \end{bmatrix}. \tag{16.3}$$

This linear algebraic equation is an **eigenvector-eigenvalue** problem. The matrix A is given, and v and λ are unknowns. This equation may have a solution – an eigenvalue-eigenvector pair (λ, v) – in one of two ways. Either $v = \mathbf{0}$ or, alternatively, the eigenvalue λ satisfies the **characteristic polynomial** of A,

$$\lambda^2 - \beta\lambda + \gamma = 0, \tag{16.4}$$

where $\beta = \text{tr}A$ and $\gamma = \det A$. Substituting $v = \mathbf{0}$ into Eq. 16.2 gives $x(t) = \mathbf{0}$, the steady state of Eq. 16.1. When $v \neq \mathbf{0}$, the dynamical behavior of $x(t)$ solving $\dot{x} = Ax$ depends on whether Eq. 16.4 has 2, 1 or 0 real-valued solutions. Using the quadratic formula we find that λ is given by[1]

$$\lambda = \frac{\beta \pm \sqrt{\beta^2 - 4\gamma}}{2} = \frac{\beta \pm \sqrt{\delta}}{2}, \tag{16.5}$$

where $\delta = \text{disc}A = \beta^2 - 4\gamma$ is the **discriminant** of the matrix A. The value(s) of λ that solve Eq. 16.5 are the **eigenvalues** of the matrix A. We have assumed $\gamma = \det A \neq 0$, so $\lambda = 0$ is not a solution of Eq. 16.4. Further assuming that $\delta \neq 0$, the eigenvalues are distinct; we will write them as $\lambda_1 \neq \lambda_2$. There are several qualitatively different cases to consider.

When $\beta^2 > 4\gamma$, the distinct eigenvalues are real valued and, without loss of generality, we may assume $\lambda_1 < \lambda_2$. In this case, there are three qualitatively different types of solutions depending on the signs of λ_1 and λ_2: **stable node**, **unstable node** and **saddle** (shown below).

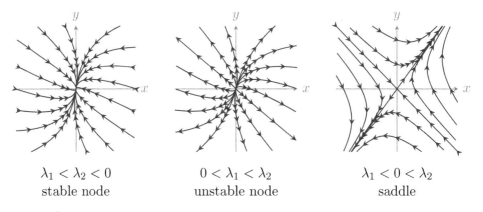

$\lambda_1 < \lambda_2 < 0$ $0 < \lambda_1 < \lambda_2$ $\lambda_1 < 0 < \lambda_2$
stable node   unstable node   saddle

When $\beta^2 < 4\gamma$, the eigenvalues of A solving Eq. 16.4 are a complex conjugate pair, $\lambda = \mu \pm i\omega$, corresponding to a **stable spiral**, **center**, or **unstable spiral** depending on the sign of the μ.

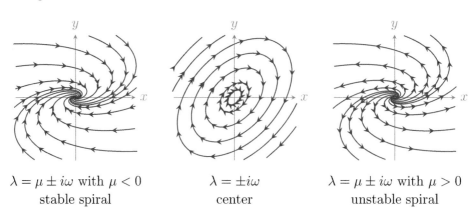

$\lambda = \mu \pm i\omega$ with $\mu < 0$ $\lambda = \pm i\omega$ $\lambda = \mu \pm i\omega$ with $\mu > 0$
stable spiral   center   unstable spiral

Consistent with the direction of flow near the origin, the stable node and stable spiral are referred to as **sinks** and the unstable node and unstable spiral are **sources**.

The Trace-Determinant Plane

Fig. 16.1 shows that the five qualitative different types of steady states illustrated above may be organized according to the value of $\beta = \mathrm{tr}A$ and $\gamma = \det A$. In this **trace-determinant plane**, the stable steady states (sinks) are located in the second quadrant (blue) where $\beta < 0$ and $\gamma > 0$. Saddles are located in both quadrants three and four ($\gamma < 0$). The spirals are located above the parabola where $\gamma > \beta^2/4$ and eigenvalues are complex valued. Nodes and saddles are located below the parabola where the eigenvalues are real (have no imaginary part).

The following two sections show the analytical form of these qualitatively different flows in 2d linear systems and briefly describe how these solutions may be derived. The case of real and distinct eigenvalues (Section 16.2) and complex conjugate eigenvalues (Section 16.3) will be described in turn.

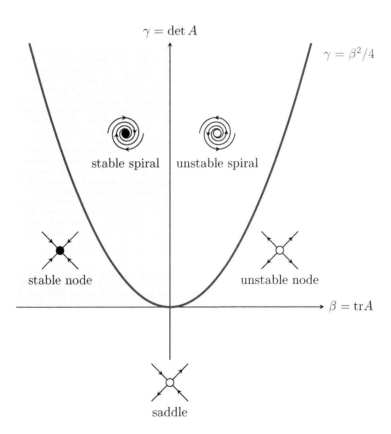

Figure 16.1   The trace-determinant plane organizes the major categories of steady states to 2d linear systems.

16.2   Real and Distinct Eigenvalues – Saddles and Nodes

Nodes and saddles are located below the parabola of the trace-determinant plane ($\gamma < \beta^2/4$, see Fig. 16.1). This region corresponds to eigenvalues that are real and distinct. On the horizontal axis (where $\gamma = 0$) the eigenvalues are β and 0. Elsewhere below the parabola the eigenvalues are real, distinct and *nonzero*. In this case, the general solution of Eq. 16.1 takes the form

$$x(t) = c_1 v_1 e^{\lambda_1 t} + c_2 v_2 e^{\lambda_2 t} \tag{16.6}$$

where the c_i are arbitrary constants and the (λ_i, v_i) are the eigenvalue-eigenvector pairs solving $A v_i = \lambda_i v_i$ ($i = 1, 2$). In expanded form,

$$\begin{bmatrix} x(t) \\ y(t) \end{bmatrix} = c_1 \begin{bmatrix} v_1^x \\ v_1^y \end{bmatrix} e^{\lambda_1 t} + c_2 \begin{bmatrix} v_2^x \\ v_2^y \end{bmatrix} e^{\lambda_2 t}, \tag{16.7}$$

where the λ_i are the roots of the characteristic polynomial (Eq. 16.4) and the eigenvectors associated to each eigenvalue solve

$$\begin{bmatrix} a & b \\ c & d \end{bmatrix} \underbrace{\begin{bmatrix} v_i^x \\ v_i^y \end{bmatrix}}_{v_i} = \lambda_i \underbrace{\begin{bmatrix} v_i^x \\ v_i^y \end{bmatrix}}_{v_i} \quad \Leftrightarrow \quad \begin{aligned} a v_i^x + b v_i^y &= \lambda_i v_i^x \\ c v_i^x + d v_i^y &= \lambda_i v_i^y, \end{aligned} \tag{16.8}$$

where v_1^x, v_1^y, v_2^x and v_2^y are unknowns. We consider two cases.

(1)   When $b \neq 0$, Eq. 16.8 implies

$$v_i^y = \frac{\lambda_i - a}{b} v_i^x$$

and choosing $v_i^x = 1$ for $i = 1, 2$ yields the eigenvectors

$$b \neq 0: \quad v_1 = \begin{bmatrix} v_1^x \\ v_1^y \end{bmatrix} = \begin{bmatrix} 1 \\ \dfrac{\lambda_1 - a}{b} \end{bmatrix} \quad v_2 = \begin{bmatrix} v_2^x \\ v_2^y \end{bmatrix} = \begin{bmatrix} 1 \\ \dfrac{\lambda_2 - a}{b} \end{bmatrix}. \tag{16.9}$$

(2)   When $b = 0$, the eigenvalues are a and d ($a \neq d$ because our assumption that $\gamma = \det A \neq 0$ implies $\delta \neq 0$, see Eq. 16.5). In this case, Eq. 16.8 implies that either $\lambda_i = a$ or $v_i^x = 0$; consequently, the eigenvectors are

$$b = 0: \quad v_d = \begin{bmatrix} v_1^x \\ v_1^y \end{bmatrix} = \begin{bmatrix} 0 \\ 1 \end{bmatrix} \quad v_a = \begin{bmatrix} v_1^x \\ v_1^y \end{bmatrix} = \begin{bmatrix} 1 \\ \dfrac{c}{a - d} \end{bmatrix}$$

where, for $\lambda_i = a$, we set $v_i^x = 1$, and $v_i^y = v_i^x c/(\lambda_i - d)$. To ensure $\lambda_1 < \lambda_2$, assign $v_1 = v_a$ and $v_2 = v_d$ when $a < d$; otherwise, $v_1 = v_d$ and $v_2 = v_a$.

Below we present and solve linear systems whose solutions are a saddle, stable node and unstable node.

Saddle: $\lambda_1 < 0 < \lambda_2$

Consider the following specific example of a 2d linear system of ODEs,

$$\frac{d}{dt}\begin{bmatrix} x \\ y \end{bmatrix} = \underbrace{\begin{bmatrix} -2 & 3 \\ 4 & -1 \end{bmatrix}}_{A} \begin{bmatrix} x \\ y \end{bmatrix}. \tag{16.10}$$

The trace and determinant of the constant coefficient matrix A are

$$\beta = \mathrm{tr}A = \mathrm{tr}\begin{bmatrix} -2 & 3 \\ 4 & -1 \end{bmatrix} = -2 + -1 = -3$$

$$\gamma = \det A = \det\begin{bmatrix} -2 & 3 \\ 4 & -1 \end{bmatrix} = -2(-1) - 3 \cdot 4 = 2 - 12 = -10.$$

The discriminant of A is $\delta = \mathrm{disc}A = \beta^2 - 4\gamma = (-3)^2 - 4(-10) = 49$. Using Eq. 16.5, we calculate the eigenvalues of A,

$$\lambda = \frac{-3 \pm \sqrt{3^2 - 4 \cdot (-10)}}{2} = \frac{-3 \pm 7}{2} = \frac{-10}{2} \text{ or } \frac{4}{2} = -5 \text{ or } 2.$$

One is negative ($\lambda_1 = -5$) and the other positive ($\lambda_2 = 2$), so the fixed point at the origin is a saddle. The general solution of Eq. 16.10 is

$$\begin{bmatrix} x(t) \\ y(t) \end{bmatrix} = c_1 \begin{bmatrix} 1 \\ -1 \end{bmatrix} e^{-5t} + c_2 \begin{bmatrix} 1 \\ 4/3 \end{bmatrix} e^{2t}, \tag{16.11}$$

where we have used

$$v_1^y = \frac{\lambda_1 - a}{b} = \frac{-5 - (-2)}{3} = -1 \quad \text{and} \quad v_2^y = \frac{\lambda_2 - a}{b} = \frac{2 - (-2)}{3} = 4/3.$$

Fig. 16.2 shows solutions near the steady state at the origin. The direction of flow indicates that this steady state is unstable, as is always the case for a saddle.

Stable Node: $\lambda_1 < \lambda_2 < 0$

Another example of a linear 2d system is

$$\frac{d}{dt}\begin{bmatrix} x \\ y \end{bmatrix} = \underbrace{\begin{bmatrix} -1 & 0.1 \\ 0.3 & -0.5 \end{bmatrix}}_{A} \begin{bmatrix} x \\ y \end{bmatrix}. \tag{16.12}$$

In this case, the trace, determinant and discriminant of A are

$$\beta = \mathrm{tr}A = -1 + (-0.5) = -1.5$$
$$\gamma = \det A = -1(-0.5) - 0.1 \cdot 0.3 = 0.5 - 0.03 = 0.47$$
$$\delta = \mathrm{disc}A = \beta^2 - 4\gamma = (-1.5)^2 - 4(0.47) = 0.37.$$

The eigenvalues of A are real, distinct and negative,

$$\lambda = \frac{\beta \pm \sqrt{\delta}}{2} = \frac{-1.5 \pm \sqrt{0.37}}{2} = \frac{-1.5 \pm 0.6083}{2} \Rightarrow \lambda_1 \approx -1.0541 \text{ and } \lambda_2 \approx -0.4459.$$

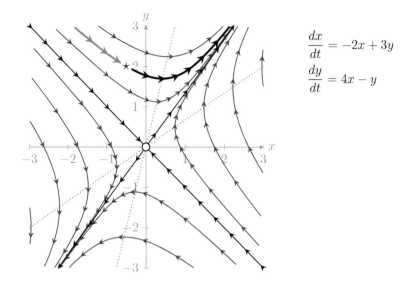

$$\frac{dx}{dt} = -2x + 3y$$

$$\frac{dy}{dt} = 4x - y$$

Figure 16.2 Solutions of a two-dimensional linear system with real and distinct eigenvalues of opposite sign ($\lambda_1 < 0 < \lambda_2$). The steady state at the origin is a saddle point (see Eq. 16.11). Solution with initial value $(-0.5, 2)$ (indicated by $\star$) is shown for $t > 0$ (black) and $t < 0$ (gray). Parameters: $\beta = -3$, $\gamma = -10$, $\delta = 49$, $\lambda_1 = -5$, $v_1^y = -1$, $\lambda_2 = 2$, $v_2^y = 1.33$.

The general solution to Eq. 16.12 is

$$\begin{bmatrix} x(t) \\ y(t) \end{bmatrix} = c_1 \begin{bmatrix} 1 \\ -0.54 \end{bmatrix} e^{-1.05t} + c_2 \begin{bmatrix} 1 \\ 5.54 \end{bmatrix} e^{-0.45t} \tag{16.13}$$

where we have used $v_i^x = 1$ and $v_i^y = (\lambda_i - a)/b$,

$$v_1^y = \frac{-1.05 - (-1)}{0.1} = -0.54 \quad \text{and} \quad v_2^y = \frac{-0.45 - (-1)}{0.1} = 5.54 \,.$$

Fig. 16.3 shows solutions directed toward the origin. The fixed point at the origin is a stable node.

Unstable Node: $0 < \lambda_1 < \lambda_2$

Our third example with real and distinct eigenvalues is

$$\frac{d}{dt} \begin{bmatrix} x \\ y \end{bmatrix} = \underbrace{\begin{bmatrix} 1 & -2 \\ 1 & 5 \end{bmatrix}}_{A} \begin{bmatrix} x \\ y \end{bmatrix} .$$

In this case,

$$\beta = \text{tr} A = 1 + 5 = 6$$
$$\gamma = \det A = 1 \cdot 5 - (-2)1 = 5 + 2 = 7$$
$$\delta = \text{disc} A = \beta^2 - 4\gamma = (6)^2 - 4(7) = 36 - 28 = 8$$

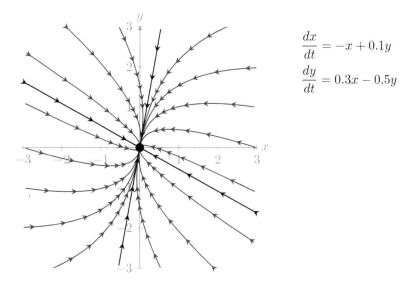

$$\frac{dx}{dt} = -x + 0.1y$$

$$\frac{dy}{dt} = 0.3x - 0.5y$$

Figure 16.3 Solutions of a two-dimensional linear system with real and distinct eigenvalues that are both negative ($\lambda_1 < \lambda_2 < 0$). The fixed point at the origin is a stable node. Parameters: $\beta = -1.5$, $\gamma = 0.47$, $\delta = 0.37$, $\lambda_2 = -0.45$, $v_2^y = 5.54$, $\lambda_1 = -1.05$, $v_1^y = -0.54$.

and the eigenvalues are both positive,

$$\lambda = \frac{\beta \pm \sqrt{\delta}}{2} = \frac{6 \pm \sqrt{8}}{2} = 3 \pm \sqrt{2} \implies \lambda_1 \approx 1.59 \text{ and } \lambda_2 \approx 4.41 \,.$$

The general solution is given by Eq. 16.7 where $v_i^y = (\lambda_i - a)/b$,

$$v_1^y = \frac{3 - \sqrt{2} - 1}{-2} = -1 + \frac{\sqrt{2}}{2} \approx -0.29$$

$$v_2^y = \frac{3 + \sqrt{2} - 1}{-2} = -1 - \frac{\sqrt{2}}{2} \approx -1.71 \,.$$

Fig. 16.4 shows representative solutions. The fixed point at the origin is unstable.

16.3   Complex Conjugate Eigenvalues – Spirals

When the discriminant $\delta = \text{disc}A = \beta^2 - 4\gamma$ that appears under the square root sign in Eq. 16.5 is negative, solutions of the characteristic polynomial are the complex numbers

$$\lambda = \frac{\beta \pm i\sqrt{|\delta|}}{2} = \frac{\beta}{2} \pm i\frac{\sqrt{|\delta|}}{2}\,, \tag{16.14}$$

where $i = \sqrt{-1}$ and $|\delta| = 4\gamma - \beta^2 > 0$. For notational convenience, let us write the two eigenvalues as follows

$$\lambda_\pm = \mu \pm i\omega \quad \text{where} \quad \mu = \frac{\beta}{2} \quad \text{and} \quad \omega = \frac{\sqrt{|\delta|}}{2} = \frac{4\gamma - \beta^2}{2}\,. \tag{16.15}$$

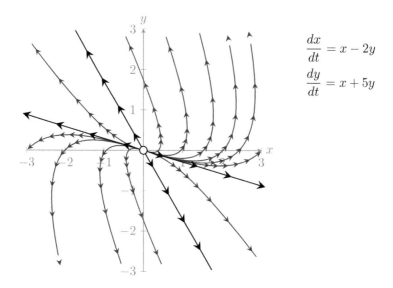

$$\frac{dx}{dt} = x - 2y$$

$$\frac{dy}{dt} = x + 5y$$

Figure 16.4   Solutions of a two-dimensional linear system with real and distinct eigenvalues that are both positive $(0 < \lambda_1 < \lambda_2)$. The fixed point at the origin is an unstable node. Parameters: $\beta = 6$, $\gamma = 7$, $\delta = 8$, $\lambda_2 = 4.41$, $v_2^y = -1.71$, $\lambda_1 = 1.59$, $v_1^y = -0.29$.

When $b \neq 0$, the general solution of $\dot{x} = Ax$ is

$$\begin{bmatrix} x(t) \\ y(t) \end{bmatrix} = c_1 e^{\mu t} \begin{bmatrix} \cos \omega t \\ v_\mu \cos \omega t - v_\omega \sin \omega t \end{bmatrix} + c_2 e^{\mu t} \begin{bmatrix} \sin \omega t \\ v_\omega \cos \omega t + v_\mu \sin \omega t \end{bmatrix} \tag{16.16}$$

where $v_\mu = (\mu - a)/b$, $v_\omega = \omega/b$, and the constants $c_1 = x_0$ and $c_2 = (y_0 - x_0 v_\mu)/v_\omega$ correspond to initial value (x_0, y_0).

As an example linear system with complex eigenvalues, consider

$$\frac{d}{dt} \begin{bmatrix} x \\ y \end{bmatrix} = \begin{bmatrix} 3 & -2 \\ 2 & 1 \end{bmatrix} \begin{bmatrix} x \\ y \end{bmatrix}. \tag{16.17}$$

The trace and determinant of the constant coefficient matrix A are

$$\beta = \text{tr}A = \text{tr} \begin{bmatrix} 3 & -2 \\ 2 & 1 \end{bmatrix} = 3 + 1 = 4$$

$$\gamma = \det A = \det \begin{bmatrix} 3 & -2 \\ 2 & 1 \end{bmatrix} = 3(1) - (-2)2 = 3 + 4 = 7,$$

and $\delta = \text{disc}A = \beta^2 - 4\gamma = (4)^2 - 4(7) = 16 - 28 = -12$. The eigenvalues are complex,

$$\lambda_\pm = \frac{4 \pm \sqrt{-12}}{2} = \frac{4}{2} \pm i\frac{\sqrt{|-12|}}{2} = 2 \pm i\sqrt{3}. \tag{16.18}$$

The general solution of Eq. 16.17 is

$$\begin{bmatrix} x(t) \\ y(t) \end{bmatrix} = c_1 e^{2t} \begin{bmatrix} \cos \sqrt{3}t \\ \frac{1+\sqrt{3}}{2} \cos \sqrt{3}t - \frac{\sqrt{3}}{2} \sin \sqrt{3}t \end{bmatrix} + c_2 e^{2t} \begin{bmatrix} \sin \sqrt{3}t \\ \frac{\sqrt{3}}{2} \cos \sqrt{3}t + \frac{1+\sqrt{3}}{2} \sin \sqrt{3}t \end{bmatrix}$$

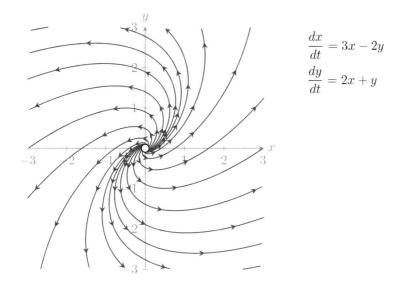

$$\frac{dx}{dt} = 3x - 2y$$

$$\frac{dy}{dt} = 2x + y$$

Figure 16.5   Solutions of a two-dimensional linear system with complex eigenvalues with positive real part ($\lambda = \mu \pm i\omega$ where $\mu > 0$). The fixed point at the origin is an unstable spiral. Parameters: $\beta = 4, \gamma = 7, \delta = -12, \mu = 2, \omega = 1.73$.

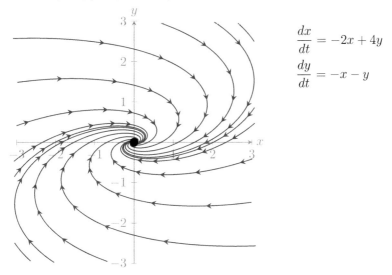

$$\frac{dx}{dt} = -2x + 4y$$

$$\frac{dy}{dt} = -x - y$$

Figure 16.6   Solutions of a two-dimensional linear system with complex eigenvalues with negative real part ($\lambda = \mu \pm i\omega$ where $\mu < 0$). The fixed point at the origin is a stable spiral. Parameters: $\beta = -3, \gamma = 6, \delta = -15, \mu = -1.5, \omega = 1.94$. *Question*: What are the eigenvalues associated with this linear system? What is the general solution?

where we have used $\mu = 2, \omega = \sqrt{3}$, and

$$v_\mu = \frac{\mu - a}{b} = \frac{\sqrt{3} - (-1)}{2} \approx 1.37 \quad \text{and} \quad v_\omega = \frac{\omega}{b} = \frac{\sqrt{3}}{2} \approx 0.87.$$

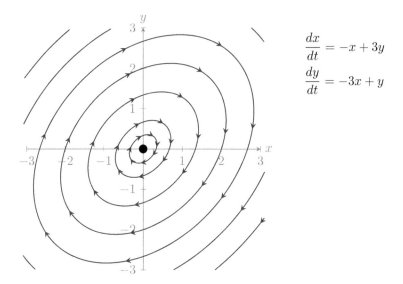

$$\frac{dx}{dt} = -x + 3y$$

$$\frac{dy}{dt} = -3x + y$$

Figure 16.7   Solutions of a two-dimensional linear system with pure imaginary eigenvalues ($\lambda = \pm i\omega$). The fixed point at the origin is a center. Parameters: $\beta = 0$, $\gamma = 8$, $\delta = -32$, $\mu = 0$, $\omega = 2.83$. *Question*: What is the general solution of this linear system?

Fig. 16.5 shows representative unstable spiral solutions for this linear system ($\mu > 0$ in Eq. 16.17). For comparison, Fig. 16.6 shows a stable spiral ($\mu < 0$) and Fig. 16.7 shows a center ($\mu = 0$).

16.4   Criterion for Stability

For a linear system with a constant coefficient matrix that has nonzero determinant ($\det A \neq 0$ in Eq. 16.1), we have observed five qualitatively different solutions.

When eigenvalues of A are real and distinct ($\lambda_1 \neq \lambda_2$; stable node, unstable node, saddle), the general solution is given by Eq. 16.6. This solution is the superposition of two solutions of the form $x_i = c_i v_i e^{\lambda_i t}$ where (λ_i, v_i) is an eigenvalue-eigenvector pair associated to A. The steady state is stable when both real-valued eigenvalues are negative ($\lambda_i < 0$).

When the eigenvalues of the constant matrix A are a complex conjugate pair ($\lambda_\pm = \mu \pm i\omega$; stable spiral, unstable spiral), the general solution is the superposition of two oscillatory solutions (Eq. 16.16). Both of these solutions include the factor $e^{\mu t}$ where $\mu = \beta/2$ is the real part of the complex conjugate pair of eigenvalues (Eq. 16.15). The steady state at the origin is stable when the complex conjugate pair of eigenvalues has negative real part ($\mu < 0$).

In both cases, the stability of the fixed point at the origin of a 2d linear system is determined by the sign of real parts of the eigenvalues of A. If we write $\Re(\lambda)$ for the real part of an eigenvalue λ, the condition for stability can be written

$$(0,0) \text{ is } \begin{cases} \text{stable} & \text{if } \Re(\lambda_i) < 0 \text{ for } i = 1,2 \\ \text{unstable} & \text{otherwise.} \end{cases} \tag{16.19}$$

This criterion is equivalent to the trace-determinant rule presented in Eq. 15.17, because the real parts of λ_1 and λ_2 are both negative precisely when $\beta = \text{tr}A = a+b < 0$ and $\gamma = \det A = ad - bc > 0$.

16.5   Further Reading and Discussion

There are many resources for further exploration of linear systems of ODEs. Puri (2009) and Strogatz (2014, Chapters 5 and 6) are highly recommended. See also Glendinning (1994), O'Malley (1997) and Thompson and Stewart (2002).

What Does $Ax = \lambda v$ Have to do with $\dot{x} = Ax$?

In Section 16.1 we stated that $\dot{x} = Ax$ is solved by $x(t) = ve^{\lambda t}$ when $Ax = \lambda v$. To see this, substitute the candidate solution into the ODE, as follows

$$\frac{d}{dt}\left[ve^{\lambda t}\right] = A\left[ve^{\lambda t}\right]$$

$$v\frac{d}{dt}\left[e^{\lambda t}\right] = Ave^{\lambda t}$$

$$ve^{\lambda t}\lambda = Ave^{\lambda t}$$

$$v\lambda = Av \implies Av = \lambda v .$$

We conclude that $\dot{x} = Ax$ has solution $x(t) = ve^{\lambda t}$ for any vector v and scalar λ that satisfy the eigenvector-eigenvalue problem $Av = \lambda v$. The trivial solution $v = 0$ corresponds to the steady state at the origin ($x(t) = 0$). To find nontrivial solutions, rearrange and factor as follows,

$$Av - \lambda v = 0 \quad \Leftrightarrow \quad \begin{bmatrix} a & b \\ c & d \end{bmatrix}\begin{bmatrix} v_x \\ v_y \end{bmatrix} - \lambda \begin{bmatrix} v_x \\ v_y \end{bmatrix} = \begin{bmatrix} 0 \\ 0 \end{bmatrix}$$

$$Av - \lambda Iv = 0 \quad \Leftrightarrow \quad \begin{bmatrix} a & b \\ c & d \end{bmatrix}\begin{bmatrix} v_x \\ v_y \end{bmatrix} - \lambda \begin{bmatrix} 1 & 0 \\ 0 & 1 \end{bmatrix}\begin{bmatrix} v_x \\ v_y \end{bmatrix} = \begin{bmatrix} 0 \\ 0 \end{bmatrix}$$

$$(A - \lambda I)v = 0 \quad \Leftrightarrow \quad \left(\begin{bmatrix} a & b \\ c & d \end{bmatrix} - \lambda \begin{bmatrix} 1 & 0 \\ 0 & 1 \end{bmatrix}\right)\begin{bmatrix} v_x \\ v_y \end{bmatrix} = \begin{bmatrix} 0 \\ 0 \end{bmatrix} .$$

Performing the matrix subtraction we obtain,

$$\begin{bmatrix} a - \lambda & b \\ c & d - \lambda \end{bmatrix}\begin{bmatrix} v_x \\ v_y \end{bmatrix} = \begin{bmatrix} 0 \\ 0 \end{bmatrix} . \tag{16.20}$$

Nontrivial solutions to Eq. 16.20 require λ to satisfy $\det (A - \lambda I) = 0$, that is,

$$\det \begin{bmatrix} a - \lambda & b \\ c & d - \lambda \end{bmatrix} = 0$$

$$(a - \lambda)(d - \lambda) - bc = 0$$

$$\vdots$$

$$\lambda^2 - \underbrace{(a+d)}_{\beta \,=\, \mathrm{tr}A}\,\lambda + \underbrace{ad - bc}_{\gamma \,=\, \det A} = 0. \tag{16.21}$$

In this way we obtain the characteristic polynomial of the matrix A (Eq. 16.4).

Satisfying the Initial Condition

The constants c_1 and c_2 appearing in the general solution to a linear system of ODEs are arbitrary, in the sense that every choice of these parameters leads to a solution. To find a solution that is associated to a given initial value, one may substitute $t = 0$, $x(0) = x_0$ and $y(0) = y_0$, into the general solution and solve for c_1 and c_2.

In the case of real and distinct eigenvalues (Eq. 16.7), substituting an initial value of (x_0, y_0) implies that

$$\begin{bmatrix} x_0 \\ y_0 \end{bmatrix} = c_1 \begin{bmatrix} v_1^x \\ v_1^y \end{bmatrix} + c_2 \begin{bmatrix} v_2^x \\ v_2^y \end{bmatrix} \quad \Leftrightarrow \quad \begin{matrix} x_0 = c_1 v_1^x + c_2 v_2^x \\ y_0 = c_1 v_1^y + c_2 v_2^y. \end{matrix}$$

This linear algebra problem has two equations with two unknowns (c_1 and c_2). By rewriting the system as follows,

$$\begin{bmatrix} v_1^x & v_2^x \\ v_1^y & v_2^y \end{bmatrix} \begin{bmatrix} c_1 \\ c_2 \end{bmatrix} = \begin{bmatrix} x_0 \\ y_0 \end{bmatrix},$$

the constants are found using Cramer's rule,

$$c_1 = \frac{\det \begin{bmatrix} x_0 & v_2^x \\ y_0 & v_2^y \end{bmatrix}}{\det \begin{bmatrix} v_1^x & v_2^x \\ v_1^y & v_2^y \end{bmatrix}} = \frac{x_0 v_2^y - v_2^x y_0}{v_1^x v_2^y - v_2^x v_1^y} \qquad c_2 = \frac{\det \begin{bmatrix} v_1^x & x_0 \\ v_1^y & y_0 \end{bmatrix}}{\det \begin{bmatrix} v_1^x & v_2^x \\ v_1^y & v_2^y \end{bmatrix}} = \frac{v_1^x y_0 - x_0 v_1^y}{v_1^x v_2^y - v_2^x v_1^y}.$$

In the case of complex conjugate eigenvalues (Eq. 16.16), the initial value of (x_0, y_0) requires

$$\begin{bmatrix} x_0 \\ y_0 \end{bmatrix} = c_1 \begin{bmatrix} 1 \\ v_\mu \end{bmatrix} + c_2 \begin{bmatrix} 0 \\ v_\omega \end{bmatrix}$$

where we have used $\cos 0 = 1$ and $\sin 0 = 0$. The initial value is satisfied by choosing $c_1 = x_0$ and $c_2 = (y_0 - x_0 v_\mu)/v_\omega$.

Straight Line Solutions

In the case of real and distinct eigenvalues, the general solution to $\dot{x} = Ax$ is the sum of two solutions, one associated to each eigenvalue λ_i (Eq. 16.6). Setting $c_2 = 0$ in Eq. 16.7, it is evident that

$$x(t) = c_1 v_1 e^{\lambda_1 t} \quad \Leftrightarrow \quad \begin{bmatrix} x(t) \\ y(t) \end{bmatrix} = c_1 \begin{bmatrix} v_1^x \\ v_1^y \end{bmatrix} e^{\lambda_1 t} \tag{16.22}$$

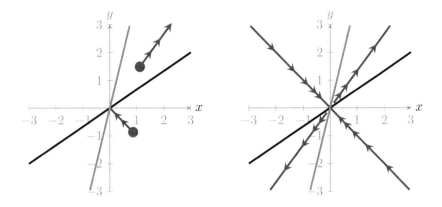

Figure 16.8 Two linearly independent straight line solutions to Eq. 16.10 given by
Eqs. 16.23 and 16.24. One solution moves toward the origin; the other moves away. Filled
circles are the points $(1, -1)$ and $(1, 4/3)$.

is a nontrivial solution to $\dot{x} = Ax$. The ratio of the two components of this solution
is the constant $y(t)/x(t) = (c_1 v_1^y e^{\lambda_1 t})/(c_1 v_1^x e^{\lambda_1 t}) = v_1^y/v_1^x$. For $c_1 \neq 0$, the trajectory is a
straight line (ray) in the phase plane that leads either toward or away from the origin
(negative or positive λ_1, respectively). This single-parameter family of **straight line
solutions** passes through the origin with slope v_1^y/v_1^x. Setting $c_1 = 0$ with $c_2 \neq 0$ gives
another straight line solution of slope v_2^y/v_2^x.

Fig. 16.8 (left) shows straight line solutions to Eq. 16.10. One corresponds to the
general solution (Eq. 16.11) with $c_1 = 1$ and $c_2 = 0$,

$$\begin{bmatrix} x(t) \\ y(t) \end{bmatrix} = \begin{bmatrix} 1 \\ 4/3 \end{bmatrix} e^{2t}. \tag{16.23}$$

The initial value is $(1, 4/3)$ and, because e^{2t} is an increasing function of time, the trajectory moves away from the origin. Setting $c_1 = 0$ and $c_2 = 1$ in the general solution
gives another straight line solution,

$$\begin{bmatrix} x(t) \\ y(t) \end{bmatrix} = \begin{bmatrix} 1 \\ -1 \end{bmatrix} e^{-5t}, \tag{16.24}$$

that begins at $(1, -1)$ and moves toward the origin.

Straight line solutions of linear systems are a distinct concept from nullclines of
linear systems (see Fig. 16.8, right). The symbols in the trace-determinant plane chosen
to represent the stable node, unstable node, and saddle are caricatures of straight line
solutions (see Fig. 16.1).

Derivation of General Solution for Complex Conjugate Eigenvalues

The 2d linear system given by $\dot{x} = Ax$ has no straight line solutions when the discriminant $\delta = \mathrm{disc} A = \beta^2 - 4\gamma$ is negative. In this case, the eigenvalues are a complex
conjugate pair (Eq. 16.14) and the general solution involves trigonometric functions

(Eq. 16.16). If one is comfortable with complex numbers, this general solution can be derived by seeking solutions of the form $x(t) = v e^{\lambda t}$ where λ is complex, and the eigenvector v is potentially complex. Using the eigenvalue with positive imaginary part ($\lambda = \mu + i\omega$), Eq. 16.9 provides the following complex-valued eigenvector,[2]

$$v = \begin{bmatrix} 1 \\ \dfrac{\lambda - a}{b} \end{bmatrix} = \begin{bmatrix} 1 \\ \dfrac{\mu + i\omega - a}{b} \end{bmatrix} = \begin{bmatrix} 1 \\ \dfrac{\mu - a}{b} + i\dfrac{\omega}{b} \end{bmatrix} = \begin{bmatrix} 1 \\ v_\mu + i v_\omega \end{bmatrix}$$

where in the last equality we define $v_\mu = (\mu - a)/b$ and $v_\omega = \omega/b$. Thus, we have a complex-valued solution given by

$$x(t) = \begin{bmatrix} 1 \\ v_\mu + i v_\omega \end{bmatrix} e^{(\mu + i\omega)t} .$$

Using $e^{(\mu + i\omega)t} = e^{\mu t} e^{i\omega t}$ and Euler's formula, $e^{i\omega t} = \cos \omega t + i \sin \omega t$,

$$x(t) = e^{\mu t} \begin{bmatrix} 1 \\ v_\mu + i v_\omega \end{bmatrix} [\cos \omega t + i \sin \omega t] .$$

Separating the real and imaginary parts of the solution,

$$x(t) = e^{\mu t} \underbrace{\begin{bmatrix} \cos \omega t \\ v_\mu \cos \omega t - v_\omega \sin \omega t \end{bmatrix}}_{x_\mu} + i e^{\mu t} \underbrace{\begin{bmatrix} \sin \omega t \\ v_\omega \cos \omega t + v_\mu \sin \omega t \end{bmatrix}}_{x_\omega} .$$

Because the real and imaginary parts of this complex-valued solution ($x = x_\mu + i x_\omega$) are linearly independent real-valued functions that satisfy $\dot{x} = Ax$, the general real-valued solution is $x = c_1 x_\mu + c_2 x_\omega$ (Eq. 16.16).

Repeated Eigenvalues: $\lambda_1 = \lambda_2$

When $\delta = \text{disc}A = 0$ the matrix A has real and repeated eigenvalues ($\lambda_1 = \lambda_2 = \beta/2$, see Eq. 16.5). In this case the general solution takes the form

$$x = c_1 v_1 e^{\lambda t} + c_2 (t v_1 + v_2) e^{\lambda t} \tag{16.25}$$

where v_1 and v_2 satisfy $Av_1 = \lambda v_1$ and $Av_2 = \lambda v_2 + v_1$, that is,

$$(A - \lambda I) v_1 = 0 \tag{16.26}$$
$$(A - \lambda I) v_2 = v_1 . \tag{16.27}$$

For example, consider the following linear system of ODEs,

$$\frac{d}{dt} \begin{bmatrix} x \\ y \end{bmatrix} = \underbrace{\begin{bmatrix} -2 & -1 \\ 1 & -4 \end{bmatrix}}_{A} \begin{bmatrix} x \\ y \end{bmatrix} . \tag{16.28}$$

In the processes of solving for the eigenvalues of A we find

$$\det (A - \lambda I) = \det \begin{bmatrix} -2 - \lambda & -1 \\ 1 & -4 - \lambda \end{bmatrix} = \cdots = (\lambda + 3)^2 .$$

We conclude that $\lambda = -3$ is a repeated eigenvalue. Using Eq. 16.26 to solve for v_1 gives

$$\begin{bmatrix} -2-\lambda & -1 \\ 1 & -4-\lambda \end{bmatrix}\begin{bmatrix} v_1^x \\ v_1^y \end{bmatrix} = \begin{bmatrix} 1 & -1 \\ 1 & -1 \end{bmatrix}\begin{bmatrix} v_1^x \\ v_1^y \end{bmatrix} = \begin{bmatrix} 0 \\ 0 \end{bmatrix} \Rightarrow v_1^x - v_1^y = 0.$$

This is an underdetermined system. To proceed, we set $v_1^x = 1$ to yield $v_1^y = 1$ and the eigenvector

$$v_1 = \begin{bmatrix} 1 \\ 1 \end{bmatrix}.$$

Using Eq. 16.27 to solve for v_2 gives

$$\begin{bmatrix} 1 & -1 \\ 1 & -1 \end{bmatrix}\begin{bmatrix} v_2^x \\ v_2^y \end{bmatrix} = \begin{bmatrix} 1 \\ 1 \end{bmatrix} \Rightarrow v_2^x - v_2^y = 1.$$

Using $v_2^x = 1$ we get $v_2^y = 0$, so

$$v_2 = \begin{bmatrix} 1 \\ 0 \end{bmatrix}.$$

The general solution given by Eq. 16.25 is thus,

$$\begin{bmatrix} x(t) \\ y(t) \end{bmatrix} = c_1 \begin{bmatrix} 1 \\ 1 \end{bmatrix} e^{-3t} + c_2 \left(t\begin{bmatrix} 1 \\ 1 \end{bmatrix} + \begin{bmatrix} 1 \\ 0 \end{bmatrix} \right) e^{-3t}, \tag{16.29}$$

that is, $x(t) = c_1 e^{-3t} + c_2 (t+1) e^{-3t}$ and $y(t) = c_1 e^{-3t} + c_2 t e^{-3t}$. Fig. 16.9 shows representative trajectories near this stable *improper* node. There is one straight line solution obtained by setting $c_2 = 0$ in Eq. 16.29 (it has slope 1).

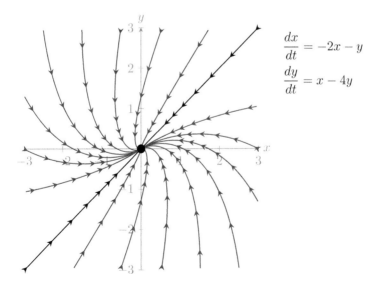

$$\frac{dx}{dt} = -2x - y$$

$$\frac{dy}{dt} = x - 4y$$

Figure 16.9 Solutions of a two-dimensional linear system with a repeated real eigenvalue that is negative ($\lambda_1 = \lambda_2 = -3$). The fixed point at the origin is an *improper* stable node. The general solution is given by Eq. 16.25 with $v_1 = (1,1)$ and $v_2 = (1,0)$.

Supplemental Problems

Problem 16.1   Calculate the trace, determinant, eigenvalues and eigenvectors of the following 2×2 matrices.

(a) $\begin{bmatrix} 0 & -1 \\ 1 & 0 \end{bmatrix}$ (b) $\begin{bmatrix} 1 & -1 \\ 1 & 1 \end{bmatrix}$ (c) $\begin{bmatrix} -2 & 1 \\ -3 & 1 \end{bmatrix}$ (d) $\begin{bmatrix} 5 & 1 \\ 0 & 7 \end{bmatrix}$

Problem 16.2   Show that for a 2×2 matrix A, the eigenvalues sum to $\operatorname{tr}A$, while the product of the eigenvalues is $\det A$. Is this true for a 3×3 matrix?

Problem 16.3   Show that the constants in Eq. 16.11 that correspond to $(x_0, y_0) = (-0.5, 2.0)$ are $c_1 = -8/7$ and $c_2 = 9/14$.

Problem 16.4   Find the general solution of the 2d linear system

$$\frac{dx}{dt} = 3x - y$$

$$\frac{dy}{dt} = 6x - 4y.$$

Sketch a graph of the flow in the (x, y)-plane based on your calculation of the eigenvalues λ_1 and λ_2 and the associated eigenvectors v_1 and v_2.

Problem 16.5   Sketch the phase portrait for each of the following systems of linear equations. Use the methods of this chapter to analytically classify the equilibrium at the origin $(0, 0)$. Check your work using an online 2d linear system solver.

(a) $\dfrac{dx}{dt} = 2x + y$

$\dfrac{dy}{dt} = x + 2y$

(b) $\dfrac{dx}{dt} = -4x - 2y$

$\dfrac{dy}{dt} = 3x - y$

(c) $\dfrac{dx}{dt} = 2x + y$

$\dfrac{dy}{dt} = x - 2y$

(d) $\dfrac{dx}{dt} = x - 4y$

$\dfrac{dy}{dt} = x + y$

Problem 16.6   Show that the eigenvalues of $A = \begin{bmatrix} a & 0 \\ c & d \end{bmatrix}$ are a and d.

Problem 16.7   Show that Eq. 16.29 solves Eq. 16.28.

Problem 16.8   The diagrams below are complex planes with imaginary and real axes denoted by $\Im$ and $\Re$, respectively. The dots represent eigenvalues of 2d linear systems, e.g., top left is a complex conjugate pair with $\Re(\lambda_i) > 0$. Determine the proper location for each diagram on the trace-determinant plane (Fig. 16.1).

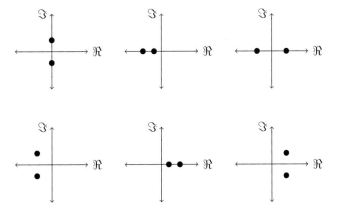

Solutions

Figure 16.6 This linear system is a stable spiral, so we expect complex conjugate eigenvalues with negative real part. In matrix form, the linear system is

$$\frac{d}{dt}\begin{bmatrix} x \\ y \end{bmatrix} = \begin{bmatrix} -2 & 4 \\ -1 & -1 \end{bmatrix} \begin{bmatrix} x \\ y \end{bmatrix}.$$

The trace and determinant of the constant coefficient matrix are $\beta = \mathrm{tr}A = -2 + (-1) = -3$ and $\gamma = \det A = -2(-1) - (4)(-1) = 2 + 4 = 6$, so the discriminant $\delta = \beta^2 - 4\gamma = (-3)^2 - 4(6) = 9 - 24 = -15$. The complex conjugate eigenvalues are

$$\lambda_\pm = \frac{-3 \pm \sqrt{-15}}{2} = -\frac{3}{2} \pm i\frac{\sqrt{15}}{2}.$$

The general solution is given by Eq. 16.16,

$$\begin{bmatrix} x(t) \\ y(t) \end{bmatrix} = c_1 e^{-\frac{3}{2}t}\begin{bmatrix} \cos\frac{\sqrt{15}}{2}t \\ \frac{1}{8}\cos\frac{\sqrt{15}}{2}t - \frac{\sqrt{15}}{8}\sin\frac{\sqrt{15}}{2}t \end{bmatrix} + c_2 e^{-\frac{3}{2}t}\begin{bmatrix} \sin\frac{\sqrt{15}}{2}t \\ \frac{\sqrt{15}}{8}\cos\frac{\sqrt{15}}{2}t + \frac{1}{8}\sin\frac{\sqrt{15}}{2}t \end{bmatrix}$$

where we have used $\mu = -3/2$, $\omega = \sqrt{15}/2$, and

$$v_\mu = \frac{\mu - a}{b} = \frac{-3/2 - (-2)}{4} = \frac{1}{8} \quad \text{and} \quad v_\omega = \frac{\omega}{b} = \frac{\sqrt{15}/2}{4} = \frac{\sqrt{15}}{8}.$$

Figure 16.7 This linear system is a center, so we expect pure imaginary eigenvalues. In matrix form, the linear system is

$$\frac{d}{dt}\begin{bmatrix} x \\ y \end{bmatrix} = \begin{bmatrix} -1 & 3 \\ -3 & 1 \end{bmatrix} \begin{bmatrix} x \\ y \end{bmatrix},$$

so $\beta = 0$, $\gamma = 8$, $\delta = \beta^2 - 4\gamma = -32$. The eigenvalues are $\lambda_\pm = \pm i2\sqrt{2}$. Using Eq. 16.16, $\mu = 0$ and $\omega = 2\sqrt{2}$, $v_\mu = (\mu - a)/b = 1/3$ and $v_\omega = \omega/b = 2\sqrt{2}/3$,

$$\begin{bmatrix} x(t) \\ y(t) \end{bmatrix} = c_1 \begin{bmatrix} \cos 2\sqrt{2}t \\ \frac{1}{3}\cos 2\sqrt{2}t - \frac{2\sqrt{2}}{3}\sin 2\sqrt{2}t \end{bmatrix} + c_2 \begin{bmatrix} \sin 2\sqrt{2}t \\ \frac{2\sqrt{2}}{3}\cos 2\sqrt{2}t + \frac{1}{3}\sin 2\sqrt{2}t \end{bmatrix}.$$

Problem 16.1 (a) $\beta = \mathrm{tr}A = 0$, $\gamma = \det A = 1$, $\delta = -4$, $\lambda = \pm i$, (b) $\beta = 2$, $\gamma = 2$, $\delta = -4$, $\lambda = 1 \pm i$, (c) $\beta = -1$, $\gamma = 1$, $\delta = -3$, $\lambda = -\frac{1}{2} \pm i\frac{\sqrt{3}}{2}$, (d) $\beta = 12$, $\gamma = 35$, $\delta = 4$, $\lambda = 5$ or 7.

Problem 16.2 Beginning with the formula for the eigenvalues of A (Eq. 16.5), the sum is

$$\lambda_1 + \lambda_2 = \frac{\beta - \sqrt{\delta}}{2} + \frac{\beta + \sqrt{\delta}}{2} = \beta = \mathrm{tr}A.$$

The product of the eigenvalues is

$$\lambda_1 \lambda_2 = \left(\frac{\beta - \sqrt{\delta}}{2}\right)\left(\frac{\beta + \sqrt{\delta}}{2}\right) = \frac{\beta^2 + \delta}{4} = \gamma = \det A$$

because $\delta = \beta^2 - 4\gamma$.

Problem 16.3 Setting $t = 0$ in Eq. 16.11 we obtain

$$\begin{bmatrix} -\frac{1}{2} \\ 2 \end{bmatrix} = c_1 \begin{bmatrix} 1 \\ -1 \end{bmatrix} + c_2 \begin{bmatrix} 1 \\ 4/3 \end{bmatrix} \quad \Leftrightarrow \quad \begin{matrix} -\frac{1}{2} = c_1 + c_2 \\ 2 = -c_1 + \frac{4}{3}c_2. \end{matrix}$$

Solving these two equations simultaneously gives $c_1 = -8/7$ and $c_2 = 9/14$.

Problem 16.6 When $b = 0$, the characteristic polynomial (Eq. 16.4) is $\lambda^2 - (a + d)\lambda + ad = (\lambda - a)(\lambda - d)$ so the eigenvalues are a and d.

Problem 16.7 The candidate solution is $x(t) = c_1 e^{-3t} + c_2 (t + 1) e^{-3t}$ and $y(t) = c_1 e^{-3t} + c_2 t e^{-3t}$. Differentiating gives

$$x'(t) = -3c_1 e^{-3t} - 3c_2 (t + 1) e^{-3t} + c_2 e^{-3t} = -3c_1 e^{-3t} - c_2 (3t + 2) e^{-3t}$$
$$y'(t) = -3c_1 e^{-3t} - 3c_2 t e^{-3t} + c_2 e^{-3t} = -3c_1 e^{-3t} - c_2 (3t - 1) e^{-3t}.$$

Then substitute these derivatives and the candidate solution into Eq. 16.28, as follows

$$-3c_1 e^{-3t} - c_2 (3t + 2) e^{-3t} = -2[c_1 e^{-3t} + c_2 (t + 1) e^{-3t}] - [c_1 e^{-3t} + c_2 t e^{-3t}]$$
$$-3c_1 e^{-3t} - c_2 (3t - 1) e^{-3t} = [c_1 e^{-3t} + c_2 (t + 1) e^{-3t}] - 4[c_1 e^{-3t} + c_2 t e^{-3t}].$$

The terms involving c_1 balance,

$$-3e^{-3t} = -2e^{-3t} - e^{-3t}$$
$$-3e^{-3t} = e^{-3t} - 4e^{-3t}.$$

Terms involving c_2 also balance,

$$- (3t + 2)\, e^{-3t} = -2\,(t + 1)\, e^{-3t} - t e^{-3t}$$
$$- (3t - 1)\, e^{-3t} = (t + 1)\, e^{-3t} - 4t e^{-3t} \,.$$

Problem 16.8

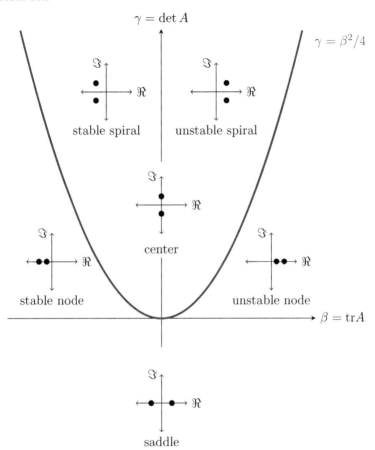

Notes

1. The quadratic formula is $(-b \pm \sqrt{b^2 - 4ac})/2$ and here $b = -\beta$ and $a = 1$ and $c = \gamma$. Note that the β is positive in Eq. 16.5 because it appears in Eq. 16.4 with a negative sign.
2. This expression uses $b \neq 0$, which is the case, because when $b = 0$ the discriminant of A is not negative and the eigenvalues of A, given by a and d, are not complex valued.

PART V
Oscillations and Bursting

17 Type II Excitability and Oscillations (Hopf Bifurcation)

The Fitzhugh-Nagumo membrane model is a minimal ODE system that exhibits bistability, excitability and oscillations. Because the model is two dimensional and the rates of change are polynomial functions of the state, it is straightforward to use phase plane and linear stability analysis to understand the dynamics of the Fitzhugh-Nagumo model. Importantly, this analysis does not necessarily assume separation of time scales. Summary bifurcation diagrams of periodic solutions show how loss of stability of a fixed point via a Hopf bifurcation may lead to repetitive spiking.

17.1 Fitzhugh-Nagumo Model

The Fitzhugh-Nagumo model of membrane excitability is the following two-dimensional ODE system,

$$\dot{x} = x - \frac{x^3}{3} - y + J \tag{17.1a}$$

$$\dot{y} = \epsilon (x + a - by) \tag{17.1b}$$

where a, b and ϵ are positive constants. The Fitzhugh-Nagumo model is simpler than the Hodgkin-Huxley model (Chapter 13), because it is a 2d (as opposed to 4d) ODE system. The Fitzhugh-Nagumo nullclines are similar to those of the Morris-Lecar model (Chapter 14). The dependent variable x in Fitzhugh-Nagumo is analogous to membrane voltage, and the ODE for x is analogous to the Morris-Lecar current balance equation (Eq. 14.2). The variable y in Fitzhugh-Nagumo is a **recovery variable** that encapsulates the dynamics of slow negative feedback. The Fitzhugh-Nagumo model exhibits behavior similar to the Morris-Lecar model, including excitability, oscillations, depolarization block, and anodal break excitation. Observe that the right sides of Eq. 17.1 are simple polynomial functions of x and y. This makes the Fitzhugh-Nagumo model a good starting point for mathematical analysis that elucidates the dynamic phenomena of excitable membranes.

Excitability

Fig. 17.1 shows solutions of the Fitzhugh-Nagumo ODEs with $a = 0.7$, $b = 0.8$, $J = 0$ and $\epsilon = 0.2$ (Eq. 17.1). In both panels, the solid curve(s) are solutions for initial values $y(0) = -0.62$ and $x(0)$ in the range -0.6 to -0.4. Observe that the action potential amplitude is *graded*, that is, the peak value of x achieved depends on the initial value $x(0)$. When plotted in the (x, y)-plane, the trajectories (blue curves) show that the action potential threshold is *soft*, i.e., there is no specific value leading to an all-or-none response (similar to the Morris-Lecar model shown in Figs. 14.3 and 14.7). As $t \to \infty$, all of the solutions asymptotically approach the steady state $(x_{ss}, y_{ss}) = (-1.20, -0.62)$ indicated by the filled circle.[1] The graded action potential amplitude

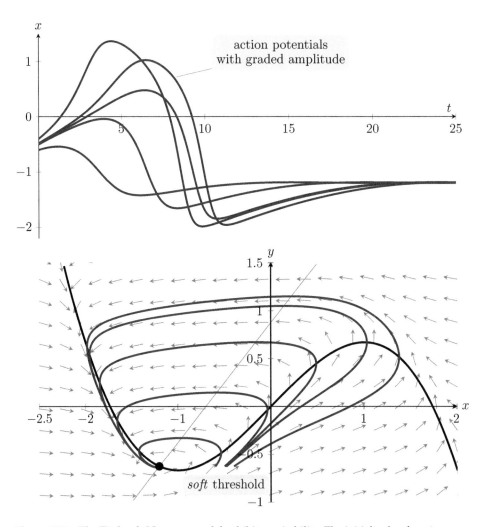

Figure 17.1 The Fitzhugh-Nagumo model exhibits excitability. The initial value for x is either -0.7 (subthreshold, dashed) or -0.4 (superthreshold, solid). Parameters: $a = 0.7$, $b = 0.8$, $J = 0$ and $\epsilon = 0.2$. Initial values for $y(0) = -0.62$. For $x(0) = -0.6, -0.5, -0.49, -0.48, -0.4$.

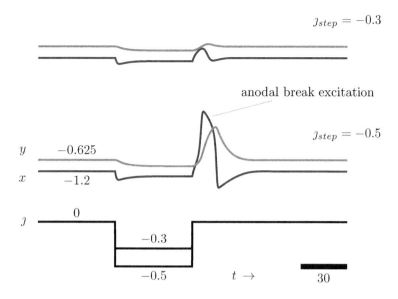

Figure 17.2   Fitzhugh-Nagumo model exhibits responses analogous to post-anodal break excitation in the Hodgkin-Huxley model (Fig. 13.10). Parameters: $\epsilon = 0.1$ and as in Fig. 17.1.

and soft threshold of Fig. 17.1 are similar to action potentials observed in the squid giant axon behavior that Hodgkin and Huxley referred to as **type II excitability** (to be contrasted with type I excitability in Chapter 18).

Anodal Break Excitation

Fig. 17.2 shows the Fitzhugh-Nagumo model exhibiting anodal break excitation. In these simulations, the parameter J (analogous to applied current) begins at 0 and is decreased to -0.3 or to -0.5 for 50 time units. In the latter case, the termination of the pulse results in a single action potential. The phase plane of Fig. 17.3 helps us understand this response. When $J = 0$, the model is at the stable fixed point (steady state). The negative value of J shifts the x nullcline downward (lower values of y). In both cases ($J = -0.3$ and -0.5), the trajectory approaches the new steady state with lower value of x and y, but only in the latter case is the slow variable y sufficiently recovered (moved far enough downward) to achieve a full post-inhibitory response.

Oscillations

Fig. 17.4 shows the response of the Fitzhugh-Nagumo model when $J(t)$ takes various positive values indicated by J_{step}. The step to 0.3 evokes an action potential. The step to 0.7 evokes repetitive action potentials (oscillations). A step to 1.5 is overstimulating (no oscillations). This sequence of responses (single action potential, oscillations, depolarization block) follows the same pattern observed in both the Morris-Lecar (Fig. 14.4) and Hodgkin-Huxley (Fig. 13.12) models.

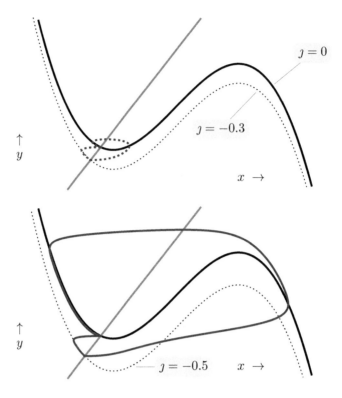

Figure 17.3   Phase planes corresponding to Fig. 17.2. *Question*: If the parameter ϵ were decreased (thereby separating the time scales of x and y even more), would the $\jmath = -0.3$ post-inhibitory response become smaller or larger?

17.2   Phase Plane Analysis of Resting Steady State

Phase plane analysis of the Fitzhugh-Nagumo model (Eq. 17.1) begins by finding algebraic equations for the x and y nullclines. Solving $0 = x - x^3/3 - y + \jmath$ and $0 = \epsilon(x + a - by)$ for y gives,

$$x \text{ nullcline:} \quad y = x - x^3/3 + \jmath \tag{17.2a}$$

$$y \text{ nullcline:} \quad y = \frac{x + a}{b}. \tag{17.2b}$$

Observe that the y nullcline is a line with slope $1/b$ and vertical intercept a/b. Because $b > 0$, the slope of the y nullcline is positive (see Fig. 17.5). The equation that defines the x nullcline is the cubic function $y(x) = x - x^3/3 + \jmath$ with local minimum and maximum (often called knees) located at $x = -1$ and $+1$ (vertical dotted lines). To see this, set the derivative $y'(x)$ to zero,

$$y'(x) = 1 - x^2 = 0 \implies x = \pm 1.$$

Substituting $x = \pm 1$ into $y(x)$ gives the knee coordinates (see Fig. 17.5). The Jacobian of Eq. 17.1 is

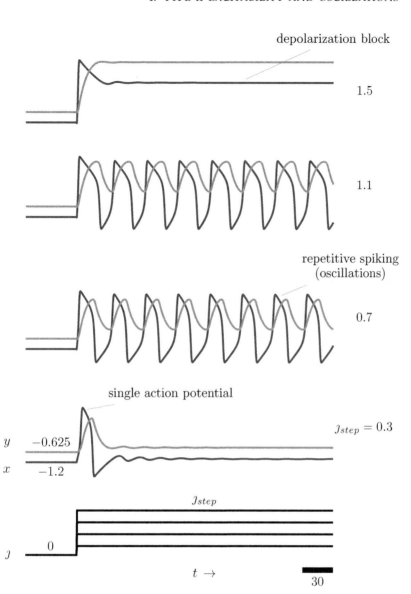

Figure 17.4 Fitzhugh-Nagumo model response to steps in the parameter J (analogous to applied current in the Hodgkin-Huxley model).

$$J(x,y) = \begin{bmatrix} \partial f/\partial x & \partial f/\partial y \\ \partial g/\partial x & \partial g/\partial y \end{bmatrix} = \begin{bmatrix} 1-x^2 & -1 \\ \epsilon & -\epsilon b \end{bmatrix}, \tag{17.3}$$

where $f(x,y) = x - x^3/3 - y + J$ and $g(x,y) = \epsilon(x + a - by)$. Observe that the Jacobian is a function of x but not y. The trace and determinant are

$$\mathrm{tr}J = 1 - x^2 - \epsilon b \tag{17.4}$$

and

$$\det J = \epsilon[1 - b(1 - x^2)]. \tag{17.5}$$

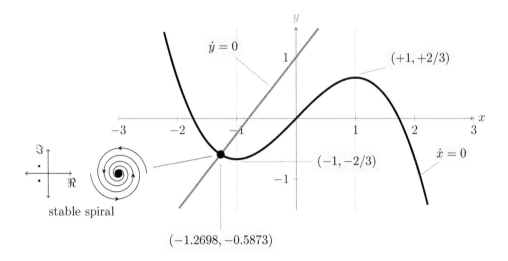

$$(-1.2698, -0.5873)$$

Figure 17.5   Phase plane for the Fitzhugh-Nagumo model (Eq. 17.1). Parameters: $a = 0.8$, $b = 0.8$, $\epsilon = 0.8$ and $\jmath = 0$. The y-axis intercept is $a/b = 1$. The steady state is located at $(x_{ss}, y_{ss}) = (-1.2698, -0.5873)$.

Fig. 17.5 plots the Fitzhugh-Nagumo nullclines using $a = b = \epsilon = 0.8$. There is one intersection located to the left of the x nullcline's left knee ($x_{ss} < -1$). The determinant of the Jacobian evaluated at this steady state is positive, because $x_{ss} < -1 \Rightarrow b(1-x_{ss}^2)$ $< 0 \Rightarrow \det J_{ss} > 0$. Substituting $\epsilon b = 0.64$ into Eq. 17.4, we see that $\mathrm{tr} J = 1 - x_{ss}^2 - \epsilon b = 0.36 - x_{ss}^2$. The trace of the Jacobian evaluated at the steady state is negative, because $x_{ss} < -1 \Rightarrow x_{ss}^2 > 1 \Rightarrow \mathrm{tr} J_{ss} = 0.36 - x_{ss}^2 < 0$. This analysis shows that the steady state is stable (Fig. 17.5, filled circle).

To classify the resting steady state as a *node* or *spiral*, we need to evaluate J_{ss}, and this requires calculation of x_{ss}. The steady state (x_{ss}, y_{ss}) is on both nullclines, so an equation for x_{ss} is obtained by equating the right sides of Eq. 17.2,

$$\frac{x_{ss} + a}{b} = x_{ss} - \frac{x_{ss}^3}{3} + \jmath \Rightarrow 0 = x_{ss}^3 + \frac{3(1-b)}{b}x_{ss} + \frac{3(a - b\jmath)}{b}.$$

Substituting $a = b = \epsilon = 0.8$ and $\jmath = 0$ gives

$$0 = x_{ss}^3 + \frac{3}{4}x_{ss} + 3.$$

An analytical solution of this cubic is unhelpful, but the Matlab command `roots([1 0 0.75 3])` gives the numerical solution $0.6349 \pm 1.3998i$ and -1.2698. Using the real-valued root, $x_{ss} = -1.2698$, we numerically evaluate the elements of the Jacobian matrix using $\epsilon = b = 0.8$ and $1 - x_{ss}^2 = -0.6124$,

$$J_{ss} = \begin{bmatrix} -0.6124 & -1 \\ 0.8 & -0.64 \end{bmatrix}. \tag{17.6}$$

Further numerical calculation gives $\mathrm{tr} J = -1.2524$, $\det J = 1.1919$, $\mathrm{disc} J = -3.1992$, and eigenvalues $\lambda = -0.6262 \pm 0.8943i$. Complex conjugate eigenvalues with negative

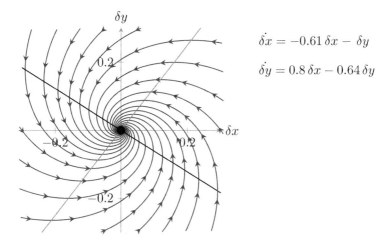

$$\dot{\delta x} = -0.61\,\delta x - \delta y$$

$$\dot{\delta y} = 0.8\,\delta x - 0.64\,\delta y$$

Figure 17.6   Solutions of the linearized Fitzhugh-Nagumo model. The deviations are $\delta x = x - x_{ss}$ and $\delta y = y - y_{ss}$ for the steady state in Fig. 17.5.

real part indicates the resting steady state is a stable spiral. Fig. 17.6 shows solutions to the linearization of the Fitzhugh-Nagumo model at this steady state.

17.3   Loss of Stability with Increasing J (Depolarization)

In the Fitzhugh-Nagumo model, the parameter J is analogous to applied current. The linear stability analysis of the previous section assumed the resting steady state (x_{ss}, y_{ss}) consistent with $J = 0$. Fig. 17.7 (top) shows that when J is increased, the x nullcline shifts vertically, because J is an additive constant in Eq. 17.2a. This changes the intersection of the nullclines and causes both x_{ss} and y_{ss} to increase. Observe that the eigenvalues of the Jacobian matrix also change, because the element of J that originates from $\partial f/\partial x$ is $1 - x_{ss}^2$. Fig. 17.7 (bottom) shows a complex plane that illustrates how the real and imaginary parts of the eigenvectors associated to J_{ss} change with increasing J and x_{ss}. As J increases from 0.7 to 0.8, the real part of this complex conjugate pair of eigenvalues changes sign (negative to positive). This corresponds to a transition from stable to unstable spirals in the linearized system. The critical value of the bifurcation parameter J is the value leading to *pure* imaginary eigenvalues (i.e., eigenvalues with zero real part). In the nonlinear system, the stability of the fixed point has also changed; this new type of bifurcation is referred to as a **supercritical Hopf**.

Bifurcation Diagram Showing Steady States and Periodic Solutions

Fig. 17.8 shows a bifurcation diagram that summarizes the Fitzhugh-Nagumo model response to increasing J (applied current). The top panel shows that the stable (solid blue) and unstable (dotted) steady states x_{ss} are a monotone increasing function of J. Loss of stability occurs at two Hopf bifurcations (blue squares). For values of J

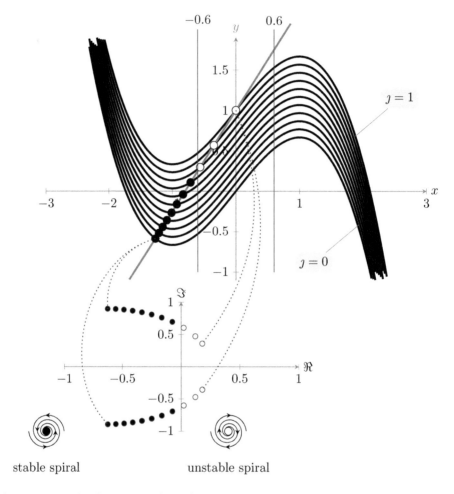

Figure 17.7 Fitzhugh-Nagumo phase plane with ten x nullclines using $0 \leq J \leq 1$. Other parameters as in Fig. 17.5.

between the two critical values (denoted J_* and J_{**}), the Fitzhugh-Nagumo model exhibits a periodic steady state (limit cycle oscillations). These oscillatory solutions are stable and the middle panels of Fig. 17.8 show how their shape changes as a function of J. The bifurcation diagram displays these periodic solutions by plotting the minimum and maximum value of x that occurs during the oscillation (black curves). Observe that the unstable steady state x_{ss} is always less than the maximum of the oscillation and greater than the minimum (middle row, open circles). Why?[2]

17.4   Analysis of Hopf Bifurcations

In the Fitzhugh-Nagumo model, depolarization-induced repetitive spiking and depolarization block occur because the 2d system passes through two different supercritical

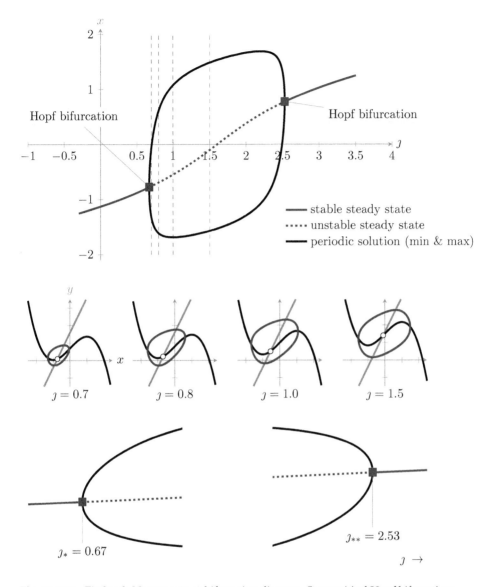

Figure 17.8 Fitzhugh-Nagumo (x, J) bifurcation diagram. Supercritical Hopf bifurcations occur at the critical values $J_* = 0.67$ and $J_{**} = 2.53$. Parameters: $a = 0.8$, $b = 0.5$, and $\epsilon = 0.8$. *Question*: Which Hopf bifurcation corresponds to depolarization block?

Hopf bifurcations as J is increased. Hopf bifurcations may be explored using the following 2d nonlinear system,[3]

$$\dot{x} = \mu x - y + \sigma x(x^2 + y^2) \tag{17.7a}$$
$$\dot{y} = x + \mu y + \sigma y(x^2 + y^2), \tag{17.7b}$$

where μ is the bifurcation parameter and $\sigma = -1$.[4] The origin is a steady state, because substituting $x_{ss} = y_{ss} = 0$ yields $\dot{x} = \dot{y} = 0$. In fact, for all values of the bifurcation parameter μ, the origin is the only fixed point (see Problem 17.3). The Jacobian of Eq. 17.7 is given by

$$J(x,y) = \begin{bmatrix} \mu + \sigma(3x^2 + y^2) & -1 + 2\sigma xy \\ 1 + 2\sigma xy & \mu + \sigma(x^2 + 3y^2) \end{bmatrix}.$$

Evaluating this Jacobian at the origin gives

$$J_{ss} = \begin{bmatrix} \mu & -1 \\ 1 & \mu \end{bmatrix}$$

with $\mathrm{tr}J_{ss} = 2\mu$ and $\det J_{ss} = 1 + \mu^2$. The determinant is positive, so the steady state will be stable provided $\mathrm{tr}J_{ss} = 2\mu < 0 \Leftrightarrow \mu < 0$.

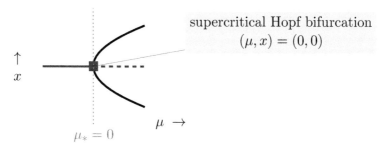

supercritical Hopf bifurcation
$(\mu, x) = (0,0)$

As illustrated above, there is a *loss of stability* of the steady state at $\mu = 0$ (the critical value for the bifurcation parameter). For $\mu < 0$, the steady state at $(x_{ss}, y_{ss}) = (0,0)$ is stable. For $\mu > 0$, this steady state is unstable. Compare this to the Hopf bifurcations in Fig. 17.8, where the critical values of J are $J_* = 0.67$ and $J_{**} = 2.53$, respectively.[5]

To explore the *periodic* solutions of Eq. 17.7 (black curve above), we will make a change of coordinate system from Cartesian (x, y) to polar (r, θ) (see Problem 17.6). This process uses the standard trigonometric relations:

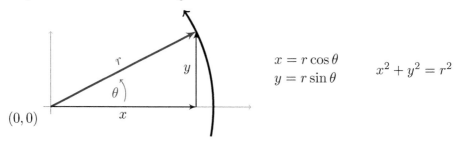

$$x = r \cos \theta$$
$$y = r \sin \theta$$

$$x^2 + y^2 = r^2$$

where the curved black arrow is the periodic solution. The *rotor* viewpoint of a Hopf bifurcation that results from this change of coordinates is

$$\dot{r} = \mu r + \sigma r^3 \qquad\qquad (17.8a)$$
$$\dot{\theta} = 1. \qquad\qquad (17.8b)$$

Observe that the radius is a nonnegative quantity ($r \geq 0$), the angle θ of rotation is periodic ($\theta + 2\pi n = \theta$ for integer n), and the ODEs are uncoupled. Eq. 17.8b is solved by $\theta(t) = t + \theta(0)$, that is, a constant rate of rotation. Below we analyze the radial coordinate (Eq. 17.8a) using the methods presented in Chapters 3 and 6.

Supercritical Hopf Bifurcation ($\sigma = -1$)

We will first consider the $\sigma = -1$ case of Eq. 17.8a,

$$\dot{r} = \mu r - r^3 \qquad r \geq 0. \tag{17.9}$$

Steady states of Eq. 17.9 solve $0 = \mu r_{ss} - r_{ss}^3 \Rightarrow 0 = \mu r_{ss} - r_{ss}^3 \Rightarrow 0 = (\mu - r_{ss}^2) r_{ss}$, that is,

$$r_{ss} = \begin{cases} 0 & \mu \leq 0 \\ 0, \sqrt{\mu} & \mu > 0. \end{cases}$$

Observe the number of steady states changes when the bifurcation parameter is at the critical value of $\mu = 0$.

Stability analysis of these steady states proceeds in the usual way. Fig. 17.9 (left column) shows phase diagrams for various values of the bifurcation parameter μ. The slope with which $\mu r - r^3$ crosses the horizontal axis shows that the steady state

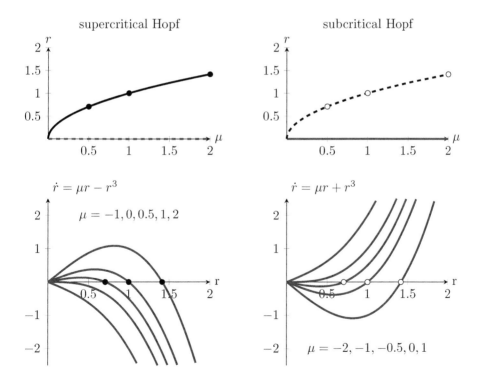

Figure 17.9 Diagrams for the supercritical and subcritical Hopf bifurcations (Eq. 17.8a). *Question*: For what values of μ is the fixed point at $r_{ss} = 0$ stable?

$r_{ss} = \sqrt{\mu}$ is stable for $\mu > 0$. This can be shown analytically by evaluating the derivative $[\mu r - r^3]' = \mu - 3r^2$ at $r_{ss} = \sqrt{\mu}$ with $\mu > 0$ to obtain $\mu - 3r_{ss}^2 = \mu - 3\mu = -2\mu < 0$ (stable). This Hopf bifurcation is referred to as *supercritical*, because the (stable) limit cycle oscillation is located *after* ($\mu > 0$) the fixed point loses stability.

Subcritical Hopf Bifurcations

The subcritical Hopf bifurcation is exhibited by Eq. 17.8a with $\sigma = 1$,

$$\dot{r} = \mu r + r^3 \qquad r \geq 0.$$

As shown below and in Fig. 17.9 (right column), the limit cycles associated with a subcritical Hopf bifurcation are *unstable*.

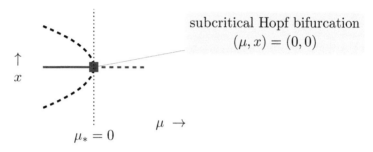

This Hopf bifurcation is referred to as *subcritical*, because the (unstable) limit cycle oscillation is located *prior* ($\mu < 0$) to the fixed point losing stability.

Stable and Unstable Limit Cycles

Fig. 17.10 shows direction fields in the Cartesian coordinates for both types of Hopf bifurcation (Eq. 17.7). Stable limit cycles are shown solid black, while unstable limit cycles are shown dashed black. Observe that the radial flow is always directed toward a stable limit cycle or stable fixed point (attractors) and away from an unstable limit cycle or fixed point (repellors). The outermost columns show caricatures of the phase portraits focusing on the fixed points and periodic solutions (both stable and unstable).

The Condition for Stability in the Fitzhugh-Nagumo Model

In the fast x/slow y limit, the Hopf bifurcation in Fig. 17.8 occurs at the value of J that leads to intersection of the x and y nullclines at the x nullcline knee. In the general case without separation of time scales, the Hopf bifurcation in Fig. 17.8 occurs when the x and y nullclines intersect somewhere between the two x nullcline knees. For what value of $x_{ss}(J)$ does loss of stability occur? Since the bifurcation is a Hopf, the transition is from stable spiral to unstable spiral; thus, in the region of interest, $\det J_{ss} > 0$ (see Fig. 16.1). Furthermore, J_{ss} has complex conjugate eigenvalues $\lambda = \mu \pm i\omega$. The condition for stability is $\Re(\lambda) < 0$, that is, $\mu < 0$. Recall that the trace of a matrix is the sum of its eigenvalues ($\operatorname{tr} J_{ss} = \lambda_1 + \lambda_2$). The sum of the complex

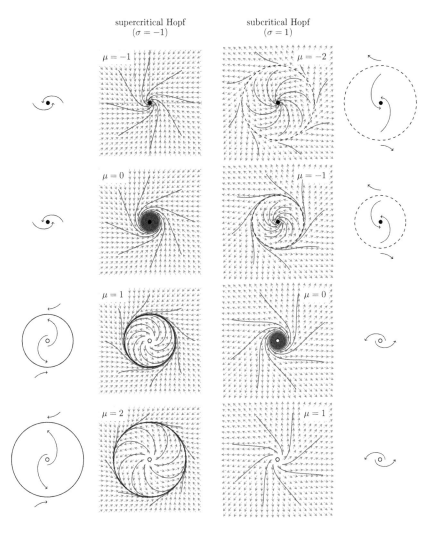

supercritical Hopf
($\sigma = -1$)

subcritical Hopf
($\sigma = 1$)

Figure 17.10   Direction fields and representative trajectories for the Hopf bifurcations exhibited by Eq. 17.7 with $\sigma = -1$ (supercritical) and $+1$ (subcritical). *Question*: Is the limit cycle observed for $\sigma = 1$ and $\mu = -1$ stable or unstable? How would you modify Eq. 17.7 to make the flow clockwise?

conjugate eigenvalues is $\mu - i\omega + \mu + i\omega = 2\mu$. This means that the stability condition for the spiral is

$$\mathrm{tr}J_{ss} = 1 - x_{ss}^2 - \epsilon b < 0 \;\Rightarrow\; x_{ss}^2 > 1 - \epsilon b \;\Rightarrow\; |x_{ss}| > \sqrt{1 - \epsilon b}.$$

In Fig. 17.8 we have used $\epsilon = b = 0.8$, so $\epsilon b = 0.64$ and the spiral is stable when

$$|x_{ss}| > \sqrt{0.36} = 0.6.$$

Among other things, this tells us that in the Fitzhugh-Nagumo model, depolarization block for $x_{ss}(J) > 0.6$ occurs for precisely the same reason as cessation of spiking when $x_{ss}(J) < -0.6$ (Fig. 17.7, vertical blue lines). As mentioned above, when time scales for

x and y are separated, the Hopf bifurcation occurs very close to the x nullcline knee, because $\sqrt{1 - \epsilon b} \to 1$ as $\epsilon \to 0$.

17.5   Limit Cycle Fold Bifurcation

In physiological models, the state variables remain finite. Consequently, the phase diagrams and direction fields for the subcritical Hopf bifurcations that show r forever increasing cannot be the entire story. Often these solutions move toward a stable limit cycle (attractor) that surrounds the unstable limit cycle (repellor). This phenomenon is exhibited by the following ODE system,

$$\dot{r} = vr - r(r^2 - 1)^2 \tag{17.10a}$$
$$\dot{\theta} = 1. \tag{17.10b}$$

This is another rotor model for which the angle θ increases at constant rate (compare Eq. 17.8). The dynamics of the radius r may be read off the phase diagrams and summary bifurcation plot in Fig. 17.11. Alternatively, you may analyze the system analytically. Setting the left side of Eq. 17.10a to zero gives

$$0 = vr_{ss} - r_{ss}(r_{ss}^2 - 1)^2 .$$

The factorization $r_{ss}[v - (r_{ss}^2 - 1)^2]$ reveals that the steady states include $r_{ss} = 0$ and solutions of $v = (r_{ss}^2 - 1)^2$. When v is negative, this auxiliary equation has no solution, so we conclude that $r_{ss} = 0$ is the only steady state. When the bifurcation parameter is nonnegative ($v \geq 0$),

$$v = (r_{ss}^2 - 1)^2 \;\Rightarrow\; \sqrt{v} = r_{ss}^2 - 1 \;\Rightarrow\; r_{ss}^2 = 1 + \sqrt{v} \;\Rightarrow\; r_{ss} = \sqrt{1 \pm \sqrt{v}},$$

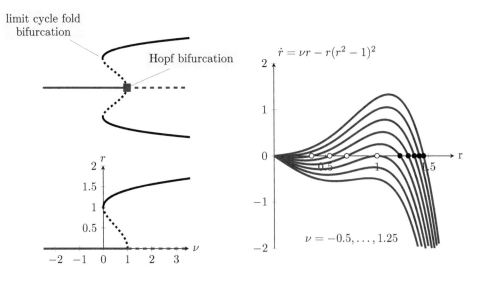

limit cycle fold bifurcation

Hopf bifurcation

$$\dot{r} = vr - r(r^2 - 1)^2$$

$v = -0.5, \ldots, 1.25$

Figure 17.11   Polar coordinate r, θ rotor model of a limit cycle fold bifurcation (top left, see Eq. 17.10). *Question*: Is the Hopf bifurcation supercritical or subcritical?

where it is understood that complex-valued solutions are nonphysical. To be clear, the real-valued solutions are,

$$
r_{ss} = \begin{cases} 0 & v \le 0 \\ 0, \sqrt{1-\sqrt{v}}, \sqrt{1+\sqrt{v}} & 0 < v < 1 \\ 0, \sqrt{1+\sqrt{v}} & 1 \le v. \end{cases}
$$

Observe the number of steady states changes when the bifurcation parameter is at the critical value of $v = 1$ (the subcritical Hopf bifurcation) and $v = 0$ (the limit cycle fold bifurcation). The reader is invited to perform the stability analysis, thereby confirming the solid and dashed lines of Fig. 17.12 (see Problem 17.5).

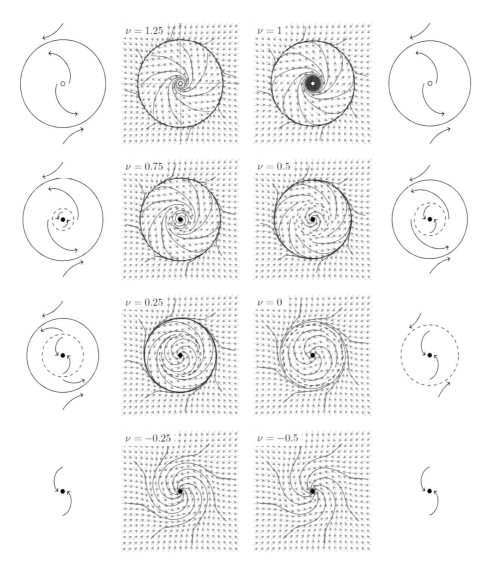

Figure 17.12   Direction fields for a subcritical Hopf and limit cycle fold bifurcation (compare to Eq. 17.11). Stable (solid) and unstable (dashed) limit cycles are shown black.

Subcritical Hopf/Limit Cycle Fold Bursting

The rotor model of Eq. 17.11 is bistable for $0 < \nu < 1$, that is, between the values leading to the subcritical Hopf and limit cycle fold bifurcations. To see the significance of this, extend the rotor model as follows,

$$\dot{r} = sr - r(r^2 - 1)^2 \tag{17.11a}$$
$$\dot{\theta} = 1 \tag{17.11b}$$
$$\text{new:} \quad \dot{s} = \epsilon(a + br^2 - s). \tag{17.11c}$$

The bottom panels of Fig. 17.13 show numerically calculated solutions to this ODE system using the parameters shown in the caption. This solution repeatedly transitions

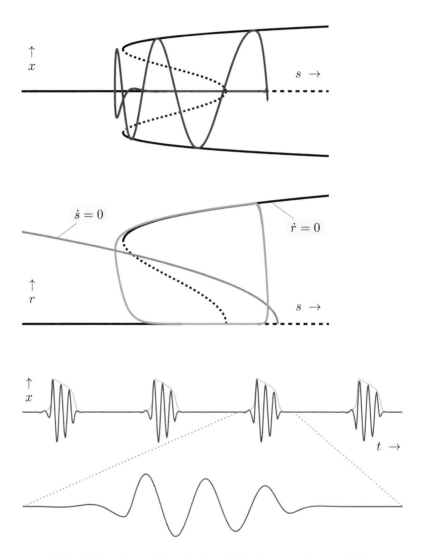

Figure 17.13 Elliptical bursting exhibited by Eq. 17.11 when s is a slow variable. Parameters: $\epsilon = 0.05$, $a = 1.5$, $b = -2$.

between an *active phase* for which there are oscillations, and a *silent phase* without oscillations. Such dynamic behavior is referred to as *bursting*.

To understand the transitions between active and silent phases, compare this system (Eq. 17.11) to the previous rotor model (Eq. 17.10). Observe that the dynamic variable s enters into the ODE for r in precisely the same way as the bifurcation parameter v had done. In both cases, the ODE for θ is uncoupled from the rest of the system. Here the dynamics of s are a function of both s and r. To analyze the dynamics, plot the r and s nullclines. From our work in Section 17.5, we see that the r nullcline is given by the real-valued solutions to $r_{ss} = \sqrt{1 \pm \sqrt{s}}$. Using Eq. 17.11c we find the nullcline for s is

$$s \text{ nullcline}: \quad s = a + br^2.$$

The s and r nullclines are shown in Fig. 17.13 (middle panel) using $a = 3/2$ and $b = -2$. A periodic solution of this model obtained when $\epsilon = 0.005$ is shown in blue. The limit cycle oscillation takes a familiar form, consistent with the time scale separation between r (fast) and s (slow). When the x component of the solution (given by $x = r\cos\theta$) is plotted in the (x, s) phase plane, the distinction between the active (x oscillating) and silent ($x \approx 0$) phases is emphasized (blue curves). This complicated trajectory can be understood by comparing it to the (x, s) bifurcation diagram (black curves) calculated using Eqs. 17.11a and 17.11b viewing s as a bifurcation parameter. Fast/slow analysis of this kind gives insight into the dynamics of bursting in neurons (action potential generation, a phase of inactivity, repeat). The particular bursting pattern shown in Fig. 17.13 is referred to as *subcritical Hopf/limit cycle fold* bursting by Izhikevich (2000), because these are the bifurcations that occur in the fast r, θ subsystem when s is viewed as a bifurcation parameter. This type of bursting is also referred to as *elliptical* bursting, because the envelope of the spikes is elliptical in shape (Rinzel, 1987; Rinzel and Ermentrout, 1989).

17.6   Further Reading and Discussion

A good reading on the Fitzhugh-Nagumo model and excitable systems is Segel and Edelstein-Keshet (2013, Chapter 11). For more on supercritical and subcritical Hopf bifurcations see Glendinning (1994, pp. 225–243), Jordan and Smith (1999, pp. 437–439) and Strogatz (2014). For more on the geometrical approach to neuronal dynamics see these pedagogical articles (Terman, 2005; Borisyuk and Rinzel, 2005; Friedman et al., 2005) and monographs (Izhikevich, 2007; Ermentrout and Terman, 2010).

Subcritical Hopf and Limit Cycle Fold in the Morris-Lecar Model

In Chapter 14, the emergence of oscillations in the Morris-Lecar model was discussed under the V fast/w slow assumption. Using the phase plane and linear stability analysis tools introduced in Chapters 15 and 16, we will revisit the emergence of oscillations in the Morris-Lecar model without assuming separation of time scales.

Fig. 17.14 shows a numerically computed (I_{app}, V) bifurcation diagram for this more general case of the Morris-Lecar model. Observe that there is a critical value of

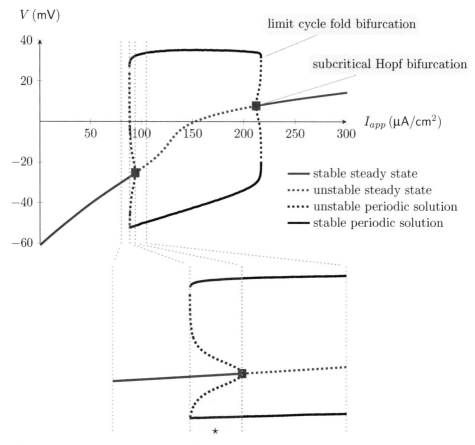

Figure 17.14   Morris-Lecar (I_{app}, V) bifurcation diagram showing supercritical Hopf and limit cycle fold bifurcations. Parameters: $a = 0.8$, $b = 0.5$, and $\epsilon = 0.8$. *Question*: Which branch of steady states is high conductance? Note bistability for the value of I_{app} indicated by the star ($\star$).

applied current at which unstable limit cycle oscillations emerge through a subcritical Hopf bifurcation. For I_{app} slightly less depolarized, there is a limit cycle fold bifurcation at which stable and unstable oscillations coalesce.

The loss of stability at the subcritical Hopf bifurcation in the Morris-Lecar model can be analyzed in much the same way as we approached the supercritical Hopf bifurcation in the Fitzhugh-Nagumo model. Following Rinzel and Ermentrout (1989), write the Morris-Lecar system as follows,

$$\dot{V} = f(V, w) = -\frac{1}{C}\left[I_{app} - I_{ion}(V, w)\right]$$

$$\dot{w} = g(V, w) = -\frac{w - w_\infty(V)}{\tau_w(V)}.$$

A steady state (V_{ss}, w_{ss}) satisfies

$$I_{app} = I_{ion}(V_{ss}, w_{ss}) \quad \text{and} \quad w_{ss} = w_\infty(V_{ss}).$$

By definition, the Jacobian of this system is

$$J(V,w) = \begin{bmatrix} \partial f/\partial V & \partial f/\partial w \\ \partial g/\partial V & \partial g/\partial w \end{bmatrix}.$$

Calculating partial derivatives such as $\partial g/\partial w = -1/\tau_w(V)$, and evaluating these at steady state, observe that the Jacobian for the linearized Morris-Lecar model takes the form,

$$J_{ss} = \begin{bmatrix} -\dfrac{1}{C}\dfrac{\partial I_{ion}}{\partial V} & -\dfrac{1}{C}\dfrac{\partial I_{ion}}{\partial w} \\ \dfrac{1}{\tau_w}\dfrac{dw_\infty}{dV} & -\dfrac{1}{\tau_w} \end{bmatrix}_{ss}. \tag{17.12}$$

Near the Hopf bifurcation, solutions of the linearized system spiral around the fixed point (V_{ss}, w_{ss}); consequently, we know J_{ss} has complex conjugate eigenvalues $\lambda = \mu \pm i\omega$. The condition for stability of the spiral is $\mu < 0$ (real part of eigenvalues negative). Because the sum of the eigenvalues is $\mathrm{tr}J_{ss} = \lambda_1 + \lambda_2 = 2\mu$, the condition for stability of a spiral is $[-C^{-1}\partial I_{ion}/\partial V - 1/\tau_w]_{ss} < 0$ or, equivalently,

$$\left[\frac{1}{C}\frac{\partial I_{ion}}{\partial V} + \frac{1}{\tau_w} \right]_{ss} > 0. \tag{17.13}$$

The second term in this expression is positive, because $\tau_w(V) > 0$ for all V including V_{ss}. Thus, a sufficient condition for an unstable spiral associated to the Hopf bifurcation is

$$\frac{1}{C}\frac{\partial I_{ion}}{\partial V}\bigg|_{ss} > 0 \Rightarrow \text{stable spiral}.$$

That is to say, the spiral will certainly be stable if the steady state is located outside the knees of the voltage nullcline. A necessary condition for the spiral to be unstable is

$$\text{unstable spiral} \Rightarrow \frac{1}{C}\frac{\partial I_{ion}}{\partial V}\bigg|_{ss} < 0.$$

That is, for an unstable spiral, depolarization from the steady state (V_{ss}, w_{ss}) must recruit inward current.

Supplemental Problems

Problem 17.1   Consider Fitzhugh-Nagumo equations given by Eq. 17.1. Show that oscillations in the V fast/w slow version of this system arise via a Hopf bifurcation. Is the Hopf supercritical or subcritical? As $_J$ increases, does y_{ss} increase or decrease? For what range of values of x_{ss} is the steady state unstable?

Problem 17.2   Near the rest point, the slopes of the nullclines in Fig. 17.5 are equal to those of the linearized system (Fig. 17.6). Why is this the case?

Problem 17.3   Using Cartesian rather than polar coordinates, show that the origin is the *only* steady state of Eq. 17.7.

Problem 17.4   Observe that the bifurcation diagrams for the supercritical and subcritical Hopf bifurcations (as on pp. 306 and 308) resemble the supercritical and subcritical pitchfork bifurcations (Figs. 6.5 and 6.6). How are these diagrams similar? What are the distinctions between Hopf and pitchfork bifurcations that these diagrams do not emphasize?

Problem 17.5   Perform stability analysis on the fixed points and periodic solutions of the limit cycle fold bifurcation system (Eq. 17.10).

Problem 17.6   Using the trigonometric relations on p. 306, transform Eq. 17.7 into Eq. 17.8. Hint: Differentiate with respect to time.

Problem 17.7   Consider the supercritical Hopf bifurcation of Fig. 17.10 with $\mu = 0$. The dark blue near the origin indicates that the asymptotic approach of $r \to 0$ as $t \to \infty$ is slow. What is responsible for this?

Problem 17.8   Observe the bistability in Fig. 17.14 for the value of I_{app} indicated by $\star$. Which direction field(s) in Fig. 17.12 are qualitatively similar to the (V, w) direction field for this value of applied current?

Problem 17.9   Using the parameters of Fig. 17.14, show that the determinant of Eq. 17.12 is positive. This is a necessary, but not sufficient, condition for stability of (V_{ss}, w_{ss}).

Problem 17.10   Recall that τ_w is proportional to the scale parameter $\bar{\tau}_m$. Observe that Eq. 17.13 $\to$ $C^{-1}\partial I_{ion}/\partial V|_{ss}$ $=$ 0 as $\bar{\tau}_m$ $\to$ ∞. What does this signify?

Solutions

Figure 17.3   The response would be larger, because the trajectory (directed more horizontally) could escape the lower knee of the x nullcline.

Figure 17.8   The critical value J_{**}.

Figure 17.9   For the supercritical Hopf, the fixed point at $r_{ss} = 0$ is stable for $\mu \leq 0$. For the subcritical Hopf, the fixed point at $r_{ss} = 0$ is stable for $\mu > 0$.

Figure 17.10   This limit cycle is unstable.

Figure 17.11   Subcritical.

Problem 17.3   Steady states of Eq. 17.7 are given by simultaneous solution of

$$0 = \mu x_{ss} - y_{ss} + \sigma x_{ss}(x_{ss}^2 + y_{ss}^2)$$
$$0 = x_{ss} + \mu y_{ss} + \sigma y_{ss}(x_{ss}^2 + y_{ss}^2).$$

Write this system in the form

$$y_{ss} = x_{ss}[\mu + \sigma(x_{ss}^2 + y_{ss}^2)]$$
$$0 = x_{ss} + y_{ss}[\mu + \sigma(x_{ss}^2 + y_{ss}^2)]$$

and substitute the first equation for y_{ss} into the equation for x_{ss} to obtain

$$0 = x_{ss} + x_{ss}[\mu + \sigma(x_{ss}^2 + y_{ss}^2)]^2 \;\Rightarrow\; 0 = x_{ss}(1 + [\mu + \sigma(x_{ss}^2 + y_{ss}^2)]^2).$$

Because the factor in parentheses is greater than or equal to 1, the steady state $x_{ss} = 0$, and this implies $y_{ss} = 0$. Thus, for all μ, Eq. 17.7 has one steady state at the origin $(0,0)$.

Problem 17.5   Differentiating the right side of Eq. 17.10 with respect to r gives $[vr - r(r^2 - 1)^2]' = v - (r^2 - 1)^2 - 2r^2(r^2 - 1)$. Evaluating at $r_{ss} = 0$ gives $v - 1$; thus, this fixed point is stable for $v < 1$. Evaluating at $r_{ss} = \sqrt{1 - \sqrt{v}}$ for $0 < v < 1$ gives $2(1 - \sqrt{v})\sqrt{v} > 0$ (unstable). Evaluating at $r_{ss} = \sqrt{1 + \sqrt{v}}$ for $v > 0$ gives $-2(1 + \sqrt{v})\sqrt{v} < 0$ (stable).

Problem 17.6   Differentiating $x = r\cos\theta$ and $y = r\sin\theta$ gives $\dot{x} = \dot{r}\cos\theta - \dot{\theta}r\sin\theta$ and $\dot{y} = \dot{r}\sin\theta + \dot{\theta}r\cos\theta$. Substituting into Eq. 17.7 gives

$$\dot{r}\cos\theta - \dot{\theta}r\sin\theta = \mu r\cos\theta - r\sin\theta + \sigma r^3\cos\theta$$
$$\dot{r}\sin\theta + \dot{\theta}r\cos\theta = r\cos\theta + \mu r\sin\theta + \sigma r^3\sin\theta.$$

Equating terms for the factors of $\cos\theta$ (or $\sin\theta$) and θ gives Eq. 17.8.

Problem 17.8 $v = 0.25, 0.5$ and 0.75.

Problem 17.9   With parameters as in Fig. 17.14, $\partial I_{ion}/\partial V|_{ss} > 0$. To see this, exchange the vertical and horizontal axes and remember that $I_{ion}(V_{ss}, w_{ss}) = I_{app}$ where $w_{ss} = w_\infty(V_{ss})$. The determinant is

$$\det J_{ss} = \left[\frac{1}{C}\frac{\partial I_{ion}}{\partial V} \cdot \frac{1}{\tau_w} + \frac{1}{C}\frac{\partial I_{ion}}{\partial w} \cdot \frac{1}{\tau_w}\frac{dw_\infty}{dV}\right]_{ss}.$$

Because $C > 0$, $\tau_w(V_{ss}) > 0$, and the gating function w_∞ is a monotone increasing function of voltage $(dw_\infty/dV > 0)$, both terms of the Jacobian are positive; thus, $\det J_{ss} > 0$.

Notes

1. $y_{ss} = -0.625$ for the Fitzhugh-Nagumo model at rest $(J = 0)$. Evidently, y should not be interpreted as a fraction of open channels (a nonnegative quantity), as we did with w (the fraction of open K_V channels) in the Morris-Lecar model.
2. In the (x, y) phase plane, the unstable steady state (x_{ss}, y_{ss}) is located within the closed curve of the limit cycle; consequently $x(t)$ takes values both greater and less than x_{ss}.
3. The normal form for a 2d Hopf bifurcation.

4. The constant $\sigma = -1$ is introduced in Eq. 17.7 so we may easily compare our analysis to the case when $\sigma = 1$. These two cases correspond to supercritical and subcritical Hopf bifurcations, respectively.

5. The left-right orientation of the Hopf bifurcation on the bifurcation diagram is not relevant to its classification. After all, this orientation can be reversed by replacing $_J$ with $-_J$ in Eq. 17.1a.

18 Type I Excitability and Oscillations (SNIC and SHO Bifurcations)

The Hopf bifurcation is important to cellular neurophysiology because this bifurcation is the explanation for the emergence of periodic solutions (repetitive action potentials) *in some cases*. However, the Hopf bifurcation is not the only way that depolarization of excitable membranes may lead to repetitive spiking. Two others include the saddle-node on an invariant circle (SNIC) bifurcation and the saddle homoclinic orbit (SHO) bifurcation.

18.1 Saddle-Node on an Invariant Circle

Consider the following modification of the Fitzhugh-Nagumo model,

$$\dot{x} = x - x^3/3 - y + J \tag{18.1a}$$

new: $\quad \dot{y} = \epsilon(x + a - by - y^3),$ $\tag{18.1b}$

in which the equation for the recovery variable y is a cubic function (compare to Eq. 17.1). The nullclines of this modified Fitzhugh-Nagumo model are

$$x \text{ nullcline:} \quad y = x - x^3/3 + J \tag{18.2a}$$

new: $\quad y \text{ nullcline:} \quad x = y^3 + by - a.$ $\tag{18.2b}$

Fig. 18.1 (bottom panel) plots these nullclines and the direction field for Eq. 18.1 in the (x, y)-plane using the parameters in the caption. Observe that the nullclines intersect *three times*; thus, there are three fixed points. One of these is stable, and the other two are unstable. Using XPPAUT to numerically calculate eigenvalues for these singular points, we find

(x_{ss}, y_{ss})	λ_1	λ_2	Classification
$(-1.18, -1.13)$	-1.61	-0.80	stable node
$(-0.44, -0.91)$	-1.08	0.54	saddle
$(0.49, -0.05)$	$0.34 - i0.56$	$0.34 + i0.56$	unstable spiral

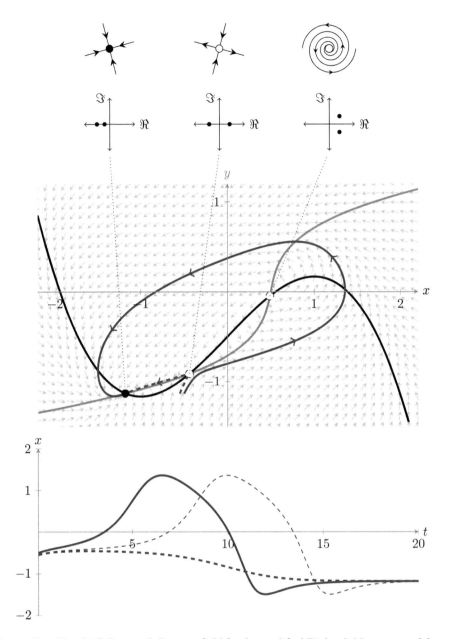

Figure 18.1   Top: Nullclines and direction field for the modified Fitzhugh-Nagumo model (Eq. 18.1). Parameters: $\epsilon = 0.5$, $a = -0.5$, $b = 0.2$, $\jmath = -0.5$. Bottom: Superthreshold (solid) and subthreshold (dotted) responses are obtained using slightly different initial values (see text).

Fig. 18.1 also plots three trajectories with different initial conditions. The superthreshold responses use $x(0) = -0.50$ and $x(0) = -0.54$, while the subthreshold response uses $x(0) = -0.55$ (all use $y(0) = -1.13$). Responses that are just slightly superthreshold remain full sized, although the peak response may be delayed (thin blue curve). The threshold is *sharp*, in contrast to the soft thresholds of Chapter 17.

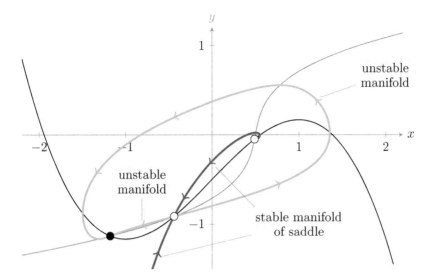

Figure 18.2   Nullclines for the modified Fitzhugh-Nagumo model (Eq. 18.1) are shown black and gray. The stable manifold of the saddle is shown red. The unstable manifold of the saddle is shown green. Parameters: $\epsilon = 0.5$, $a = -0.5$, $b = 0.2$, $j = -0.5$. *Question*: How many heteroclinic trajectories are shown?

This sharp threshold is a feature of **type I excitability** that can be contrasted with the soft threshold of type II excitability (Fig. 17.1). The explanation for this phenomenon is illustrated in Fig. 18.2, which plots several trajectories that either begin or end at the saddle point (green and red curves). These special solutions limit on the saddle point in the direction of the eigenvectors associated to the system linearized at the saddle. The **unstable manifold of the saddle** (green) has two branches that are similar to the subthreshold and superthreshold responses of Fig. 18.1. One of these is a **heteroclinic** trajectory that is directed leftward and connects the saddle point to the stable node (short path). The other branch of the unstable manifold of the saddle is a heteroclinic trajectory that is directed rightward at first, but which takes the longer route to the stable node that loops around the unstable spiral.

The **stable manifold of the saddle** (red) also has two branches. The first branch is a heteroclinic connecting the unstable spiral and the saddle. The second branch limits on the saddle point from below. The sharp threshold phenomenon associated with type I excitability occurs because solutions that begin to the right of the stable manifold of the saddle (and below/outside the unstable manifold) must remain outside the unstable manifold of the saddle as the flow carries the solution to the rest point (stable node).

As the parameter j is increased from -0.5 to -0.3, a saddle-node bifurcation occurs that gives rise to a limit cycle oscillation (Fig. 18.3, black). Because this saddle-node bifurcation occurs on the **invariant set** that is topologically a circle, this type of bifurcation is referred to as a **saddle-node on an invariant circle** or SNIC bifurcation. It is also sometimes called a *saddle-node infinite period* bifurcation or SNIPER bifurcation.

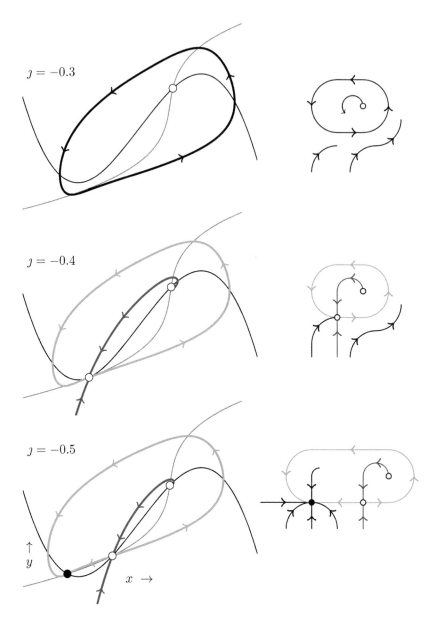

Figure 18.3   Left: Phase planes for the modified Fitzhugh-Nagumo model that are pre-critical, critical, and post-critical to a saddle-node on an invariant circle bifurcation. Parameters as in Fig. 18.2. *Question*: What is the critical value of $\jmath$? Right: Diagrammatic representation of the SNIC bifurcation.

This second name refers to the fact that the oscillation that emerges at the critical value of the bifurcation parameter is a **homoclinic** trajectory that takes an infinite amount of time to traverse (i.e., the flow begins and ends asymptotically). When post-critical, the limit cycle oscillation is very low frequency (long period) near the *ghost* of this saddle-node bifurcation. As the bifurcation parameter moves to the critical value from above, the oscillation can have an arbitrarily long period (see Fig. 18.4)

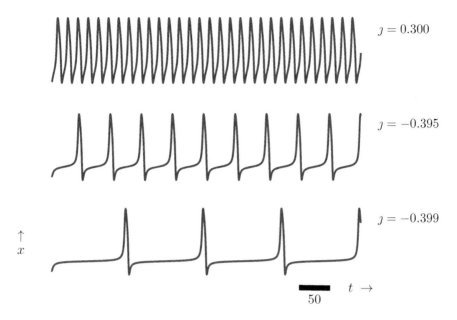

$$J = 0.300$$

$$J = -0.395$$

$$J = -0.399$$

x

$t \rightarrow$

50

Figure 18.4 Type I oscillations exhibited by the modified Fitzhugh-Nagumo model may occur with arbitrarily low frequency.

Fig. 18.5 shows the bifurcation diagram corresponding to the sequence of phase planes in Fig. 18.2. Note the three steady states (one stable, two unstable) when the bifurcation is pre-critical (−), and the stable oscillations when post-critical (+). The saddle-node bifurcation is located at the critical value of the bifurcation parameter (⋆), and this is precisely where the periodic solutions begin.

18.2   Saddle Homoclinic Bifurcation

Fig. 18.6 shows (x, y) phase planes for the modified Fitzhugh-Nagumo model with the parameter ϵ increased from 0.5 to 1.3 (less separation of time scales). Comparing Figs. 18.3 and 18.6, one may confirm that this change does not affect the shape of the nullclines, which still intersect three times at a stable node, saddle and unstable spiral (ϵ does not appear in Eq. 18.2). Of course, the relative location of the nullclines changes with J. In particular, more negative values of J lower the x nullcline and increase the distance between the stable node and saddle. However, there is a profound change in the location(s) of the stable (red) and unstable (green) manifolds of the saddle.

The three panels of Fig. 18.6 show a **saddle homoclinic orbit bifurcation**. For the critical value of J (SHO, middle panel), there is a homoclinic trajectory (orbit) that is part of the unstable manifold of the saddle (green) as well as the stable manifold (red). In the pre-critical phase plane (−, bottom), the stable and unstable manifolds of the saddle do not coincide; rather, the stable manifold includes a heteroclinic trajectory that connects the unstable spiral to the saddle. Both branches of the unstable manifold of the saddle (green) are heteroclinic orbits that limit onto the stable node.

Fig. 18.7 shows the bifurcation diagram corresponding to the saddle homoclinic bifurcation. When the bifurcation parameter is pre-critical ($J < -0.5096$), there are

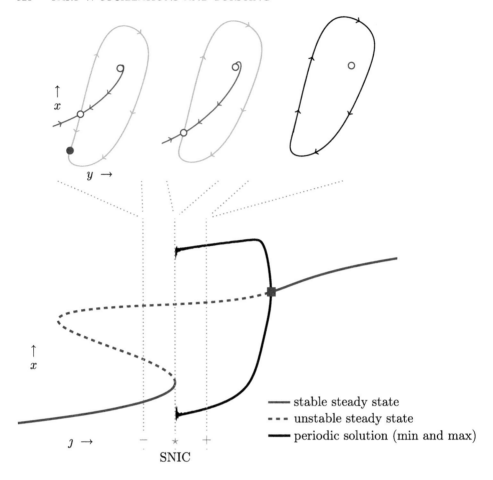

Figure 18.5   Fitzhugh-Nagumo $(x, \jmath)$ bifurcation diagram illustrating the saddle-node on an invariant circle bifurcation (SNIC).

three steady states (one stable, two unstable, see vertical line labelled −). When $\jmath$ is post-critical, the three steady states remain, but stable limit cycle oscillations are also observed (vertical line labelled +). This bifurcation diagram indicates that, as $\jmath$ decreases, the periodic orbits terminate on an unstable steady state located at ⋆ (the phase planes show this is the saddle). Although not our primary interest at this time, the diagram also shows another SHO bifurcation, a saddle-node bifurcation, and two Hopf bifurcations. Comparing Figs. 18.7 and 18.5, one sees that the adjustment made to the parameter ϵ had no effect on the steady state current-voltage relation, which is consistent with the lack of effect on the nullclines mentioned above.

18.3   Square-Wave Bursting

One salient feature of Fig. 18.7 is the bistability observed when $\jmath$ is post-critical to the SHO bifurcation, but pre-critical to the saddle-node bifurcation. In this parameter

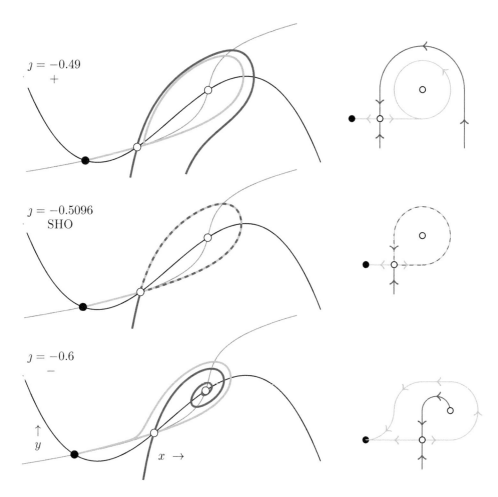

Figure 18.6 Phase planes for the modified Fitzhugh-Nagumo model that are pre-critical (−), critical (SHO) and post-critical (+) of a saddle homoclinic bifurcation. Parameters: $a = -0.5, b = 0.2, \epsilon = 1.3$.

regime, the unstable fixed points (blue dashed curve) serve as a threshold separating the stable rest point (solid blue) and the stable limit cycle oscillations (black). As in Fig. 17.13, bistability allows the introduction of a slow variable to create a model that exhibits bursting, as follows:

$$\text{fast subsystem:} \begin{cases} \dot{x} = x - x^3/3 - y + z \\ \dot{y} = \epsilon(x + a - by - y^3) \end{cases} \tag{18.3a}$$

$$\text{slow variable:} \quad \dot{z} = \sigma[c(z-h)^3 + d - x]. \tag{18.3b}$$

Observe that the equation for x has been modified; the dynamic variable z takes the place of $\jmath$. The x and y nullclines of this modified Fitzhugh-Nagumo model are given by Eq. 18.2 with the replacement $z \leftrightarrow \jmath$. The z nullcline is

$$z \text{ nullcline:} \quad x = c(z-h)^3 + d. \tag{18.4}$$

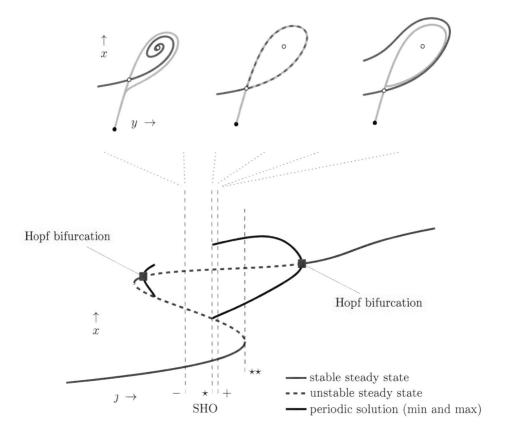

Figure 18.7   The (x, J) bifurcation diagram for the modified Fitzhugh-Nagumo model includes a saddle homoclinic bifurcation (SHO). *Question*: Can you sketch a phase portrait corresponding to the saddle-node bifurcation located at ⋆⋆?

Fig. 18.8 plots this z nullcline (gray) on top of the (z, x) bifurcation diagram for the fast subsystem (black, Eq. 18.3a). Observe that the parameters c, d and h have been chosen so that the z nullcline intersects *unstable* fixed points in the bistable regime of the fast (x, y) subsystem. The periodic solutions that terminate on the SHO bifurcation are above the z nullcline, while the saddle-node bifurcation is below the z nullcline. This configuration leads to bursting that is similar to the subHopf/fold cycle bursting of Chapter 17, in that the bistability and hysteresis of the fast subsystem is playing an important role, but different in that the bifurcations at the beginning and end of the active phase are a saddle-node and a SHO (compare Fig. 18.8 to 17.13).

The active phases of the bursts shown in Fig. 18.8 have four spikes each. For comparison, Fig. 18.9 shows a similar calculation using five-fold smaller value of σ. Because the time scales of the (x, y) fast subsystem and the slow variable z have been further separated, there are more spikes per burst (about 15 in Fig. 18.9).

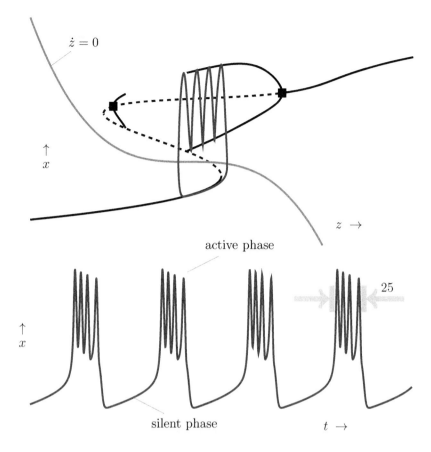

Figure 18.8 Square-wave bursting results from the relationship of the $(x, \jmath)$ bifurcation diagram for the fast subsystem and the nullcline for the slow variable z. Parameters: $\epsilon = 1.3$, $a = -0.5, b = 0.2, c = -20, d = -0.6, h = -0.5$ and $\sigma = 0.005$.

Fig. 18.10 was obtained from Fig. 18.9 by making an adjustment of one of the parameters for the z nullcline (d decreased from -0.6 to -1.1). The full three variable system that results is no longer oscillatory, but it is excitable. One might describe the neuron as *primed to burst* in response to depolarization. The filled black circle at the intersection of the z nullcline and the branch of hyperpolarized fixed points of the (x, y) subsystem is a stable fixed point. From this rest state, a rapid increase in x (voltage) has one of two outcomes depending on whether x is (or is not) increased beyond the unstable steady state that exists in the fast (x, y) subsystem for the resting value of z. A superthreshold depolarization causes x to enter the basin of attraction of the periodic solutions. This corresponds to a burst of action potentials. During the burst, the slow variable z decreases until the burst terminates via a saddle homoclinic bifurcation. After termination of spiking, the voltage drops to the only attractor available when z is precritical to the SHO. These fixed points are slightly more hyperpolarized than the

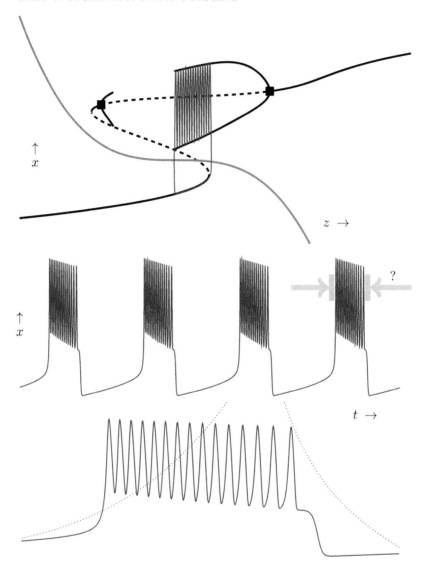

Figure 18.9 Square-wave bursting. *Question*: The only difference in parameters between this figure and Fig. 18.8 is a five-fold change in σ. Was this change in σ a decrease or an increase? In Fig. 18.8 the active phase had duration 25. What is the duration of the active phase here?

rest point. The slow variable z then increases to its resting value and, as it does so, x follows the stable fixed points associated to each value of z.

18.4   Calcium-Activated Potassium Currents as Slow Variable

In both elliptical and square-wave bursting, the nullcline for the slow variable is positioned so that, during the active phase, the slow variable moves in the direction that is

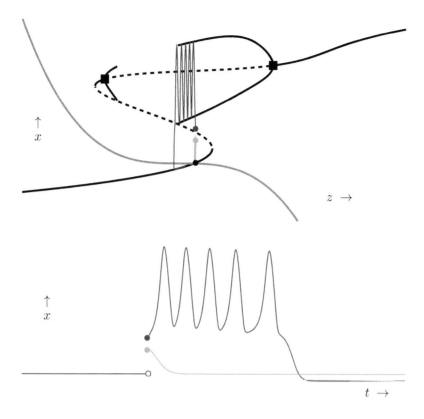

Figure 18.10   Bifurcation diagram for the fast subsystem (black) and slow variable z (gray) for Eq. 18.3 with $d = -1.1$. The full three variable system is excitable and primed to burst in response to depolarization.

hyperpolarizing (leading to cessation of spiking). Similarly, during the silent phase, the slow variable moves in the direction that is depolarizing (eventually initiating spiking).

What physiological mechanisms might correspond to the dynamics of the slow variable z? One possibility is that z represents the activation gating variable for an inward current, for example, a persistent sodium current that supplements Na_V and K_V currents responsible for the periodic solution of the fast subsystem. The bifurcation diagram for such a neuron would look like Figs. 18.8 and 18.9. Increasing z would correspond to the recruitment of inward current, and the (stable) steady state voltage would be an increasing function of z. For bursting to occur, the persistent sodium current would be required to slowly increase when the membrane was hyperpolarized, and slowly decrease when the membrane was depolarized (repetitive spiking).

Rhythmic bursting may also occur via slow activation of an outward current during the active phase. If s were the activation gating variable for such an outward current, the bifurcation diagram of the fast subsystem would look something like:

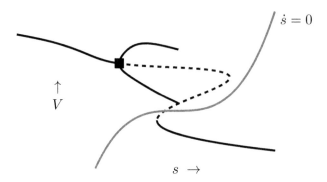

because increasing s is hyperpolarizing ($\partial V/\partial s < 0$ on stable branches of the steady state current-voltage relation).

Many neurons express high-voltage activated calcium channels in addition to the transient $\mathrm{Na_V}$ current that is primarily responsible for the upstroke of the action potential. As a consequence, repetitive action potentials increase the calcium influx rate. Elevated intracellular calcium may recruit outward (hyperpolarizing) calcium-activated potassium current (I_{KCa}) eventually terminating spiking. This was the physiological mechanism for electrical bursting in insulin-secreting pancreatic β-cells hypothesized by Chay and Keizer (1983, 1985). Though the story is now known to be more complex (Bertram and Sherman, 2005), the Chay-Keizer model is of interest for the physiological plausibility of the slow variable dynamics,[1]

$$\text{fast:} \begin{cases} C\dot{V} = I_{app} - I_{\mathrm{Ca_V}}(V) - I_{\mathrm{K_V}}(V,w) - I_{\mathrm{KCa}}(V,c) - I_L(V) \\ \dot{w} = -[w - w_\infty(V)]/\tau_w(V) \end{cases} \tag{18.5a}$$

$$\text{slow:} \quad \dot{c} = \sigma\left[-\alpha I_{\mathrm{Ca_V}}(V) - kc\right]. \tag{18.5b}$$

The current balance equation includes four ionic membrane currents. Action potentials are mediated by an inward regenerative $\mathrm{Ca_V}$ current and delayed rectifying outward $\mathrm{K_V}$ current; the gating variables m_∞ and w_∞ are increasing sigmoids as in the Morris-Lecar model. The calcium-activated potassium current is given by

$$I_{\mathrm{KCa}} = \underbrace{\bar{g}_{\mathrm{KCa}}\frac{c^n}{c^n + \kappa^n}}_{g_{\mathrm{KCa}}}(V - E_{\mathrm{K}}).$$

Observe the Hill-type factor with coefficient n that could reflect rapid sequential binding steps necessary for regulation of the $\mathrm{K_{Ca}}$ current by intracellular calcium. For low calcium concentration, the potassium conductance is small ($g_{\mathrm{KCa}} \to 0$ as $c \to 0$). For high calcium concentration, the potassium conductance achieves its maximum ($g_{\mathrm{KCa}} \to \bar{g}_{\mathrm{KCa}}$ as $c \to \infty$).

The calcium influx term in Eq. 18.5b is $-\alpha I_{\mathrm{Ca_V}}(V)$ where α and σ are positive parameters. The negative sign converts an inward current $I_{\mathrm{Ca_V}} < 0$ into a positive material flux ($\dot{c} > 0$). The calcium concentration c is a slow variable in part because $\sigma = 0.01$ is a small dimensionless parameter. This parameter accounts for buffering of intracellular calcium by high concentration, low affinity calcium binding proteins that are ubiquitous in the cell cytoplasm (e.g., calmodulin). That is, only a small fraction of the calcium that enters the cell is *free* intracellular calcium; most (99%) becomes bound to intracellular proteins.

In the Chay-Keizer model, the calcium influx rate during repetitive spiking is greater than the efflux rate kc. This increases the intracellular calcium concentration that activates K_{Ca} channels, increases outward current, and eventually hyperpolarizes the membrane enough to terminate spiking.

A similar mechanism of slow negative feedback mediated through intracellular calcium may be responsible for **spike frequency adaptation** of *regular spiking* (as opposed to *fast spiking*) pyramidal neurons in neocortex and hippocampus (Wang, 1998; Ermentrout, 1998b; Liu and Wang, 2001; Nowak et al., 2003). When sufficiently stimulated to respond repetitively, these neurons exhibit a gradual decrease in spike frequency mediated by calcium influx and subsequent activation of outward K_{Ca} current. In some neurons, this mechanism leads to a distinctive hyperpolarization after every action potential, in which case the responsible K_{Ca} currents are sometimes denoted I_{AHP} (**a**fter-**h**yper**p**olarization).

18.5   Further Reading and Discussion

Good papers to read on bursting include Hindmarsh and Rose (1984), Chay and Keizer (1985) and Rinzel and Ermentrout (1989). An excellent edited volume focusing on bursting and rhythm generation in the central nervous system is Coombes and Bressloff (2005). For further discussion on classification of bursting mechanisms see Rinzel (1987) and Izhikevich (2000).

Comparison of Elliptical and Square-Wave Bursting

The square-wave burster of Fig. 18.9 works because (1) there is a hysteresis loop generated by bistability in the fast (x, y) subsystem (stable fixed point, unstable fixed point, stable periodic orbit) and (2) the nullcline for the slow variable z is located between the two branches of stable fixed points of the fast subsystem. Elliptical bursting is mechanistically similar (Fig. 17.13), but the bistability in the fast subsystem involves a stable fixed point, unstable periodic orbit, and stable periodic orbit.

For both square-wave and elliptical bursters, the period of the bursting oscillation depends on the time scale of the slow variable. If the slow variable is made even slower, the bursting period increases due to an increase in the time spent in both active and silent phases. The proportion of the oscillation period that is active (the *duty cycle*) does not change much.

The bistability and hysteresis loop in the fast subsystems of square-wave and elliptical bursters leads to similar phase resetting behavior (not shown). In both cases, if one stimulates in a superthreshold manner during the silent phase, the solution moves upward beyond the unstable fixed point and into the attracting stable periodic orbit. The slow variable, which was increasing during the silent phase, immediately begins to decrease. Thus, premature ending of the silent phase shortens the subsequent active phase (Tabak et al., 2001). This means that the number of spikes observed in a burst that is evoked by stimulation during the silent phase will be fewer than that observed during rhythmic bursting.

Although square-wave and elliptical bursters both involve bistability and hysteresis, the properties of the respective fast subsystems are sufficiently different that these types of bursting are easily distinguished. Signatures of a square-wave burster include the following. (1) The minimum of voltage oscillations is greater than the voltage during the silent phase. (2) The minimum of the oscillation decreases during the active phase. (3) Oscillations gradually slow (decrease in frequency) during the active phase. Signatures of elliptical bursting include the following. (1) The minimum of voltage oscillations is less than the voltage during the silent phase. (2) The oscillation magnitude first increases and then decreases during the active phase. (3) The oscillation frequency does not change much during the active phase.

Local Versus Global Bifurcations

This chapter introduced two new bifurcations: the saddle-node on an invariant circle (SNIC) and the saddle homoclinic (SHO). These two bifurcations, as well as the limit cycle fold bifurcation, are examples of *global bifurcations*. The bifurcations we encountered in previous chapters were all *local bifurcations*:

> **local bifurcations:**   saddle-node, transcritical, pitchfork, Hopf, ...
> **global bifurcations:**   limit cycle fold, SNIC, SHO, ...

Local bifurcations are those that may be analyzed by "zooming in" to the bifurcation point where the number and/or stability of fixed points changes. More formally, a bifurcation is *local* when the Jacobian evaluated at the steady state associated to the critical value of the bifurcation parameter has an eigenvalue with zero real part. For example, the Hopf bifurcation analyzed in Fig. 17.7 has a Jacobian with pure imaginary eigenvalues when $J = -0.6$. Because $\Re(\lambda_i) = 0$ for this critical value of J, this is a *local* bifurcation. The saddle-node bifurcation,

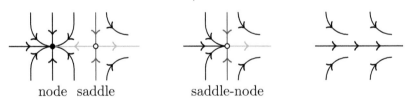

is a local bifurcation, because the critical saddle-node has two real eigenvalues, one of which is zero (in the illustration above $\lambda_1 < 0 = \lambda_2$).

Analysis of the saddle homoclinic and SNIC bifurcations requires more mathematical machinery, because these global bifurcations involve the overall flow in the phase plane. In both cases, there is a critical homoclinic trajectory with a distinctive form. In the saddle homoclinic bifurcation, neither eigenvalue is zero at the critical value of the bifurcation parameter; rather, $\lambda_1 < 0 < \lambda_2$. In the saddle-node on an invariant circle bifurcation, the critical saddle-node has a zero eigenvalue, but the *invariant circle* is an aspect of the overall flow throughout the phase plane, so the bifurcation is global. Is the heteroclinic bifurcation illustrated below a local or global bifurcation?

Comparison of Type I and Type II Dynamics

The type I excitability and oscillations discussed in this chapter may be contrasted with the type II excitability and oscillations (Chapter 17). In type I membranes, the resting neuron is subthreshold to a global bifurcation (SNIC or SHO). In type II membranes, the resting neuron is subthreshold to a local bifurcation (Hopf). This fundamental difference leads to distinct response properties when the neuron is stimulated either briefly (resulting in excitation) or in a prolonged manner (resulting in oscillations).

type	bifurcation	– oscillation – amplitude	period	– excitability – amplitude	threshold
I	SHO/SNIC	finite	infinite	all-or-none	hard
II	HB	zero	finite	graded	soft

The input/output properties of resting neurons also depend on whether they are subthreshold to a global (type I) or local (type II) bifurcation. Type I neurons respond to synaptic input as leaky *integrators*, while type II neurons are *resonators* (see Izhikevich, 2000 for discussion).

What Can Happen in the Phase Plane? What Cannot?

Beginning with Figs. 18.2 and 18.6, the flow of solutions in the phase plane was sometimes illustrated without nullclines. Rather, fixed points, limit cycles, and homoclinic and heteroclinic trajectories were emphasized (e.g., stable and unstable manifolds of saddle points). Neuroscientists with an understanding of dynamics often communicate to one another by sketching diagrams such as these. To my knowledge, there is no agreed upon standard for these drawings. For example, all three diagrams below represent a phase plane subthreshold to a SNIC bifurcation:

 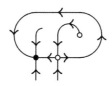 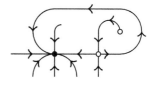

The rightmost diagram is the most elaborate. There is an unstable spiral, saddle, and stable node. The flow near the node indicates that the eigenvalue associated to vertical movement is more negative than the eigenvalue associated to horizontal movement ($\lambda_{vert} < \lambda_{horiz} < 0$). The leftmost diagram is minimal. It does not make explicit the fact that the unstable source is a spiral, but this "untwisting" is consistent with a topological, rather than geometric, presentation of the phase portrait. For another example, here are three sketches of the critical phase portrait of a SNIC:

Each shows the defining feature of a SNIC, a saddle-node homoclinic trajectory (i.e., a saddle-node loop).

Here are phase portraits pre-critical, critical and post-critical to a saddle homoclinic bifurcation (Fig. 18.6):

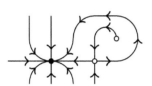 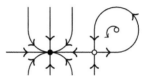 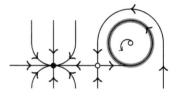

When sketching a phase portrait, the following is recommended.

(1)   Include all fixed points and periodic orbits. (You may wish to indicate stability using filled and open circles for fixed points, solid and dashed closed curves for limit cycles.)

(2)   Draw the stable and unstable manifolds of any saddle points.

(3)   Ensure every trajectory, including homoclinics and heteroclinics, moves away from a repellor and toward an attractor. (Solutions that grow without bound are okay.)

(4)   Require solutions moving nearby an unstable manifold to approach it tangentially.

(5)   Ensure the direction field that is implicit in your sketch changes *continuously*. Abrupt changes are not allowed, as these would indicate something important is missing from the diagram.

(6)   When drawing trajectories, always consider the time-reversed flow where attractors become repellors, sources become sinks, and vice versa. The above rules must still be satisfied.

As you practice drawing phase portraits according to these rules, you will become convinced of certain necessities that are well known to dynamicists. For example, a stable limit cycle must have something unstable within. This can be an unstable fixed point (usually a spiral) or an unstable limit cycle.

You may also convince yourself that the critical phase portraits we have shown in this chapter must exist if there is to be a continuous transition from the pre-critical to post-critical phase portraits. For example, the critical heteroclinic orbit on p. 332 is required if the phase portrait on the left is to change continuously to the phase portrait on the right. A similar argument can be made about the critical homoclinic (saddle loop) below:

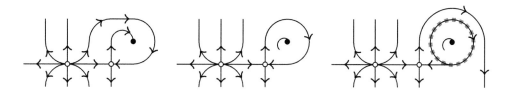

Supplemental Problems

Problem 18.1 Consider the fast/slow version of Eq. 18.1 obtained as $\epsilon \to 0$. Treating J as a bifurcation parameter, show that oscillations are not observed, but rather bistability. For what critical value(s) of J does a saddle-node bifurcation occur?

Problem 18.2 Consider the 2d ODE system

$$\dot{x} = x^2 - y$$
$$\dot{y} = x - y + \mu$$

with bifurcation parameter μ. What is the critical value of μ? Confirm that the bifurcation is local. Hint: Fig. 15.5 shows the subcritical phase plane.

Problem 18.3 Are bifurcations of one-dimensional systems local or global?

Problem 18.4 The saddle-node bifurcation illustrated on p. 332 implicitly assumes that the eigenvalues of the saddle-node are $\lambda_1 < 0 = \lambda_2$. Sketch three diagrams that represent a saddle-node bifurcation in which the saddle-node has eigenvalues $\lambda_1 = 0 < \lambda_2$.

Problem 18.5 In type I membranes, oscillations emerge with zero frequency (infinite period). In the case of the SHO, for values of the bifurcation parameter μ that are greater than, but close to, the critical value μ_*, the frequency is proportional to

$$f \propto \frac{1}{|\ln(\mu - \mu_*)|} \qquad 0 < \mu - \mu_* \ll 1.$$

For the SNIC bifurcation the frequency of oscillation is proportional to

$$f \propto \sqrt{\mu - \mu_*} \qquad 0 < \mu - \mu_* \ll 1.$$

Show that $f \to 0$ as $\mu \to \mu_*$ in both cases. What determines the oscillation frequency of post-critical type II membranes?

Problem 18.6 What are the physical dimensions of the parameter α that appears in the Chay-Keizer model (Eq. 18.5b)?

Solutions

Figure 18.2   There are three heteroclinic trajectories. Saddle leftward to node. Saddle rightward long way around to node. Unstable spiral to saddle.

Figure 18.3   The critical value of $J_* = -0.4$.

Figure 18.7

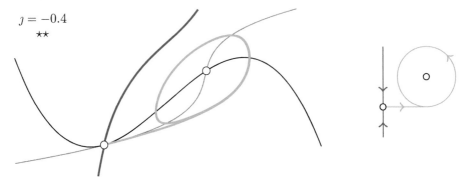

$J = -0.4$
★★

The unstable manifold of the saddle-node has only one branch (green in the middle diagram on p. 332).

Figure 18.9   A five-fold decrease, that is, $\sigma = 0.001$ in Fig. 18.9. Smaller σ slows the dynamics of z leading to more spikes in each active phase of the burst, and longer inter-burst intervals. The duration of the active phase in Fig. 18.9 is 125, five-fold longer than in Fig. 18.8.

Problem 18.2   Steady states of the ODE solve $y_{ss} = x_{ss}^2$ and $y_{ss} = x + \mu$, so

$$0 = x_{ss}^2 - x_{ss} - \mu \implies x_{ss} = \frac{1 \pm \sqrt{1 + 4\mu}}{2}.$$

The critical value of μ is the value that leads to zero discriminant, i.e., $\mu = -1/4$. For $\mu < -1/4$, there are no steady states. For $\mu > -1/4$, there are two steady states. When the Jacobian,

$$J = \begin{bmatrix} 2x & -1 \\ 1 & -1 \end{bmatrix},$$

is evaluated at the steady state associated to the critical value of μ, namely, $x_{ss} = 1/2$, we obtain

$$J_{ss} = \begin{bmatrix} 1 & -1 \\ 1 & -1 \end{bmatrix}.$$

This matrix has eigenvalues $\lambda_1 = \lambda_2 = 0$. Because J_{ss} has an eigenvalue with zero real part, the bifurcation is *local*.

Problem 18.3   Bifurcations observed in one-dimensional systems are *local*. The ODEs take the form $\dot{y} = f(y; \mu)$ and steady states solve $0 = f(y_{ss}, \mu)$. In these scalar systems, the analog of the Jacobian is $\partial f/\partial y$. Evaluating this derivative at the steady state associated to the critical value of μ yields zero for the fold, transcritical and pitchfork bifurcations. For example, $\dot{y} = f(y; \mu) = \mu y - y^2$ has a transcritical bifurcation at $\mu = 0$ for which $y_{ss} = 0$. Evaluating $\partial f/\partial y = -2y$ at this steady state gives $-2y_{ss} = 0$.

Problem 18.4 This is most easily done by reversing the arrows in each of the diagrams on p. 332, as follows:

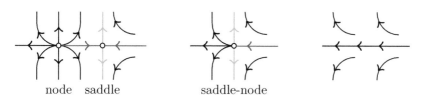

node   saddle   saddle-node

Problem 18.6 Because σ is dimensionless, the physical dimensions of αI_{Ca_V} must balance the physical dimensions of dc/dt, as follows

$$\frac{\#}{\text{length}^3 \cdot \text{time}} \ominus \frac{dc}{dt} \ominus \alpha I_{Ca_V} \ominus \frac{\#}{\text{length} \cdot \text{charge}} \cdot \frac{\text{charge/time}}{\text{length}^2}$$

where current $=$ charge/time. Thus, $\alpha \ominus \#/(\text{length} \cdot \text{charge})$. Observe that $1/\text{length} = \text{area/volume}$, and $\#/\text{charge}$ is the reciprocal of the physical dimensions of $zF \ominus \text{charge}/\#$. Thus, $\alpha = s/(vzF)$ where s and v are the surface area and volume of the cell, z is the valence of the charge carrier, and F is Faraday's constant.

Note

1. The equations of the Chay-Keizer model follow the review by Bertram and Sherman (2005). For simplicity, the ATP-sensitive potassium current denoted in their work as $I_{K(ATP)}$ is denoted I_L.

19 The Low-Threshold Calcium Spike

Some ionic currents are inactivated at resting membrane potentials. One important example is the low-threshold calcium current (I_T) that is responsible for post-inhibitory rebound bursting in thalamocortical relay neurons.

19.1 Post-Inhibitory Rebound Bursting

The intrinsic membrane properties of neurons are more complex than the previous chapters may suggest. Consider the simulated responses of the thalamocortical relay neuron presented in Fig. 19.1. When this neuron model is depolarized by superthreshold applied current, repetitive action potentials mediated by I_{Na_V} and I_{K_V} are evoked ($I_{app} = 20\,\mu A/cm^2$, top panel). Positive applied current that is subthreshold leads to a small depolarization, but no action potentials ($I_{app} = 5\,\mu A/cm^2$, middle panel). Negative applied current ($I_{app} = -4\,\mu A/cm^2$) leads to a deep hyperpolarization and – perhaps surprisingly – the termination of the hyperpolarizing current pulse is followed by a burst of five action potentials that lasts about 10 ms (bottom panel).

In thalamocortical relay neurons, this phenomenon of **post-inhibitory rebound bursting** is mediated by T-type *low-voltage activated* calcium channels (T for transient), not to be confused with the L-type (long lasting) high-voltage activated channels that mediate I_{Ca_V}. In many respects the T-current is similar to the transient inward regenerative current I_{Na_V}, and it may be modeled in a similar fashion,[1]

$$I_T = g_T\, m^3\, h\, (V - E_{Ca}),\tag{19.1}$$

where m and h are activation and inactivation gating variables with steady states that are increasing and decreasing sigmoidal functions of voltage, respectively (Fig. 19.2). However, there are important differences between I_T and I_{Na_V}. First, T-type channels are highly specific for calcium and, consequently, the reversal potential is E_{Ca}. Second, the half maxima of the activation (m_∞) and inactivation (h_∞) gating functions for I_T are about 20 mV more hyperpolarized than those of I_{Na_V} and I_{Ca_V} (Fig. 19.2). The leftward shift in m_∞ is responsible for the low threshold for activation of I_T. The leftward shift

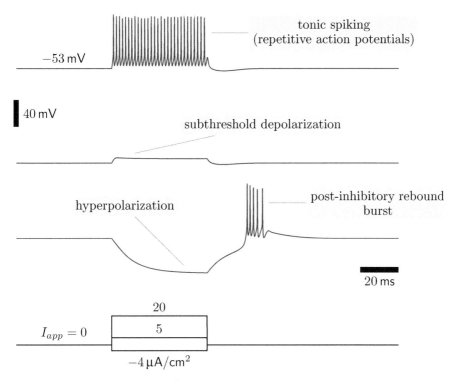

Figure 19.1 Response properties of a model thalamocortical relay neuron include tonic spiking and post-inhibitory rebound bursting. *Question:* $V_{rest} = -53$ mV for this neuron. Is the membrane resistance greater for $V > V_{rest}$ or $V < V_{rest}$?

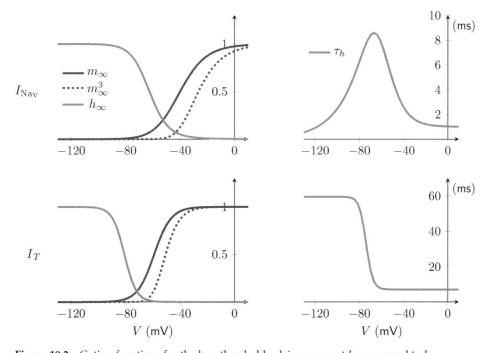

Figure 19.2 Gating functions for the low-threshold calcium current I_T compared to I_{Nav}.

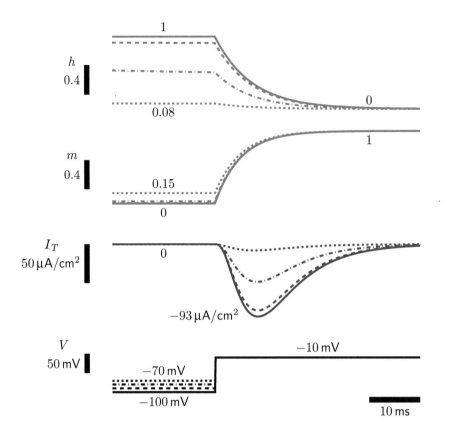

Figure 19.3   Top: Gating variables for a transient T current (Eqs. 12.11 and 12.14). Bottom: Simulated voltage clamp recordings. *Question*: Which variable, *m* or *h*, is the inactivation gating variable? The activation gating variable? Which is faster?

in h_∞ means that, unlike Na$_V$, the T current may be partly or almost entirely *inactivated* at resting membrane potentials.

To illustrate this resting inactivation of I_T, Fig. 19.3 shows simulated voltage-clamp recordings. From a holding potential of -70 mV, a depolarizing step in the command potential to -10 mV evokes little inward current (dotted lines). However, from a more hyperpolarized holding potential of -100 mV, the step to -10 mV evokes significant transient inward current that activates and then inactivates in about 50 ms (solid lines). Holding potentials in the range -70 to -100 mV lead to different amounts of evoked current because the inactivation gating variable *h* equilibrates to $h_\infty(V_{hold})$ prior to the depolarizing pulse. Fig. 19.3 (top) shows that for $V_{hold} = -70$ mV, the T current is mostly inactivated ($h = 0.08$) when the depolarizing step to -10 mV begins. Conversely, when $V_{hold} = -100$ mV, the T current is fully de-inactivated ($h = 1$) when the step begins.

Fig. 19.4 presents simulated current-clamp recordings using a model that includes action potential generating currents I_{Na_V} and I_{K_V}, a leakage current, and the low-threshold calcium current I_T (Rush and Rinzel, 1994). Observe that the number of

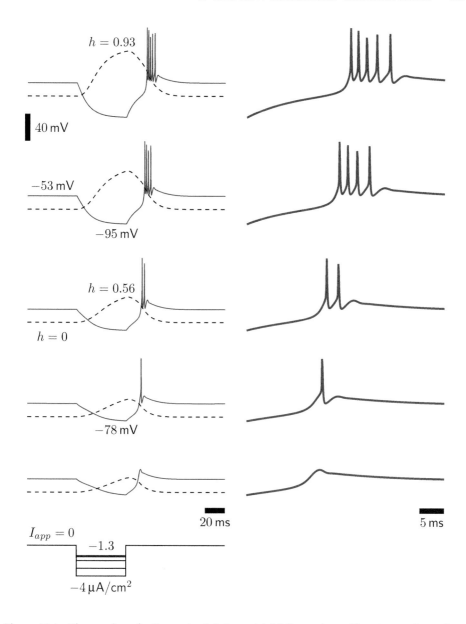

$h = 0.93$

40 mV

-53 mV

-95 mV

$h = 0.56$

$h = 0$

-78 mV

20 ms

5 ms

$I_{app} = 0$

-1.3

$-4\,\mu A/cm^2$

Figure 19.4 The number of action potentials in post-inhibitory rebound bursts may depend on the depth of preceding hyperpolarization. *Question*: Which represents more inactivation of T current, $h = 0.93$ or $h = 0.56$?

action potentials in the post-inhibitory responses depends on the depth of hyper-polarization and the level of de-inactivation upon release (larger h, more spikes). The robustness of a post-inhibitory rebound response may also depend on the duration of hyperpolarizing current pulse as compared to the de-inactivation time constant τ_h, which is about 60 ms for $V < -80$ mV (see Fig. 19.2).

19.2   Fast/Slow Analysis of Post-Inhibitory Rebound Bursting

The membrane model of post-inhibitory bursting in thalamic neurons used in Fig. 19.4 includes a minimal representation of action potential generating currents ($I_{AP} = I_{Nav} + I_{Kv}$) and the low-threshold calcium current I_T, as follows:[2]

$$\text{fast AP subsystem:} \begin{cases} C\dot{V} = I_{app} - I_{AP}(V,n) - I_T(V,h) - I_L(V) \\ \dot{n} = -[n - n_\infty(V)]/\tau_n(V) \end{cases} \tag{19.2}$$

slow I_T de-inactivation: $\dot{h} = -[h - h_\infty(V)]/\tau_h(V)$.

The black curves of Fig. 19.5 show the bifurcation structure of the fast subsystem when h, the level of de-inactivation of T current, is viewed as a bifurcation parameter. Two different values of applied current ($I_{app} = 0$ and $-2\,\mu A/cm^2$) lead to different bifurcation structure for subthreshold voltages, but the periodic solutions are similar.[3] For both values of applied current, periodic solutions arise from a SNIC bifurcation when the T current is about 30% de-inactivated ($h \approx 0.3$). The rest point of the model neuron is located at the intersection ($\star$) of the stable steady state voltages of the fast subsystem (solid black) and $h = h_\infty(V)$ (gray nullcline).

When the applied current is $-2\,\mu A/cm^2$, the stable branch of fixed points becomes hyperpolarized and an unstable branch of fixed points is located between two saddle-node bifurcations (one at $h = 0.2$, the other at a nonphysical value of h that is off the

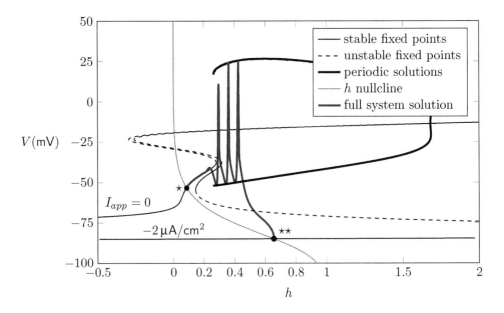

Figure 19.5 Fast/slow analysis of a post-inhibitory rebound burst that occurs after a hyperpolarizing current pulse of $I_{app} = -2\,\mu A/cm^2$. Black: Bifurcation diagram of the (V,n) fast subsystem. Gray: h nullcline. Calculation uses parameter set 1 in Rush and Rinzel (1994) and $g_T = 0.3\,mS/cm^2$. *Question*: What is the range of physical values for h? What type of bifurcation leads to termination of spiking at the end of the post-inhibitory rebound burst?

plot). For this hyperpolarizing applied current, the new rest point of the model neuron is located at $V \approx -85\,\text{mV}$ and $h = 0.65$ (★★). If the hyperpolarizing current is applied for sufficiently long time, the neuron's state will asymptotically approach the rest point ★★. When the hyperpolarizing current pulse ends, the relevant voltage nullcline instantaneously shifts upward and leftward ($I_{app} = 0$). Because V is faster than h, the trajectory moves upward into the stable periodic solutions. At these voltages, the steady state value of h is near 0; consequently, the trajectory moves leftward until the SNIC bifurcation terminates spiking. The trajectory then asymptotically approaches the stable resting point given by ★ (where h is over 90% inactivated).

19.3   Rhythmic Bursting in Response to Hyperpolarization

Fig. 19.6 shows the (V, I_{app}) bifurcation diagram for the model thalamocortical relay neuron (Eq. 19.2). This diagram shows type I dynamics in which periodic solutions arise from a SNIC bifurcation at $I_{app} = 16\,\mu\text{A}/\text{cm}^2$ and depolarization block occurs at the subcritical Hopf at $I_{app} = 97\,\mu\text{A}/\text{cm}^2$. Interestingly, the branch of fixed points also loses stability for hyperpolarizing applied current in the range -1.5 to $-0.8\,\mu\text{A}/\text{cm}^2$

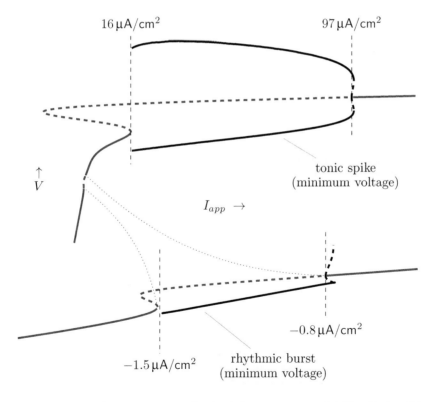

Figure 19.6 (V, I_{app}) bifurcation diagram for the thalamic neuron model. Top: Tonic spiking for positive applied current. Bottom: Rhythmic bursting mediated by the low-threshold calcium current I_T for negative applied current.

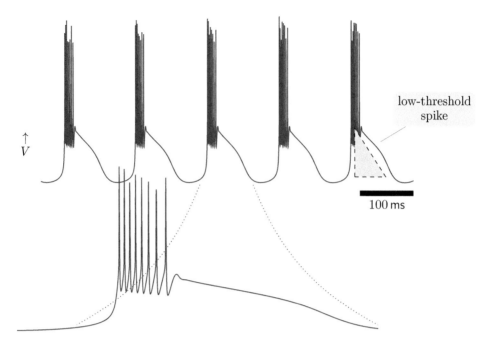

Figure 19.7   Rhythmic bursting mediated by the low-threshold calcium current I_T. The triangular waveform is the signature of the T current mediated low-threshold spike.

(see inset). In this parameter regime, stable periodic solutions emerge through a SNIC bifurcation at $-1.5\ \mu A/cm^2$ and are eliminated by a limit cycle fold bifurcation associated with a subcritical Hopf at $-0.8\ \mu A/cm^2$. Fig. 19.7 shows current-clamp simulation of the stable periodic solutions that exist for $-1.1\ \mu A/cm^2$.[4] Neuroscientists who study the thalamus refer to this type of response as *rhythmic bursting*, to distinguish it from transient post-inhibitory rebound burst responses.

19.4   Fast/Slow Analysis of Rhythmic Bursting

Fig. 19.8 shows the bifurcation diagram of the fast subsystem during rhythmic bursting, the h nullcline, and a periodic solution. Observe that the mechanism of rhythmic bursting in thalamic relay neurons is both similar and different to the square-wave bursting of the previous chapter (Fig. 18.9). In both cases, the transition from the silent phase to the active phase occurs via a saddle-node bifurcation. In both cases, the minimum silent phase voltage is more hyperpolarized than the minimum voltage during spiking. However, the transition from the active phase to silent phase occurs via a saddle homoclinic bifurcation in square-wave bursting, whereas in the model thalamic neuron spike termination via a SNIC results in a triangular waveform that is characteristic of the low-threshold calcium spike (Fig. 19.7).

Fig. 19.9 shows the dynamics of V and h when the maximum conductance of the sodium current is set to zero ($\bar{g}_{Na} = 0$). This parameter change simulates application

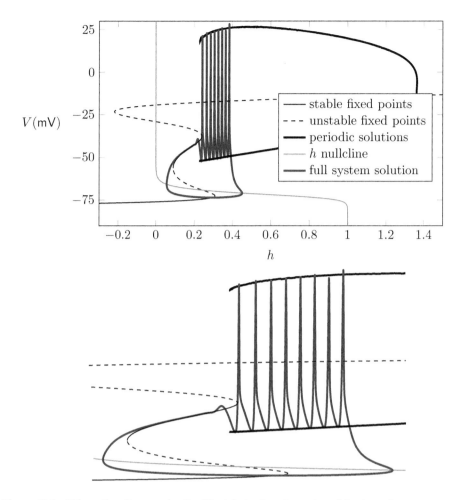

V(mV)

- stable fixed points
- - - unstable fixed points
— periodic solutions
— h nullcline
— full system solution

Figure 19.8 Bifurcation diagram for the (V, n) fast subsystem where h is viewed as a bifurcation parameter (black) and solution of the full system (blue).

of tetrodotoxin (TTX).[5] The bistability and hysteresis loop responsible for rhythmic bursting mediated by the T current is highlighted by this block of action potential generation. The greatly simplified fast subsystem has three branches of fixed points, two of which are stable, and no periodic solutions (no spike generation). The limit cycle (blue) superimposed on the (V, h) bifurcation diagram corresponds to rhythmic low-threshold calcium spikes (Fig. 19.9, middle panel). One can easily imagine adding spikes back onto the depolarized phase of the rhythmic burst labelled *active* in Fig. 19.9 (top and bottom panels).

The bottom panel of Fig. 19.9 rotates the top panel so that V is on the horizontal axis and h is on the vertical axis. This orients the h nullcline as in the plots of the T current gating functions (Fig. 19.2). Because h is a gating variable for an inward current, the stable branches (solid black) of fixed points for the fast subsystem have positive slope ($\partial V / \partial h > 0$). The relaxation oscillation shown in Fig. 19.9 (bottom) highlights a

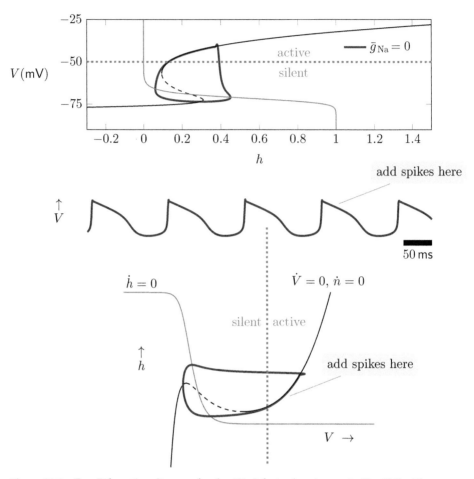

Figure 19.9 Top: Bifurcation diagram for the (V, n) fast subsystem as in Fig. 19.8 with $\bar{g}_{Na} = 0$ simulating blockade of I_{Na_V} by TTX. Middle: Periodic solution corresponding to repetitive low-threshold calcium spikes. Bottom: Diagram rotated to correspond to phase planes for a minimal model of low-threshold spike (Figs. 19.10 and 19.11).

possible mechanism of the rhythmic bursting exhibited by thalamic neurons, provided one remembers that when V is sufficiently depolarized (right of vertical dotted line) the membrane will be spiking (*active*). The *silent* phase of the rhythmic burst occurs when the voltage is not sufficiently depolarized (left of vertical dotted line).

19.5 Minimal Model of the Low-Threshold Calcium Spike

Fig. 19.9 suggests that a minimal model of the low-threshold calcium spike may be produced by eliminating the action potential generating currents (I_{K_V} and I_{Na_V}) in the thalamic relay neuron model given by Eq. 19.2 (Wang et al., 1991; Wang and Rinzel, 1992). The resulting dynamical system has only two variables,[6]

$$C\dot{V} = I_{app} - g_T\, m_\infty^2(V)\, h\, (V - E_{Ca}) - g_L(V - E_L) \tag{19.3a}$$

$$\dot{h} = -[h - h_\infty(V)]/\tau_h(V). \tag{19.3b}$$

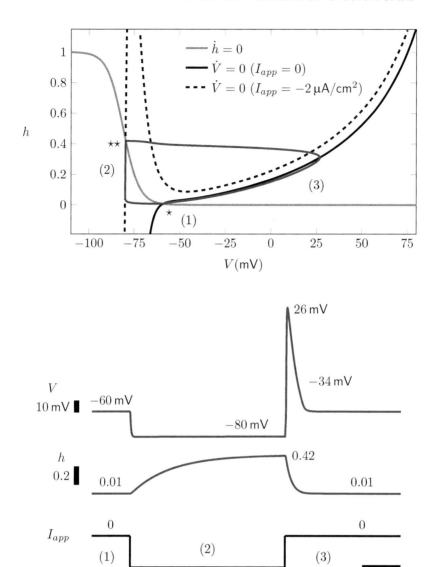

Figure 19.10   The (V, h) phase plane for a minimal model of the low-threshold calcium spike gives insight into post-inhibitory rebound bursting. *Question*: The maximum voltage of the low-threshold spike is 26 mV, while in Fig. 19.9 the maximum voltage is −40 mV. What is responsible for this difference?

Fig. 19.10 shows the (V, h) phase plane for this system. The h nullcline, given by $h_\infty(V)$ does not depend on the applied current. The voltage nullcline, which is given by the points (V, h) solving

$$V \text{ nullcline:} \quad h = \frac{I_{app} - g_L(V - E_L)}{-g_T \, m_\infty^2(V) \, (V - E_{Ca})},$$

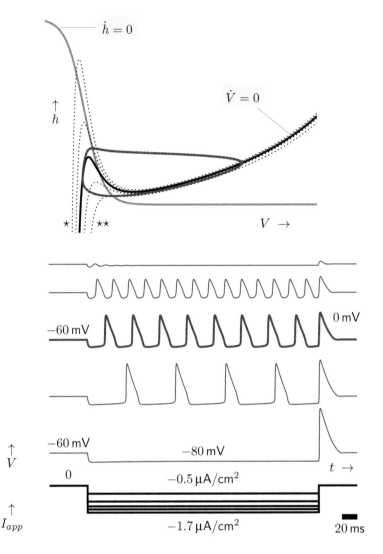

Figure 19.11   Top: The (V, h) phase plane for a minimal model rhythmic low-threshold calcium spikes. Bottom: Time course of voltage (blue) during hyperpolarizing current pulses (black). *Question*: Five V nullclines are shown for $I_{app} = -0.5, -1.0, -1.3, -1.5$ and $-1.7\,\mu A/cm^2$ (dotted curves). What value corresponds to the V nullcline labelled $\star$? Which V nullcline ($\star$ or $\star\star$) corresponds to the steady state in which the T current is most inactivated?

does depend on the applied current and is shown for $I_{app} = 0$ (solid black) and $-2\,\mu A/cm^2$ (dotted). Observe that negative applied current causes the voltage null-cline to shift *upward* because h is the gating variable for an *inward* current. This is the opposite of what occurs in the Morris-Lecar model (Fig. 14.9).

By assuming that voltage is much faster than the inactivation gate h for the low-threshold calcium current I_T, we obtain insight into why a calcium spike occurs after a hyperpolarizing current pulse.

(1) Prior to the pulse the resting membrane potential is $V = -60\,\text{mV}$ and the T current is almost entirely inactivated ($h = 0.01$, see $\star$ in Fig. 19.10). This steady state is given by the intersection of the voltage nullcline for $I_{app} = 0$ (solid black) and the h nullcline (gray).

(2) When the applied current steps down to $I_{app} = -2\,\mu\text{A/cm}^2$, the relevant voltage nullcline moves upward (Fig. 19.10, dashed curve). For this applied current, the steady state voltage (indicated by $\star\star$) is about 20 mV more hyperpolarized than rest ($\star$). The corresponding steady state value for h is 0.4, that is, the T current is 40% de-inactivated. Because V is fast compared to h, the trajectory from $\star$ to $\star\star$ begins by moving leftward toward the stable branch of the V nullcline at $-80\,\text{mV}$, then upward toward the h nullcline.

(3) When the membrane is released from hyperpolarization, the relevant nullcline is once again solid ($I_{app} = 0$). Because voltage is the faster variable, the solution first moves rightward to the voltage nullcline. Once across the voltage nullcline, the T current inactivates (h decreases) and the trajectory hugs the V nullcline until it reaches the rest state $\star$.

Fig. 19.11 shows that this minimal model may also exhibit rhythmic low-threshold spikes in response to hyperpolarization. When the applied current is $-1\,\mu\text{A/cm}^2$, the h nullcline intersects the voltage nullcline between its knees. Because voltage is a fast variable, this steady state is unstable and there is a stable periodic solution (relaxation oscillation).

19.6   Further Reading and Discussion

For more discussion of the electrophysiological properties of thalamocortical relay neurons see McCormick and Huguenard (1992), Huguenard and McCormick (1992), Zhan et al. (1999) and the monograph *Thalamocortical Assemblies* by Destexhe and Sejnowski (2001) and references therein. Although not emphasized here, the nonspecific cationic current I_h contributes to the post-inhibitory rebound and rhythmic bursting exhibited by thalamic neurons (Lüthi and McCormick, 1998). The fast/slow analysis of bursting presented in this chapter follows Rush and Rinzel (1994). The minimal model of the low-threshold calcium spike is based on Wang et al. (1991), Wang and Rinzel (1992) and Golomb et al. (1994).

Each of the mechanisms of bursting emphasized in Chapters 17–19 involves bistability, a hysteresis loop in the fast subsystem for action potential generation, and one slow variable. We have not discussed a wide variety of bursting mechanisms that involve two slow variables, for which bistability in the fast subsystem is not required (Rinzel, 1987; Izhikevich, 2000). For example, parabolic (SNIC-SNIC) bursting involves opposing influences of slow positive and slow negative feedback moving a membrane through an oscillatory region of the fast subsystem bounded by saddle-node homoclinic bifurcations (Rinzel and Ermentrout, 1989, Figs. 11 and 12).

The A Current

This chapter began with the observation that some ionic currents of excitable membranes are inactivated at resting membrane potentials. The low-threshold calcium

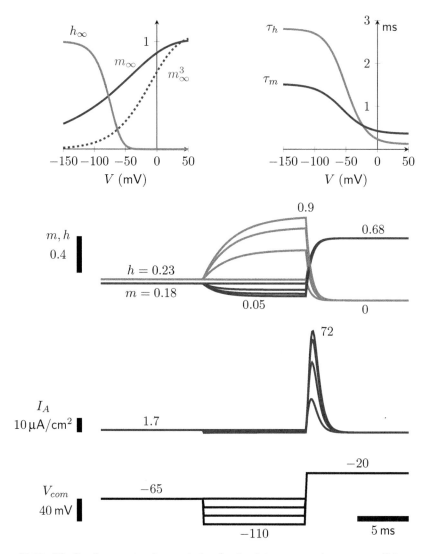

Figure 19.12   Idealized current-voltage relation for the A-type potassium current (I_A).

current I_T is one example. Another is the voltage-gated potassium current referred to as I_A. The A current is similar to the delayed rectifier potassium current I_{K_V} (e.g., both are depolarization activated and outward); however, I_A is inactivated at resting membrane potentials. Thus, the A current activates with little delay after depolarization (and resists further depolarizing) *provided the membrane is first hyperpolarized*, because this hyperpolarization de-inactivates the potassium channels that mediate the A current.

Fig. 19.12 shows activation (m_∞) and inactivation (h_∞) gating functions for an A current modeled as $I_A = \bar{g}_A m^3 h(V - E_K)$. The simulated voltage-clamp protocol shows depolarization from -65 (rest) to -20 mV results in little outward A current. Prepulses to more hyperpolarized voltages de-inactivate the A current (middle row, increasing h, gray), which increases the amount of outward current evoked upon depolarization.

Expression of the A current may increase the interval between action potentials and facilitates low frequency repetitive firing of neurons. Membrane models that include the A current may transition, as $\bar{g}_A$ is increased, from a type II membrane with periodic solutions originating via a Hopf into a type I membrane with periodic solutions originating via a SNIC (Rush and Rinzel, 1995). Low-threshold potassium currents similar to I_A promote sensitivity to phasic depolarization and suppress responses to sustained depolarization, a phenomenon referred to as **type III excitability** (Clay et al., 2008). Type III excitability is observed in medial superior olive neurons of the auditory brainstem that express low-threshold potassium currents (Meng et al., 2012). During Hodgkin and Huxley's Nobel Prize winning work, type I, II, and III behavior of squid giant axon was observed.

Solutions

Figure 19.1   The membrane resistance is greater when $V < V_{rest}$.

Figure 19.3   The activation gating variable is m. The inactivation gating variable is h. At $-10\,\mathrm{mV}$ the time constants of these two gating variables are similar.

Figure 19.4 $h = 0.56$ represents more inactivation of T current than $h = 0.93$. In the latter case, the T current is mostly de-inactivated.

Figure 19.5 h represents the fraction of low-threshold calcium channels that are not inactivated, so physical values of h are from 0 to 1. Termination of spiking at the end of the post-inhibitory rebound burst is via a SNIC bifurcation.

Figure 19.10   Fig. 19.9 includes I_{K_V} while Fig. 19.10 does not. I_{K_V} is restorative and resists the depolarizing influence of I_T.

Figure 19.11   The V nullcline labelled $\star$ was obtained using $-1.7\,\mu\mathrm{A/cm^2}$. The V nullcline that corresponds to the steady state in which the T current is most inactivated is labelled $\star\star$, because $h = 0$ represents complete inactivation, while $h = 1$ represents complete de-inactivation (zero inactivation).

Notes

1. Equations and parameters for I_T follow Golomb et al. (1994).
2. To reduce the number of equations in the fast subsystem, a linear relationship was imposed between the inactivation gating variable for I_{Na}, denoted by $\hat{m}$ below, and the activation gating variable for I_K, as follows:

$$I_{AP} = \bar{g}_{Na}\hat{m}_\infty^3(V)(0.85 - n)(V - E_{Na}) + \bar{g}_K n^4 (V - E_K).$$

This simplification does not change the bifurcation structure of the fast subsystem (Rush and Rinzel, 1994).

3. For clarity, periodic solutions are only shown for $I_{app} = 0$.
4. This simulation uses parameter set 2 in Rush and Rinzel (1994) with $g_T = 0.5\,\text{mS/cm}^2$.
5. Fig. 19.8 is not a phase plane, because there is a third dynamic variable, the gating variable n in the action potential generating currents $I_{AP}(V, n)$.
6. The exponent 2 for the activation gating function m_∞ in Eq. 19.3a, as opposed to 3 in Eq. 19.3a and the Hodgkin-Huxley Na$_V$ model, is phenomenological and should not be interpreted mechanistically.

20 Synaptic Currents

In central nervous system function, two important means of intercellular communication are chemical and electrical synapses. When chemical synaptic transmission is fast, phase plane analysis gives insight into how small networks of excitatory or inhibitory neurons will function together.

20.1 Electrical Synapses

Chapters 13–19 have emphasized the ionic currents and membrane excitability of *single* neurons. This final chapter focuses on how neurons of the central nervous system influence one another and how these interactions are modeled. Intercellular communication in the central nervous system occurs in a myriad of ways, including ionotropic and metabotropic chemical neurotransmission mediated by ligand-receptor interactions (Chapter 4), endocrine and paracrine hormonal signaling, contact-dependent signaling, and unusual second messengers such as the volatile gas nitric oxide that can easily diffuse across the cellular membranes.

Perhaps the simplest interaction between cells of the central nervous system is direct electrical coupling. When neurons or muscle cells are coupled via gap junctions, ionic current may flow from the interior of one cell to another. As diagrammed in Fig. 20.1, let us arbitrarily define current flowing from cell 1 to cell 2 as *positive* gap junctional current ($I_{gap} > 0$). Positive gap junctional current is outward from the perspective of cell 1 (hyperpolarizing) and inward from the perspective of cell 2 (depolarizing). Consequently, I_{gap} enters into the current balance equations for the two cells with opposite sign, as follows

$$C\dot{V}_1 = I_{app}^{(1)} - I_{ion}^{(1)} - I_{gap}$$
$$C\dot{V}_2 = I_{app}^{(2)} - I_{ion}^{(2)} + I_{gap},$$

where we assume the cells have the same specific capacitance. We assume the gap junctional current is passive and proportional to the difference in intracellular

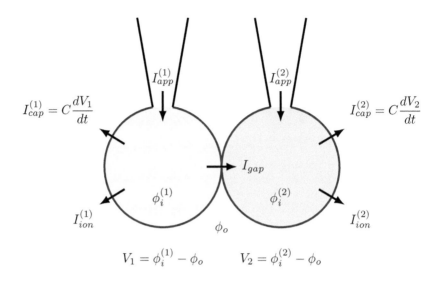

$$V_1 = \phi_i^{(1)} - \phi_o \qquad V_2 = \phi_i^{(2)} - \phi_o$$

Figure 20.1   Schematic of applied, ionic, and gap junctional membrane currents for two electrically coupled cells.

potentials, $\phi_i^{(1)} - \phi_i^{(2)}$. This voltage drop[1] is equal to the difference in transmembrane potentials, $V_1 - V_2$, so the gap junctional current is

$$I_{gap} = g_{gap} (V_1 - V_2) .$$

Equations for two passive electrically coupled cells are

$$C\dot{V}_1 = I_{app}^{(1)} - g_L (V_1 - E_L) - g_{gap} (V_1 - V_2)$$
$$C\dot{V}_2 = I_{app}^{(2)} - g_L (V_2 - E_L) - g_{gap} (V_2 - V_1) .$$

For the time being, assume no applied current for either cell, $I_{app}^{(1)} = I_{app}^{(2)} = 0$, and denote the difference in the cell membrane potentials as $\Delta V = V_1 - V_2$. In that case, we may differentiate to obtain $\dot{\Delta V} = \dot{V}_1 - \dot{V}_2$, substitute the right sides of the current balance equations, and simplify to obtain

$$C\dot{\Delta V} = - (g_L + 2g_{gap}) \, \Delta V .$$

This simple calculation shows that the voltages of two passive cells are expected to equilibrate with an exponential time constant of $\tau = C/(g_L + 2g_{gap})$. Higher gap junctional coupling conductance leads to faster relaxation. This is consistent with the intuitive notion that strong electrical coupling should result in a similar dynamical state for the two cells.

   If neuron 1 is stimulated with positive applied current, neuron 2 will be affected to a degree that depends on the gap junctional conductance. Assuming both cells have the same leakage conductance, the difference in membrane potentials at steady state is

$$\Delta V = \frac{\Delta I_{app}}{g_L + 2g_{gap}} \quad \text{where} \quad \Delta I_{app} = I_{app}^{(1)} - I_{app}^{(2)} .$$

Observe that for constant applied current the potential difference of the two cells decreases as the gap junctional coupling increases.

20.2 Electrical Synapses and Synchrony

The calculation above leads to the intuitive notion that strong electrical coupling will promote synchronization of neurons exhibiting repetitive spiking. To explore this possibility, consider two identical Morris-Lecar models (Chapter 14) coupled via a gap junctional current,

$$C\dot{V}_1 = I_{app}^{(1)} - I_{ion}(w_1, V_1) - g_{gap}(V_1 - V_2) \tag{20.1a}$$
$$\dot{w}_1 = -[w_1 - w_\infty(V_1)]/\tau_w(V_1) \tag{20.1b}$$
$$C\dot{V}_2 = I_{app}^{(2)} - I_{ion}(w_2, V_2) - I_L(V_2) - g_{gap}(V_2 - V_1) \tag{20.1c}$$
$$\dot{w}_2 = -[w_2 - w_\infty(V_2)]/\tau_w(V_2) \tag{20.1d}$$

where $I_{ion} = I_{Ca_V} + I_{K_V} + I_L$ as in Eq. 14.2. Fig. 20.2 (top) shows uncoupled Morris-Lecar models that are initially repetitively spiking out of phase. As expected, when gap junctional coupling is turned on (g_{gap} changed from 0 to $2\,\mathrm{mS/cm^2}$ at $t = 100\,\mathrm{ms}$), the repetitive spiking of these neurons synchronizes in a phase-locked fashion. Perhaps less intuitively, gap junctional coupling may also promote anti-phase locked synchro-

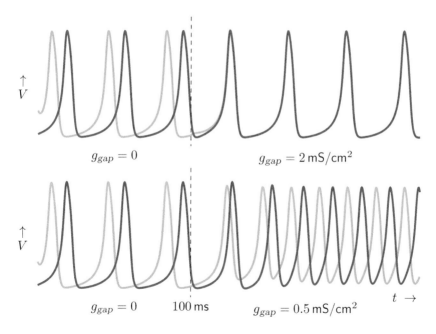

Figure 20.2   Top: Two Morris-Lecar oscillators (Eq. 20.4) that are initially out of phase are synchronized by strong gap junctional coupling. Bottom: Weaker gap junctional coupling leading to anti-phase synchronization. Adapted from Sherman and Rinzel (1992) and Rinzel (2002). Sherman, A, and Rinzel, J. 1992. Rhythmogenic effects of weak electrotonic coupling in neuronal models. *Proceedings of the National Academy of Sciences*, **89**(6), 2471–2474.

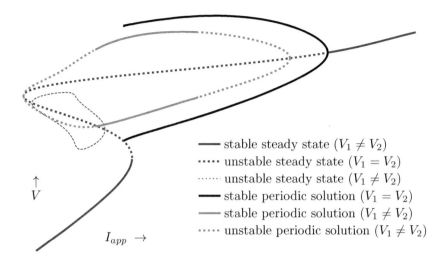

stable steady state $(V_1 \neq V_2)$
unstable steady state $(V_1 = V_2)$
unstable steady state $(V_1 \neq V_2)$
stable periodic solution $(V_1 = V_2)$
stable periodic solution $(V_1 \neq V_2)$
unstable periodic solution $(V_1 \neq V_2)$

Figure 20.3 (V_i, I_{app}) bifurcation diagram for two electrically coupled Morris-Lecar models $(g_{gap} = 1.5\,\text{mS/cm}^2,$ Eq. 20.4) indicates coexistence of stable in-phase and anti-phase synchronous solutions.

nization (bottom panel, $g_{gap} = 0.5\,\text{mS/cm}^2$). Fig. 20.3 presents a (V_i, I_{app}) bifurcation diagram for two coupled Morris-Lecar models using $g_{gap} = 1.5\,\text{mS/cm}^2$. The familiar fixed point and periodic solutions associated with type I membranes are present. These represent in-phase synchronous solutions where $V_1 = V_2$ and $w_1 = w_2$. The bifurcation diagram also indicates stable and unstable periodic solutions that represent anti-phase locked synchrony.

20.3   Chemical Synapses

Chemical synaptic transmission may be modeled by considering the dynamics of bimolecular association between a postsynaptic receptor and neurotransmitter ligand. Following Section 4.5, write

$$L + R \underset{k_-}{\overset{k_+}{\rightleftharpoons}} LR,$$

where L is a neurotransmitter, and LR and R are bound and unbound receptor. Assuming mass action kinetics, the ODE for the ligand-bound receptor concentration is

$$\frac{d[\text{LR}]}{dt} = k_+[\text{L}][\text{R}] - k_-[\text{LR}].$$

Using the conserved total receptor concentration, $r_T = [\text{R}] + [\text{LR}]$, this can be written as a single ODE that gives the bound receptor concentration response to the changing ligand concentration,

$$\frac{d[\text{LR}]}{dt} = k_+[\text{L}](r_T - [\text{LR}]) - k_-[\text{LR}]. \tag{20.2}$$

Let us denote the fraction of ligand-bound postsynaptic receptors as $s = [LR]/r_T$ and divide Eq. 20.2 by r_T to obtain,

$$\dot{s} = k_+[L](1 - s) - k_- s.$$

We shall not explicitly model the dynamics of neurotransmitter concentration, $[L]$, in the synaptic clefts associated to a particular presynaptic neuron. Nevertheless, the above reasoning motivates a phenomenological approach to chemical neurotransmission in which a dimensionless **synaptic gating variable** associated to a presynaptic neuron responds to that neuron's voltage (V_{pre}) as follows,

$$\dot{s} = \alpha_s(V_{pre})[1 - s] - \beta_s s \qquad (20.3)$$

where activation of s is voltage dependent and β_s Θ time^{-1} is the rate constant for synaptic decay. For example, we might use $\alpha_s(V) = \alpha_s^0/(1 + e^{-(V-\theta)/\sigma})$ with $\sigma > 0$, which is an increasing sigmoidal function of voltage that takes values between 0 and α_s^0. For a sustained presynaptic voltage (V_{pre}), the steady state of this synaptic gating variable,

$$s_\infty(V) = \frac{\alpha_s(V)}{\alpha_s(V) + \beta_s},$$

takes values between 0 and 1. The exponential time constant for the synaptic gating variable is $1/(\alpha_s(V) + \beta_s)$, a monotone decreasing function that takes values between $1/\beta_s$ and $1/(\alpha_s^0 + \beta_s)$. A synaptic current I_{syn} with conductance $\bar{g}_{syn}s$ and driving force $V_{post} - E_{syn}$ is included in the postsynaptic neuron's current balance equation. For example, two Morris-Lecar models reciprocally coupled via chemical synaptic transmission would be modeled using Eqs. 20.1b and 20.1d and

$$C\dot{V}_1 = I_{app}^{(1)} - I_{ion}(w_1, V_1) - g_{syn}s_2(V_1 - E_{syn}) \qquad (20.4a)$$
$$\dot{s}_1 = \alpha_s(V_1)[1 - s_1] - \beta_s s_1 \qquad (20.4b)$$
$$C\dot{V}_2 = I_{app}^{(2)} - I_{ion}(w_2, V_2) - I_L(V_2) - g_{syn}s_1(V_2 - E_{syn}) \qquad (20.4c)$$
$$\dot{s}_2 = \alpha_s(V_2)[1 - s_2] - \beta_s s_2. \qquad (20.4d)$$

Observe that the synaptic gating variable for neuron 2 scales the synaptic current in the current balance equation for neuron 1 (and vice versa), thereby coupling the cells.

20.4   Phase Plane Analysis of Instantaneously Coupled Cells

Several times in this primer we have found that by assuming a separation of time scales between various cellular processes, we put ourselves in a position to use phase plane analysis and thereby develop a geometric intuition about neuronal behavior. The remainder of this chapter does just this in the context of synaptically coupled neurons. We will assume two identical neurons with a Fitzhugh-Nagumo-like phase plane governing oscillations in the subthreshold currents expressed by these cells. When the voltage is above some fixed value, we presume that the neuron is active, i.e., spiking; and when the voltage is below the threshold the neuron will be presumed to be silent (not spiking). The influence of an active neuron will be represented by changing the position of the voltage nullcline for the neuron(s) to which it is presynaptic.

More concretely, consider the following ODEs for two synaptically coupled Fitzhugh-Nagumo-like model neurons,

$$\dot{x}_1 = x_1 - x_1^3/3 - y_1 + J + s(x_2) \tag{20.5a}$$
$$\dot{y}_1 = \epsilon[y_\infty(x) - y_1] \tag{20.5b}$$
$$\dot{x}_2 = x_2 - x_2^3/3 - y_2 + J + s(x_1) \tag{20.5c}$$
$$\dot{y}_2 = \epsilon[y_\infty(x) - y_2], \tag{20.5d}$$

where $y_\infty = \tanh(-x/\sigma)$ is an increasing sigmoidal function of x and $s(x) = s_*$ for $x > 0$ and 0 otherwise. In these equations, J determines the baseline location of the x nullcline and, thus, the intrinsic character of the two identical neurons (silent, oscillatory, or tonically active). The term $s(x)$ represents synaptic coupling with $s_* < 0$ corresponding to inhibition and $s_* > 0$ excitation.

Fig. 20.4 (left) shows the phase plane for two identical neurons that are intrinsically tonically active. When uncoupled, the neurons (whose states are denoted by circles filled red and green) eventually find their way to the stable steady state at the intersection of the nullclines for voltage (black) and the recovery variable (gray). This intersection is on the active side of the vertical dotted line, consistent with our assumption that the neurons are tonically active. Fig. 20.4 (right) shows how the phase plane changes to represent neurons with the same intrinsic properties, but which are reciprocally coupled via inhibition. When either neuron is inhibited by the other, the relevant voltage nullcline is shifted downward, because the recovery variable y in this Fitzhugh-Nagumo-like model is analogous to the activation gating variable of an

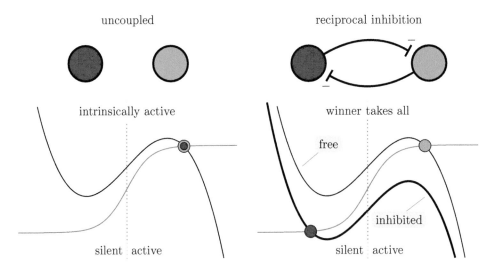

Figure 20.4   Left: Phase plane for two identical tonically active neurons that are uncoupled (left) or coupled via inhibition (right). Voltage nullclines are shown black. Slow variable nullclines are gray. When a neuron is inhibited the relevant voltage nullcline is shifted downward (thick black).

outward current. We will assume x is much faster than y, so if the green neuron is active (right of vertical dotted line), the red neuron must be on the lower *inhibited* voltage nullcline (and vice versa). Otherwise, the neurons would move about the depolarized and hyperpolarized branches of the voltage nullcline, falling off the knees leftward or rightward toward the recovery variable nullcline. After some trial and error, and a little thought, the reader will conclude that there are many *transient* locations in phase space that the two cells may occupy, such as

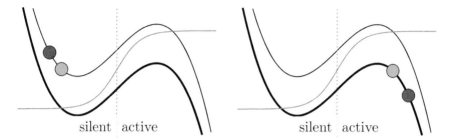

On the other hand, there are only two valid steady states. Fig. 20.4 (right) shows one of these: the green neuron is active and suppressing the activity of the red neuron, and both are located at the intersection of the y nullcline with the appropriate x nullcline. The symmetric case in which the red cell continuously suppresses the activity of the green cell is the second steady state. This type of synaptic interaction between neurons is referred to as "winner takes all" dynamics.

Fig. 20.5 shows solutions of Eq. 20.5 with parameters chosen to represent mutually inhibitory neurons. The separation of time scales between fast and slow variables is relatively strong ($\epsilon = 0.4$), but not as strong as assumed in the thought experiment of the previous paragraph ($0 < \epsilon \ll 1$). Once again the neurons are intrinsically active (note position of *free x* nullcline), but the inhibition in Fig. 20.5 ($s_0 = -1/2$) is half as strong as in Fig. 20.4 ($s_0 = -1$). Observe that the *inhibited x* nullcline in Fig. 20.5 (thick black) implies that activity in a presynaptic neuron moves the intrinsically active postsynaptic neuron into a relaxation oscillator regime. The numerically calculated stable periodic solution (blue) corresponds to red and green neurons that burst in anti-phase locked synchrony (imagine spikes during the depolarized active phases). Fig. 20.5 may confirm your intuition that mutually inhibitory neurons could synchronize but, if they did so, it would make sense for them to oscillate 180 degrees out of phase.

It is important to note that the bursting rhythm of Fig. 20.5 is not an intrinsic property of the two neurons, which are tonically active. Rather, the rhythmogenic behavior (referred to as **half-center oscillation**) is an emergent property of the inhibitory network connectivity. The mechanism leading to the rhythmic bursting is referred to by dynamicists as an **escape** mechanism, because the neurons can escape the inhibition received from each other. That is, silent phases end when the inhibited neuron falls rightward off the lower knee of the inhibited voltage nullcline. Because hyperpolarization-activated inward currents may contribute to escape from inhibition, they may function as pacemaker currents that may be modulated to control rhythmic frequency (Lüthi and McCormick, 1998).

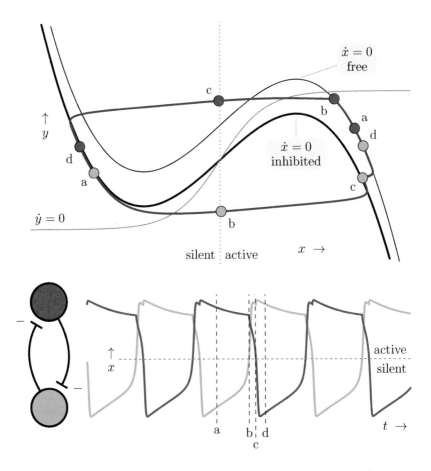

Figure 20.5   A half-center oscillator composed of two reciprocally coupled inhibitory and intrinsically active neurons (Eq. 20.5). Anti-phase synchronization occurs through an *escape* mechanism. Parameters: $\epsilon = 0.08$, $\sigma = 0.3$, $J = 0.5$, $s_0 = -0.5$.

Fig. 20.6 shows a simulation of the converse *release* mechanism. In this case, an inhibited neuron eventually finds its way to the stable *silent* steady state located at the intersection of the y nullcline and the lower knee of the inhibited x nullcline. The silent phase for this neuron ends when the active neuron *times out*, falls leftward off the upper knee of the free x nullcline, and goes silent, thereby releasing its partner.

Fig. 20.7 summarizes and contrasts the release and escape mechanisms for rhythmic bursting that may emerge from reciprocally coupled inhibitory neurons. Network-based rhythmicity of both types has been observed in small, relatively autonomous neural networks that subserve repetitive rhythmic behaviors of locomotion, respiration and mastication (Skinner et al., 1993; Daun et al., 2009). These **central pattern generators** have been extensively studied in invertebrate experimental preparations such as the stomatogastric ganglion of lobster and crab (Hudson et al., 2010) and in the vertebrate brain stem (Feldman et al., 2013) and spinal cord (Tabak et al., 2001).

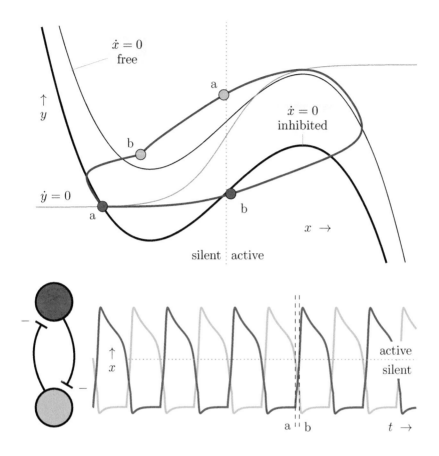

Figure 20.6   A network oscillator composed of two intrinsically bursting reciprocally coupled inhibitory neurons (Eq. 20.5). Anti-phase synchronization occurs through a *release* mechanism. Parameters as in Fig. 20.5 with $\epsilon = 0.4$, $\jmath = 0.2$, $s_0 = -1$.

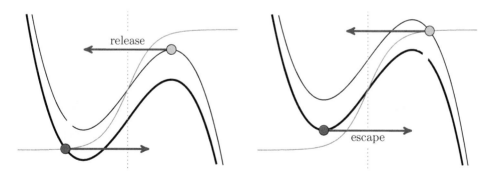

Figure 20.7   Comparison of release and escape mechanisms for network rhythmicity of reciprocally coupled inhibitory neurons.

20.5 Reciprocally Coupled Excitatory Neurons

Fig. 20.8 shows a simulation of two intrinsically oscillatory reciprocally coupled excitatory neurons. The out-of-phase solution exists but is unstable; after a few oscillations
the neurons synchronize. In the fast/slow regime a stable in-phase synchronous solution for two mutually excitatory oscillators might be expected. Imagine the state of
the two almost synchronized neurons as red and green filled circles in a Fitzhugh-
Nagumo-like phase plane. Let us assume that (1) both neurons are silent and (2) the
red neuron is slightly ahead of the green neuron (phase advanced). When the red
neuron arrives at the lower knee of the free x nullcline, it quickly jumps rightward
and becomes active. The activity of the red neuron excites the green neuron, elevating
the x nullcline relevant for the green neuron. The green neuron then quickly moves
rightward to the lower or upper branch of the excited x nullcline, whichever is closer.
Either way, the green neuron ends up closer (in phase) to the red neuron, confirming
our intuition.

But let us not forget that the above reasoning assumes excitatory neurons
that are identical, intrinsically oscillatory relaxation oscillators. Relax any of these

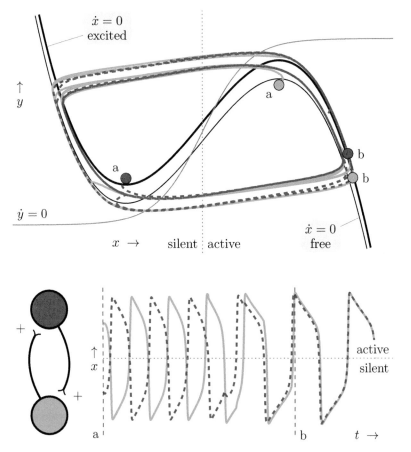

Figure 20.8   In-phase synchronization of two intrinsically oscillatory reciprocally coupled
excitatory neurons. Parameters as in Fig. 20.5 with $\epsilon = 0.1$, $J = -0.1$, $s_0 = 0.2$.

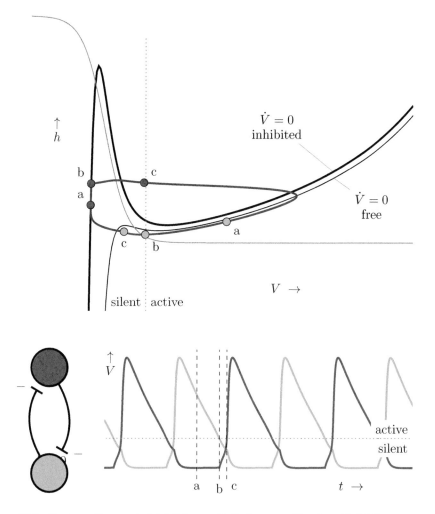

Figure 20.9   Reciprocally coupled, inhibitory cells that are capable of post-inhibitory rebound may synchronize in an anti-phase locked manner (compare Fig. 19.11).

assumptions and the emergent behavior of coupled neurons will be more complex and probably less intuitive. For example, Fig. 20.9 shows that rhythmic bursting generated by reciprocally coupled, inhibitory, non-oscillatory cells that are capable of post-inhibitory rebound may synchronize in an anti-phase locked manner (perhaps as expected). However, when the postsynaptic conductance is not instantaneous but decays slowly, the neurons may rhythmically burst synchronously with no phase difference, and different stable activity patterns may coexist (Wang and Rinzel, 1992).

20.6   Further Reading and Discussion

See Wang and Rinzel (1992) and Skinner et al. (1993) for further reading on modeling of central pattern generators. Ligand-receptor based modeling of glutamatergic synaptic transmission mediated by AMPA and NMDA receptors is discussed in Destexhe et al.

(1994). See Sherman and Rinzel (1992), De Vries et al. (1998) and Rinzel (2002) for more on in-phase and anti-phase synchronization of gap junctional couple cells. See Kopell et al. (2000), Kopell and Ermentrout (2004), Mancilla et al. (2007) and Oswald et al. (2009) for more on inhibitory neurons in cortical microcircuits. Although we have not discussed it here, there is a fruitful mathematical approach to neuronal synchronization that focuses on *weak* coupling of oscillatory neurons (Hoppensteadt and Izhikevich, 1997). The phase response curves that characterize intercellular communication in this limit are qualitatively distinct for type I and II neurons (Ermentrout and Kopell, 1991; Ermentrout, 1996; Galán et al., 2005). Ermentrout (1998a) is a highly recommended overview of mathematical modeling of neural networks.

Slow Synapses

Metabotropic synaptic transmission models can be quite elaborate when the dynamics of second messengers and cascade of intracellular signaling events are explicitly modeled. A relatively simple phenomenological approach to modeling slow synapses introduces sequential activation to Eq. 20.3 using an auxiliary gating variable. For example, an inhibitory synaptic current mediated by metabotropic $GABA_B$ receptor regulation of a potassium current could take the form (Golomb et al., 1994),

$$I_{GABA_B} = g_{GABA_B}\, s(V_{post} - E_K)$$

where

$$\dot{x} = \alpha_x(V_{pre})[1 - x] - \beta_x x\,,$$
$$\dot{s} = \alpha_s(x)[1 - s] - \beta_s s\,,$$

$\alpha_x(V) = \alpha_x^0/(1 + e^{-(V-V_\theta)/V_\sigma})$, and $\alpha_s(x) = \alpha_s^0/(1 + e^{-(x-x_\theta)/x_\sigma})$. Note that $\alpha_s(x)$ is a function of the auxiliary variable x as opposed to voltage. For $\alpha_x \gg \beta_x$, a presynaptic voltage pulse rapidly activates x, which then decays with a time constant of $1/\beta_x$. During the time when $x > x_\theta$, the synaptic gating variable increases at rate α_s; later, s decays with time constant $1/\beta_s$.

Short Term Synaptic Plasticity

Auxiliary synaptic gating variables may also be used to model short term synaptic plasticity, e.g., synaptic depression mediated by depletion of presynaptic vesicles. If a is the fraction of presynaptic vesicles available for release, one may write

$$\dot{a} = -\alpha_a(V_{pre})\,a + \beta_a(1 - a)\,, \tag{20.6}$$

where $\alpha_a(V)$ is a voltage-dependent release rate and β_a is the recovery rate. An inhibitory synaptic current mediated by ionotropic $GABA_A$ receptors that includes synaptic depression would take the form

$$I_{GABA_A} = g_{GABA_A}\, s\, a\, (V_{post} - E_{Cl})$$

where s solves Eq. 20.3 and a solves Eq. 20.6.

Conclusion

This chapter concludes with two take home messages regarding intercellular commu-nication through chemical and electrical synapses. First, phase plane analysis provides a measure of intuition regarding the effect of fast excitatory or inhibitory synapses on neural responses. Second, the functional effect of synaptic transmission may be complex and counterintuitive.

Examples of this complexity include the following. (1) The activity of inhibitory neurons that are presynaptic to neuron expressing the low-threshold calcium current I_T may result in action potentials (excitation), as these neurons may respond to hyper-polarization with a post-inhibitory rebound burst. (2) Populations of neurons that are reciprocally coupled via inhibitory synapses may show increased activity when dis-inhibited via blockage of chemical synaptic transmission. (3) The dynamics of short term plasticity and synaptic summation may fundamentally depend on the firing mode of a presynaptic neuron (burst versus tonic).

Finally, it should be emphasized that the classical neurotransmitters (e.g., gluta-mate, GABA, etc.) are not excitatory or inhibitory in their own right. The excitatory or inhibitory effect of acetylcholine, dopamine, serotonin, etc., on a postsynaptic neuron depends on the type(s) of postsynaptic receptors expressed by that cell, changes in synaptic conductance that result from ligand-gating of these receptors, and inward or outward ionic currents that result given the reversal potentials of these synaptic currents. GABA, a neurotransmitter that is usually inhibitory, may be excitatory when the intracellular chloride concentration is abnormally high, or the extracellular chlo-ride concentration is abnormally low, because GABAergic synaptic currents mediated by GABA$_A$ receptors reverse at $E_{Cl} = (RT/z_{Cl}F)\ln[Cl^-]_o/[Cl^-]_i$.

Supplemental Problems

Problem 20.1 Acetylcholine depolarizes thalamocortical relay neurons, but hyperpolarizes inhibitory thalamic interneurons. Is acetylcholine an excitatory or inhibitory neurotransmitter?

Problem 20.2 Say you had two identical excitatory neurons that were intrinsi-cally inactive (subthreshold). If excitation was strong enough to lift the excited nullcline into the oscillatory regime (below left), what would be the possible steady states of this system? What about if excitation is so strong that it lifts the excited nullcline into a position that allows for tonic activity (below right)?

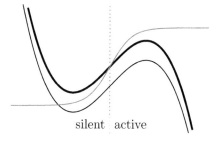

silent active

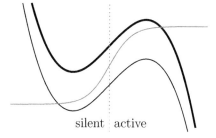

silent active

Problem 20.3   The postsynaptic conductance increase due to a presynaptic action potential is sometimes represented by a function of the form,

$$g(t) = \alpha^2 t \exp(-\alpha t). \tag{20.7}$$

In this case, the synaptic current would be

$$I_{syn} = g(t)(V_{post} - E_{syn}). \tag{20.8}$$

(a)   Confirm that the factor α^2 normalizes this function so that $\int_0^\infty g(t)\,dt = 1$.
(b)   What is the peak value of $g(t)$ and for what value of t does this peak occur?

Problem 20.4   Find the gating functions $s_\infty(V)$ and $\tau_s(V)$ that make

$$\dot{s} = [s_\infty(V) - s]/\tau_s(V)$$

equivalent to Eq. 20.3.

Problem 20.5   Derive an ODE for the level of synaptic depression, given by $d = 1 - a$, where a solves Eq. 20.6.

Problem 20.6   Consider two neurons that are capable of post-inhibitory rebound bursting. One is excitatory (E) and the other is inhibitory (I).

The phase planes characterizing the low-threshold spike are:

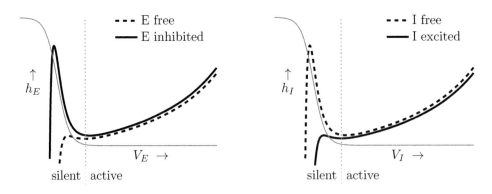

Assuming V fast/h slow and instantaneous synaptic interactions, how many stable steady states can you find for this small network?

Solutions

Problem 20.4 $s_\infty(V) = \beta/[\alpha(V) + \beta]$ and $\tau_s(V) = 1/[\alpha(V) + \beta]$.

Problem 20.5 $d = 1 - a$ so $\dot{d} = -\dot{a} \Rightarrow \dot{d} = \alpha(V)(1 - d) - \beta d$.

Note

1. Because $\phi_i^{(1)} - \phi_i^{(2)} = \phi_i^{(1)} - \phi_o - \phi_i^{(2)} + \phi_o = \phi_i^{(1)} - \phi_o - [\phi_i^{(2)} - \phi_o]$
    $= V_1 - V_2$.

Afterword

Even in literature and art, no man who bothers about originality will ever be original: whereas if you simply try to tell the truth (without caring twopence how often it has been told before) you will, nine times out of ten, become original without ever having noticed it.

— C. S. Lewis

I have written this primer on the computational biology of excitable cells for students of cellular biophysics and modeling – sophomore and junior neuroscience majors who have, in prior semesters, completed course work in introductory neurobiology and single variable calculus. I am hopeful that our slow deliberate steps have prepared you to engage textbooks, scientific monographs, and research papers at the intersection of mathematics and neurophysiology.

Our focus has been isopotential, deterministic, conductance-based modeling. These topics are the tip of an iceberg that is both gigantic (length^3) and extremely cool (temp). Many fundamentals of computational cellular and systems neuroscience have scarcely been mentioned. These include the following.

- *Spatial phenomena* such as action potential propagation, the application of compartmental modeling and cable theory to model the electronic properties of dendrites (Jack et al., 1975; Segev et al., 2003; Korogod and Tyč-Dumont, 2009; Tuckwell, 1988a), dendritic excitability (Stuart et al., 2016) and spines (Yuste, 2010), and neuronal calcium signaling (Berridge, 1998; Berridge et al., 2003).
- *Stochastic phenomena* such as membrane noise due to the finite number of ion channels in the neural membrane (DeFelice, 2012), single-channel recording and mathematical modeling of the stochastic gating of ion channels (Sakmann, 2013), and many aspects of neuronal function that are best modeled using probability theory (Laing and Lord, 2010; Bressloff, 2014).

Except for the final chapter, our focus has been single cells (no more, no less). A subsequent volume might begin with the dynamics of neuronal *populations*: local field potentials, extracellular multiunit recording, Wilson-Cowan-type modeling (Destexhe and Sejnowski, 2009), or subcellular phenomena such as localized calcium elevations, the postsynaptic density, and the biochemistry of short and long term synaptic plasticity.

◇

Depending on your interests, there are many books to recommend for continued study. To learn more about mathematical aspects of cellular biophysics and neuroscience, consider reading Sterratt et al. (2011), Izhikevich (2007), Ermentrout and Terman (2010), Koch (2004) and Koch and Segev (1998). These are similar in spirit to this primer, but written at a higher level. To explore nonlinear dynamics and bifurcations and the discipline of applied mathematics, you might try Robinson (2012), Boyce and DiPrima (2012), Glendinning (1994), Guckenheimer and Holmes (2013), Strogatz (2014), Kaplan and Glass (2012), Jordan and Smith (1999), Perko (2013), Thompson and Stewart (2002), Keener (1995), Logan (2013) and Scott (2003). If you have not done so already, I highly recommend downloading and experimenting with the software package XPPAUT (Ermentrout, 2002).

Other excellent books that might be read next are Dayan and Abbott (2001), Gabbiani and Cox (2017), Bialek (2012), Rieke et al. (1996), Gerstner and Kistler (2002), Wilson (1999), Wallisch et al. (2014), Schutter (2009) and Tuckwell (1988b, 1989). For computational and modeling approaches in physiology with less emphasis on neuroscience see Fall et al. (2002), Segel (1984), Keener and Sneyd (1998) and Lauffenburger and Linderman (1996). Accessible and general introductions to mathematical biology include Segel and Edelstein-Keshet (2013), Britton (2012), Garfinkel et al. (2017) and Edelstein-Keshet (2005). Murray (2002a,b) is a well-known advanced mathematical biology text. Textbooks and monographs on cellular biophysics, neurobiology, and neurophysiology include Aidley (1998), Gulrajani (1998), Fain (1999, 2003), Johnston and Wu (1994), Levitan and Kaczmarek (2015), Ferreira and Marshall (1985), Weiss (1996), Nicholls et al. (2011), Plonsey and Barr (2007), Gomperts et al. (2009) and, of course, Hille (2001). For readers without a neuroscience background who want to learn more, I recommend Greenspan (2007), Swanson (2012), Shepherd (1988, 2004) and Schneider (2014).

◇

I am forever indebted to those who have taught, mentored, and inspired me, especially Susan Tucker, Joel Keizer, John Rinzel, Arthur Sherman, John Tyson and Jim Keener. As a first year graduate student in the biophysics graduate group at UC Davis, Susan Tucker's courses in physical chemistry – her enthusiasm and remarkable blackboard skills – got me wondering if there existed a scientific discipline that might be described as "theoretical cell biology." When I asked this question of Richard Nuccitelli, through whose lab I was rotating, his answer was, in so many words, "No, but you might want to talk to Joel Keizer." Joel was a physical chemist by training whose research interests were moving from nonequilibrium statistical thermodynamics to mathematical aspects of cell physiology.

Joel eventually became my Ph.D. supervisor and encouraged me to apply to the 1993 CRM-UBC Summer School on Mathematical Biology. During this 4 week program, I heard lectures by, and ate meals with, Leah Edelstein-Keshet, Bard Ermentrout, Jim Keener, Nancy Kopell, Robert Miura, John Rinzel, Arthur Sherman and John Tyson, among others. These heroes of my graduate education are now colleagues and friends, to whom I dedicate this book. The concepts emphasized in the final chapters of this primer are, in essence, gentle introductions to the published work of John Rinzel's collaborators and trainees, including Artie and Bard who are mentioned above, but also David Golomb, John Rubin, Xiao-Jing Wang, Maureen Rush and David Terman.

I would also like to thank my friends and colleagues at William & Mary, especially the students and faculty associated with the Mathematical Biology Journal Club/Seminar, now in its 11th year, my past and current Ph.D. and REU students, and the teaching assistants for *APSC 351 Cellular Biophysics & Modeling*, the W&M

course that makes use of this primer. The cover art was drawn by Olivia Walch (Class of 2001). The Chair of the Department of Applied Science, Christopher Del Negro, has been consistently supportive of this writing project, and taught APSC 351 in alternate years prior to 2010. I am grateful for the encouragement, support and patience of Cambridge University Press, especially Katrina Halliday and Jenny van der Meijden. This work was supported in part by the National Science Foundation under Grant Nos. 0133132, 0443843 and 1121606.

◇

My deepest thanks are for my daughter Brenna, my wife Kristin, and my three extended families: the Smiths, the Conradis, and Peace Hill. I have never been happier or felt more loved. With Kristin by my side, each season is appealing and the future is treasured.

I began writing this book when Brenna was 6 years old. Her advice to me at that age was that a good book has many color pictures. A decade later, my daughter's perspicuous insights and succinct advice continue to impress me, and she remains my pride and joy.

References

Ahern, Christopher A, Payandeh, Jian, Bosmans, Frank, and Chanda, Baron. 2016. The hitchhiker's guide to the voltage-gated sodium channel galaxy. *Journal of General Physiology*, **147**(1), 1–24.

Aidley, David J. 1998. *The Physiology of Excitable Cells*. 4th edn. Cambridge University Press.

Andersen, OS. 2016. Introduction to biophysics week: What is biophysics? *Biophysical Journal*, **110**, E01–E03.

Bean, Bruce P. 2007. The action potential in mammalian central neurons. *Nature Reviews Neuroscience*, **8**(6), 451–465.

Berridge, Michael J. 1998. Neuronal calcium signaling. *Neuron*, **21**(1), 13–26.

Berridge, Michael J, Bootman, Martin D, and Roderick, H Llewelyn. 2003. Calcium: calcium signalling: dynamics, homeostasis and remodelling. *Nature Reviews Molecular Cell Biology*, **4**(7), 517.

Bertram, Richard, and Sherman, Arthur. 2005. Negative calcium feedback: the road from Chay-Keizer. Pages 19–48 of: Coomes, S, and Bressloff, PC (eds), *Bursting: The Genesis of Rhythm in the Nervous System*. World Scientific.

Bialek, William. 2012. *Biophysics: Searching for Principles*. Princeton University Press.

Borisyuk, A, and Rinzel, J. 2005. Understanding neuronal dynamics by geometrical dissection of minimal models. Pages 17–72 of: *Les Houches Summer School Proceedings*, vol. 80.

Bower, James, and Beeman, David. 1995. *The Book of GENESIS: Exploring Realistic Neural Models with the GEneral NEural SImulations System*. Springer.

Boyce, William E, and DiPrima, Richard C. 2012. *Elementary Differential Equations and Boundary Value Problems*. 10th edn. Wiley.

Bressloff, Paul C. 2014. *Stochastic Processes in Cell Biology*. Interdisciplinary Applied Mathematics, vol. 41. Springer.

Britton, Nicholas F. 2012. *Essential Mathematical Biology*. Springer Science & Business Media.

Bullock, Theodore H, Bennett, Michael VL, Johnston, Daniel, Josephson, Robert, Marder, Eve, and Fields, R Douglas. 2005. Neuroscience. The neuron doctrine, redux. *Science*, **310**(5749), 791–793.

Cajal, Santiago Ramón y. 1906. The structure and connexions of neurons. *Nobel Lecture*.

Catterall, William A. 2011. Voltage-gated calcium channels. *Cold Spring Harbor Perspectives in Biology*, **3**(8), a003947.

Chay, Teresa Ree, and Keizer, Joel. 1983. Minimal model for membrane oscillations in the pancreatic beta-cell. *Biophysical Journal*, **42**(2), 181–189.

Chay, TR, and Keizer, J. 1985. Theory of the effect of extracellular potassium on oscillations in the pancreatic beta-cell. *Biophysical Journal*, **48**(5), 815–827.

Clay, John R, Paydarfar, David, and Forger, Daniel B. 2008. A simple modification of the Hodgkin and Huxley equations explains type 3 excitability in squid giant axons. *Journal of the Royal Society Interface*, **5**(29), 1421–1428.

Coombes, Stephen, and Bressloff, Paul C. 2005. *Bursting: The Genesis of Rhythm in the Nervous System*. World Scientific.

Cornish-Bowden, Athel. 2012. *Fundamentals of Enzyme Kinetics*. Wiley-VCH Weinheim.

Daun, Silvia, Rubin, Jonathan E, and Rybak, Ilya A. 2009. Control of oscillation periods and phase durations in half-center central pattern generators: a comparative mechanistic analysis. *Journal of Computational Neuroscience*, **27**(1), 3.

Dayan, Peter, and Abbott, Laurence F. 2001. *Theoretical Neuroscience*. MIT Press.

DeFelice, Louis J. 2012. *Introduction to Membrane Noise*. Springer Science & Business Media.

Destexhe, A, and Sejnowski, TJ. 2001. *Thalamocortical Assemblies*. Oxford University Press.

Destexhe, Alain, and Sejnowski, Terrence J. 2009. The Wilson-Cowan model, 36 years later. *Biological Cybernetics*, **101**(1), 1–2.

Destexhe, A, Mainen, Z, and Sejnowski, T. 1994. Synthesis of models for excitable membranes, synaptic transmission and neuromodulation using a common kinetic formalism. *Journal of Computational Neuroscience*, **1**, 195–230.

Devaney, Robert L, Hirsch, Morris W, and Smale, Stephen. 2006. First-order equations. Pages 1–20 of: *Differential Equations, Dynamical Systems, and an Introduction to Chaos*. 3rd edn. Academic Press.

De Vries, Gerda, Sherman, Arthur, and Zhu, Hsiu-Rong. 1998. Diffusively coupled bursters: effects of cell heterogeneity. *Bulletin of Mathematical Biology*, **60**(6), 1167.

Dolphin, Annette C. 2006. A short history of voltage-gated calcium channels. *British Journal of Pharmacology*, **147**(S1), S56–S62.

Edelstein-Keshet, Leah. 2005. *Mathematical Models in Biology*. Society for Industrial and Applied Mathematics.

Ermentrout, Bard. 1996. Type I membranes, phase resetting curves, and synchrony. *Neural Computation*, **8**(5), 979–1001.

Ermentrout, Bard. 1998a. Neural networks as spatio-temporal pattern-forming systems. *Reports on Progress in Physics*, **61**(4), 353.

Ermentrout, G Bard. 1998b. Linearization of F-I curves by adaptation. *Neural Computation*, **10**(7), 1721–1729.

Ermentrout, Bard. 2002. *Simulating, Analyzing, and Animating Dynamical Systems: a Guide to XPPAUT for Researchers and Students*. Software, Environments, Tools, vol. 14. Society for Industrial and Applied Mathematics.

Ermentrout, G Bard, and Kopell, Nancy. 1991. Multiple pulse interactions and averaging in systems of coupled neural oscillators. *Journal of Mathematical Biology*, **29**(3), 195–217.

Ermentrout, G Bard, and Terman, David H. 2010. *Mathematical Foundations of Neuroscience*. Interdisciplinary Applied Mathematics, vol. 35. Springer Science & Business Media.

Fain, Gordon L. 1999. *Molecular and Cellular Physiology of Neurons*. Harvard University Press.

Fain, Gordon L. 2003. *Sensory Transduction*. Sinauer Associates.

Fall, CP, Marland, ES, Wagner, JM, and Tyson, JJ (eds). 2002. *Computational Cell Biology*. Springer.

Feldman, Jack L, Del Negro, Christopher A, and Gray, Paul A. 2013. Understanding the rhythm of breathing: so near, yet so far. *Annual Review of Physiology*, **75**, 423–452.

Ferreira, Hugo Gil, and Marshall, Michael W. 1985. *The Biophysical Basis of Excitability*. Cambridge University Press.

Friedman, Avner, Borisyuk, Alla, Terman, David, and Ermentrout, Bard. 2005. Tutorials in Mathematical Biosciences I: *Mathematical Neuroscience*. Springer.

Gabbiani, Fabrizio, and Cox, Steven James. 2017. *Mathematics for Neuroscientists*. Academic Press.

Galán, Roberto F, Ermentrout, G Bard, and Urban, Nathaniel N. 2005. Efficient estimation of phase-resetting curves in real neurons and its significance for neural-network modeling. *Physical Review Letters*, **94**(15), 158101.

Garfinkel, Alan, Shevtsov, Jane, and Guo, Yina. 2017. *Modeling Life: The Mathematics of Biological Systems*. Springer.

Gerstner, Wulfram, and Kistler, Werner M. 2002. *Spiking Neuron Models: Single Neurons, Populations, Plasticity*. Cambridge University Press.

Glendinning, Paul. 1994. *Stability, Instability and Chaos: An Introduction to the Theory of Nonlinear Differential Equations*. Cambridge University Press.

Glickstein, Mitch. 2006. Golgi and Cajal: The neuron doctrine and the 100th anniversary of the 1906 Nobel Prize. *Current Biology*, **16**(5), R147–R151.

Golgi, Camillo. 1906. The neuron doctrine: theory and facts. *Nobel Lecture*.

Golomb, David, Wang, Xiao-Jing, and Rinzel, John. 1994. Synchronization properties of spindle oscillations in a thalamic reticular nucleus model. *Journal of Neurophysiology*, **72**(3), 1109–1126.

Gomperts, Bastien D, Kramer, Ijsbrand M, and Tatham, Peter ER. 2009. *Signal Transduction*. 2nd edn. Academic Press.

Greenspan, Ralph J. 2007. *An Introduction to Nervous Systems*. Cold Spring Harbor Laboratory Press.

Guckenheimer, John, and Holmes, Philip. 2013. *Nonlinear Oscillations, Dynamical Systems, and Bifurcations of Vector Fields*. Applied Mathematical Sciences, vol. 42. Springer Science & Business Media.

Guillery, RW. 2005. Observations of synaptic structures: origins of the neuron doctrine and its current status. *Philosophical Transactions of the Royal Society of London Series B, Biological Sciences*, **360**(1458), 1281–1307.

Gulrajani, Ramesh M. 1998. *Bioelectricity and Biomagnetism*. Wiley.

Hansel, D, Mato, G, and Meunier, C. 1993. Phase dynamics for weakly coupled Hodgkin-Huxley neurons. *Europhysics Letters*, **23**(5), 367.

Häusser, Michael. 2000. The Hodgkin-Huxley theory of the action potential. *Nature Neuroscience*, **3**, 1165.

Hill, Archibald Vivian. 1956. Why biophysics? *Science*, **124**(3234), 1233–1237.

Hille, Bertil. 2001. *Ionic Channels of Excitable Membranes*. 3rd edn. Sinauer Associates.

Hindmarsh, James L, and Rose, RM. 1984. A model of neuronal bursting using three coupled first order differential equations. *Proceedings of the Royal Society of London, Series B*, **221**(1222), 87–102.

Hoppensteadt, Frank C, and Izhikevich, Eugene M. 1997. *Weakly Connected Neural Networks*. Applied Mathematical Sciences, vol. 126. Springer.

Hounsgaard, J, and Kiehn, O. 1989. Serotonin-induced bistability of turtle motoneurones caused by a nifedipine-sensitive calcium plateau potential. *Journal of Physiology*, **414**(1), 265–282.

Hounsgaard, J, Hultborn, H, Jespersen, B, and Kiehn, O. 1988. Bistability of alpha-motoneurones in the decerebrate cat and in the acute spinal cat after intravenous 5-hydroxytryptophan. *Journal of Physiology (London)*, **405**, 345–367.

Hudson, Amber E, Archila, Santiago, and Prinz, Astrid A. 2010. Identifiable cells in the crustacean stomatogastric ganglion. *Physiology*, **25**(5), 311–318.

Huguenard, John R, and McCormick, David A. 1992. Simulation of the currents involved in rhythmic oscillations in thalamic relay neurons. *Journal of Neurophysiology*, **68**(4), 1373–1383.

Izhikevich, Eugene M. 2000. Neural excitability, spiking and bursting. *International Journal of Bifurcation and Chaos*, **10**(6), 1171–1266.

Izhikevich, Eugene M. 2007. *Dynamical Systems in Neuroscience: The Geometry of Excitability and Bursting*. MIT Press.

Jack, James Julian Bennett, Noble, Denis, and Tsien, Richard W. 1975. *Electric Current Flow in Excitable Cells*. Clarendon Press.

Johnston, Daniel, and Wu, Samuel Miao-Sin. 1994. *Foundations of Cellular Neurophysiology*. MIT Press.

Jordan, Dominic William, and Smith, Peter. 1999. *Nonlinear Ordinary Differential Equations: An Introduction to Dynamical Systems*. 3rd edn. Oxford University Press.

Julien, Jean-Pierre, and Kriz, Jasna. 2006. Transgenic mouse models of amyotrophic lateral sclerosis. *Biochimica et Biophysica Acta (BBA) Molecular Basis of Disease*, **1762**(11), 1013–1024.

Kaplan, Daniel, and Glass, Leon. 2012. *Understanding Nonlinear Dynamics*. Springer.

Keener, James P. 1995. *Principles of Applied Mathematics*. Addison-Wesley.

Keener, James P, and Sneyd, James. 1998. *Mathematical Physiology: Cellular Physiology*. Vol. 1. Springer.

Kiehn, Ole, and Eken, Torsten. 1998. Functional role of plateau potentials in vertebrate motor neurons. *Current Opinion in Neurobiology*, **8**(6), 746–752.

Koch, Christof. 2004. *Biophysics of Computation: Information Processing in Single Neurons*. Oxford University Press.

Koch, Christof, and Segev, Idan. 1998. *Methods in Neuronal Modeling: From Ions to Networks*. MIT Press.

Kopell, Nancy, and Ermentrout, Bard. 2004. Chemical and electrical synapses perform complementary roles in the synchronization of interneuronal networks. *Proceedings of the National Academy of Sciences*, **101**(43), 15482–15487.

Kopell, N, Ermentrout, GB, Whittington, MA, and Traub, RD. 2000. Gamma rhythms and beta rhythms have different synchronization properties. *Proceedings of the National Academy of Sciences*, **97**(4), 1867–1872.

Korogod, Sergiy Mikhailovich, and Tyč-Dumont, Suzanne. 2009. *Electrical Dynamics of the Dendritic Space*. Cambridge University Press.

Laing, Carlo, and Lord, Gabriel J. 2010. *Stochastic Methods in Neuroscience*. Oxford University Press.

Lauffenburger, Douglas A, and Linderman, Jennifer. 1996. *Receptors: Models for Binding, Trafficking, and Signaling*. Oxford University Press.

Levitan, Irwin B, and Kaczmarek, Leonard K. 2015. *The Neuron: Cell and Molecular Biology*. Oxford University Press.

Lipscombe, Diane, Helton, Thomas D, and Xu, Weifeng. 2004. L-type calcium channels: the low down. *Journal of Neurophysiology*, **92**(5), 2633–2641.

Liu, YH, and Wang, Xiao-Jing. 2001. Spike-frequency adaptation of a generalized leaky integrate-and-fire model neuron. *Journal of Computational Neuroscience*, **10**(1), 25–45.

Logan, J David. 2013. *Applied Mathematics*. John Wiley & Sons.

Lüthi, A, and McCormick, DA. 1998. H-current: properties of a neuronal and network pacemaker. *Neuron*, **21**(1), 9–12.

Mancilla, Jaime G, Lewis, Timothy J, Pinto, David J, Rinzel, John, and Connors, Barry W. 2007. Synchronization of electrically coupled pairs of inhibitory interneurons in neocortex. *Journal of Neuroscience*, **27**(8), 2058–2073.

Maor, Eli. 2009. *e: The Story of a Number*. Princeton University Press.

Masia, Ricard, Krause, Daniela S, and Yellen, Gary. 2015. The inward rectifier potassium channel Kir2.1 is expressed in mouse neutrophils from bone marrow and liver. *American Journal of Physiology: Cell Physiology*, **308**(3), C264–C276.

McCormick, DA. 1998. Membrane properties and neurotransmitter actions. Pages 37–75 of: Shepherd, GM (ed), *Synaptic Organization of the Brain*. Oxford University Press.

McCormick, David A, and Huguenard, John R. 1992. A model of the electrophysiological properties of thalamocortical relay neurons. *Journal of Neurophysiology*, **68**(4), 1384–1400.

McHugh, D, Sharp, EM, Scheuer, T, and Catterall, WA. 2000. Inhibition of cardiac L-type calcium channels by protein kinase C phosphorylation of two sites in the N-terminal domain. *Proceedings of the National Academy of Sciences*, **97**(22), 12334–12338.

Meng, Xiangying, Huguet, Gemma, and Rinzel, John. 2012. Type III excitability, slope sensitivity and coincidence detection. *Discrete and Continuous Dynamical Systems, Series A*, **32**(8), 2729.

Morris, Catherine, and Lecar, Harold. 1981. Voltage oscillations in the barnacle giant muscle fiber. *Biophysical Journal*, **35**(1), 193–213.

Murray, James D. 2002a. *Mathematical Biology: An Introduction*.

Murray, James D. 2002b. *Mathematical Biology: Spatial Models and Biomedical Applications*. 3rd edn. Vol. 2. Springer.

Nauta, WJ, and Feirtag, M. 1979. The organization of the brain. *Scientific American*, **241**(3), 88–111.

Nicholls, John G, Martin, A Robert, Brown, David A, Diamond, Matthew E, Weisblat, David A, and Fuchs, Paul A. 2011. *From Neuron to Brain*. Sinauer Associates.

Nobel, Park S. 2009. *Physicochemical and Environmental Plant Physiology*. 4th edn. Academic Press.

Nonner, Wolfgang, Chen, Duan P, and Eisenberg, Bob. 1999. Progress and prospects in permeation. *Journal of General Physiology*, **113**(6), 773–782.

Nowak, Lionel G, Azouz, Rony, Sanchez-Vives, Maria V, Gray, Charles M, and McCormick, David A. 2003. Electrophysiological classes of cat primary visual cortical neurons in vivo as revealed by quantitative analyses. *Journal of Neurophysiology*, **89**(3), 1541–1566.

O'Malley, Robert E. 1997. *Thinking About Ordinary Differential Equations*. Cambridge University Press.

Oswald, Anne-Marie M, Doiron, Brent, Rinzel, John, and Reyes, Alex D. 2009. Spatial profile and differential recruitment of GABA$_B$ modulate oscillatory activity in auditory cortex. *Journal of Neuroscience*, **29**(33), 10321–10334.

Perko, Lawrence. 2013. *Differential Equations and Dynamical Systems*. Texts in Applied Mathematics, vol. 7. Springer Science & Business Media.

Perrier, Jean-François, Alaburda, Aidas, and Hounsgaard, J. 2002. Spinal plasticity mediated by postsynaptic L-type Ca_{2+} channels. *Brain Research Reviews*, **40**(1), 223–229.

Pieri, Massimo, Caioli, Silvia, Canu, Nadia, Mercuri, Nicola B, Guatteo, Ezia, and Zona, Cristina. 2013. Over-expression of N-type calcium channels in cortical neurons from a mouse model of amyotrophic lateral sclerosis. *Experimental Neurology*, **247**, 349–358.

Plonsey, Robert, and Barr, Roger C. 2007. *Bioelectricity: A Quantitative Approach.* Springer Science & Business Media.

Puri, NN. 2009. *Fundamentals of Linear Systems for Physical Scientists and Engineers.* CRC Press.

Rieke, Fred, Warland, David, van Steveninck, Rob de Ruyter, and Bialek, William. 1996. *Spikes: Exploring the Neural Code.* MIT Press.

Rinzel, John. 1987. A formal classification of bursting mechanisms in excitable systems. Pages 267–281 of: Teramoto, E, and Yumaguti, M (eds), *Mathematical Topics in Population Biology, Morphogenesis and Neurosciences.* Springer.

Rinzel, John. 2002. Intercellular communication. Pages 140–169 of: Fall, CP, Marland, ES, Wagner, JM, and Tyson, JJ (eds), *Computational Cell Biology.* Springer.

Rinzel, John, and Ermentrout, G Bart. 1989. Analysis of neural excitability and oscillations. Pages 135–169 of: Koch, C, and Segev, I (eds), *Methods in Neuronal Modeling: From Synapses to Networks.* MIT Press.

Robinson, Rex Clark. 2012. *An Introduction to Dynamical Systems: Continuous and Discrete.* American Mathematical Society.

Rubinstein, Isaak. 1990. *Electro-diffusion of Ions.* Society for Industrial and Applied Mathematics.

Rush, Maureen E, and Rinzel, John. 1994. Analysis of bursting in a thalamic neuron model. *Biological Cybernetics*, **71**(4), 281–291.

Rush, Maureen E, and Rinzel, John. 1995. The potassium A-current, low firing rates and rebound excitation in Hodgkin-Huxley models. *Bulletin of Mathematical Biology*, **57**(6), 899–929.

Sakmann, Bert. 2013. *Single-Channel Recording.* Springer Science & Business Media.

Schneider, Gerald E. 2014. *Brain Structure and Its Origins in Development and in Evolution of Behavior and the Mind.* MIT Press.

Schutter, Erik De. 2009. *Computational Modeling Methods for Neuroscientists.* MIT Press.

Schwiening, Christof J. 2012. A brief historical perspective: Hodgkin and Huxley. *Journal of Physiology (London)*, **590**(11), 2571–2575.

Scott, Alwyn. 2003. *Nonlinear Science: Emergence and Dynamics of Coherent Structures.* Oxford University Press.

Segel, Irwin H. 1993. *Enzyme Kinetics: Behavior and Analysis of Rapid Equilibrium and Steady-State Enzyme Systems.* Wiley-Interscience.

Segel, Lee A. 1984. *Modeling Dynamic Phenomena in Molecular and Cellular Biology.* Cambridge University Press.

Segel, Lee A, and Edelstein-Keshet, Leah. 2013. *A Primer on Mathematical Models in Biology.* Society for Industrial and Applied Mathematics.

Segev, Idan, Rinzel, John, and Shepherd, Gordon M. (eds). 2003. *The Theoretical Foundation of Dendritic Function: The Collected Papers of Wilfrid Rall with Commentaries.* MIT Press.

Shepherd, Gordon M. 1988. *Neurobiology.* Oxford University Press.

Shepherd, Gordon M. 2004. *The Synaptic Organization of the Brain*. Oxford University Press.

Shepherd, Gordon M. 2015. *Foundations of the Neuron Doctrine*. 2nd edn. Oxford University Press.

Sherman, Arthur, and Rinzel, John. 1992. Rhythmogenic effects of weak electrotonic coupling in neuronal models. *Proceedings of the National Academy of Sciences*, **89**(6), 2471–2474.

Sherman, SM, and Guillery, RW. 2004. Thalamus. Pages 311–359 of: Shepherd, GM (ed), *The Synaptic Organization of the Brain*. 5th edn. Oxford University Press.

Shiflet, AB, and Shiflet, GW. 2006. Systems dynamics problems with rate proportional to the amount. Pages 71–110 of: *Introduction to Computational Science: Modeling and Simulation for the Sciences*. Princeton University Press.

Shou, Wenying, Bergstrom, Carl T, Chakraborty, Arup K, and Skinner, Frances K. 2015. Theory, models and biology. *eLife*, **4**(July).

Sims, Christopher E, and Allbritton, Nancy L. 1998. Metabolism of inositol 1, 4, 5-trisphosphate and inositol 1, 3, 4, 5-tetrakisphosphate by the oocytes of Xenopus laevis. *Journal of Biological Chemistry*, **273**(7), 4052–4058.

Skinner, Frances K, Turrigiano, Gina G, and Marder, Eve. 1993. Frequency and burst duration in oscillating neurons and two-cell networks. *Biological Cybernetics*, **69**(5–6), 375–383.

Sterratt, David, Graham, Bruce, Gillies, Andrew, and Willshaw, David. 2011. *Principles of Computational Modelling in Neuroscience*. Cambridge University Press.

Stone, Jonathan. 1983. *Parallel Processing in the Visual System: The Classification of Retinal Ganglion Cells and its Impact on the Neurobiology of Vision*. 1st edn. Perspectives in Vision Research. Springer.

Strogatz, Steven H. 2014. *Nonlinear Dynamics and Chaos: With Applications to Physics, Biology, Chemistry, and Engineering*. 2nd edn. Westview Press.

Stuart, Greg, Spruston, Nelson, and Häusser, Michael. 2016. *Dendrites*. Oxford University Press.

Sutherland, Wilson A. 2009. *Introduction to Metric and Topological Spaces*. 2nd edn. Oxford University Press.

Swanson, Larry W. 2012. *Brain Architecture: Understanding the Basic Plan*. Oxford University Press.

Tabak, Joël, Rinzel, John, and O'Donovan, Michael J. 2001. The role of activity-dependent network depression in the expression and self-regulation of spontaneous activity in the developing spinal cord. *Journal of Neuroscience*, **21**(22), 8966–8978.

Terman, David. 2005. An introduction to dynamical systems and neuronal dynamics. Pages 21–68 of: Friedman, A, Borisyuk, A, Terman, D, and Ermentrout, B (eds), *Tutorials in Mathematical Biosciences I: Mathematical Neuroscience*. Springer.

Thompson, John Michael Tutill, and Stewart, H Bruce. 2002. *Nonlinear Dynamics and Chaos*. John Wiley & Sons.

Tuckwell, HC. 1988a. *Introduction to Theoretical Neurobiology: Linear Cable Theory and Dendritic Structure*. Vol. 1. Cambridge University Press.

Tuckwell, Henry C. 1988b. *Introduction to Theoretical Neurobiology: Nonlinear and Stochastic Theories*. Vol. 2. Cambridge University Press.

Tuckwell, Henry C. 1989. *Stochastic Processes in the Neurosciences*. CBMS-NSF Regional Conference Series in Applied Mathematics, vol. 56. Society for Industrial and Applied Mathematics.

Tyson, John J. 2007. Bringing cartoons to life. *Nature*, **445**(7130), 823.

Tyson, John J, Chen, Katherine C, and Novak, Bela. 2003. Sniffers, buzzers, toggles and blinkers: dynamics of regulatory and signaling pathways in the cell. *Current Opinion in Cell Biology*, **15**(2), 221–231.

Wallisch, Pascal, Lusignan, Michael E, Benayoun, Marc D, Baker, Tanya I, Dickey, Adam Seth, and Hatsopoulos, Nicholas G. 2014. *MATLAB for Neuroscientists: An Introduction to Scientific Computing in MATLAB*. Academic Press.

Wang, Xiao-Jing. 1998. Calcium coding and adaptive temporal computation in cortical pyramidal neurons. *Journal of Neurophysiology*, **79**(3), 1549–1566.

Wang, XJ, and Rinzel, J. 1992. Alternating and synchronous rhythms in reciprocally inhibitory model neurons. *Neural Computation*, **4**(1), 84–97.

Wang, Xiao-Jing, Rinzel, John, and Rogawski, Michael A. 1991. A model of the T-type calcium current and the low-threshold spike in thalamic neurons. *Journal of Neurophysiology*, **66**(3), 839–850.

Weiss, Thomas Fischer. 1996. *Cellular Biophysics*. MIT Press.

Wilson, Charles. 2008. Up and down states. *Scholarpedia Journal*, **3**(6), 1410.

Wilson, Hugh Reid. 1999. *Spikes, Decisions, and Actions: The Dynamical Foundations of Neuroscience*. Oxford University Press.

Yamazaki, D, Kito, H, Yamamoto, S, Ohya, S, Yamamura, H, Asai, K, and Imaizumi, Y. 2010. Contribution of Kir2 potassium channels to ATP-induced cell death in brain capillary endothelial cells and reconstructed HEK293 cell model. *American Journal of Physiology: Cell Physiology*, **300**(1), C75–C86.

Yuste, Rafael. 2010. *Dendritic Spines*. MIT Press.

Zhan, XJ, Cox, Charles L, Rinzel, John, and Sherman, S Murray. 1999. Current clamp and modeling studies of low-threshold calcium spikes in cells of the cat's lateral geniculate nucleus. *Journal of Neurophysiology*, **81**(5), 2360–2373.

Index

A current, 349
action potential, 216
anodal break excitation, 222, 299
antagonist, 84
applied current, 134
association reaction, *see* chemical reaction,
    bimolecular association

barnacle muscle fiber, 236
bifurcation
    fold, 99, 188
    Hopf, 304
    limit cycle fold, 310
    local versus global, 332
    parameter, 98
    pitchfork, 102
    saddle homoclinic, 323
    saddle-node on an invariant circle, 319
    subcritical Hopf, 308
    subcritical versus supercritical, 104
    supercritical Hopf, 307
    transcritical, 101
bifurcation diagram
    type II oscillations, 303
bifurcation parameter
    critical value, 98
binding curve, 70, 72
bistability, 50, 185
bursting
    Chay-Keizer model, 331
    elliptical, 312, 331
    post-inhibitory rebound, 338
        fast/slow analysis, 342
    rhythmic, 343
        fast/slow analysis, 344
    slow variable, 328
    square-wave, 324, 331
    subcritical Hopf-limit cycle fold, 312

calcium
    intracellular, 16
calcium-activated potassium current, 328
calcium channel
    L-type, 171
capacitative current, 136
characteristic time, 88
Chay-Keizer model, *see* bursting, Chay-Keizer
    model, 331

chemical flux, 60
chemical potential, 122
chemical reaction, 59
    bimolecular association, 65
    EC_{50}, 68, 85
    equilibrium constant, 64, 67
    equilibrium relation, 67
    isomerization, 61
    order, 60
    sequential binding, 69, 70
conductance, 138
conservation of charge, 135
conserved quantity, 62, 67, 74
constitutive relation, 17
    battery, 144
    capacitor, 145
    Nernst-Planck, 154, 160
    resistor, 144
current balance equation, 135
    units, 146
current-clamp recording, 199, 217
current-voltage relation, 133
    hyperpolarization-activated cation current,
        177
    inward rectifying potassium current, 171
    L-type calcium current, 171
    qualitative features, 179
    regenerative, 180
    restorative, 180
    subtraction, 181
    voltage-gated potassium current, 177

depolarization, 133
detailed balance, 53
driving force, 174

EC_{50}, *see* chemical reaction, EC_{50}
eigenvalues, 276
    saddle, 279
    stable node, 279
    stable spiral, 282
    unstable node, 280
    unstable spiral, 282
electrical potential, *see* membrane potential
electrocardiogram, 90
equilibria, *see* steady state
excitability
    fast/slow analysis, 242

type I, 319–331
    sharp threshold, 320
type I versus type II, 333
type II, 297–304
exponential
    relaxation, 18, 64
    time constant, 64
extensive versus intensive quantities, 25

Fitzhugh-Nagumo model, 297
fixed point, *see* steady state
function
    biphasic, 83
    Gaussian, 83
    Hill, *see* Hill function
    hyperbolic, 82
    qualitative similarity, 84
    rational, 91
    scaling, 82
    shifting, 82
    versus relation, 81

gap junctional coupling, 353
gating variable, 205
    differential equations, 211
global balance, 53
Goldman-Hodgkin-Katz
    current equation, 154–161
        limiting conductance, 157
        physical dimensions, 155
        rectification, 157
    voltage equation, 125

Hill function, 72, 84
Hodgkin-Huxley model, 216, 219
    oscillations, 224
    repetitive spiking, 224
hyperpolarization, 133
hyperpolarization-activated cation current, 177

initial value, 17
injected current, *see* applied current
intensive quantities, *see* extensive versus
        intensive
interval notation, 85
ionic channel, 117
ionic concentration, 119
ionic current, 138
    current-voltage relation, *see* current-voltage
        relation
    inward versus outward, 132, 134
    regenerative, 185
    reversal potential, *see* reversal potential

ionic membrane current
    constitutive relation, 137

Kirchoff's current law, 135

leakage current, 138
ligand binding, 65
limit cycle, 308
linear systems, 275
    straight line solutions, 286
logistic equation, 57
low-threshold calcium current, 340
low-threshold calcium spike, 346

mass action kinetics, 59
material balance equation, 16
membrane bistability, 188
    bifurcation diagram, 190
    conductance, 195
    exponential time constant, 196
    linearization, 195
    mediated by I_{Ca_V}, 185
    mediated by I_{Kir}, 191
membrane excitability, 221
membrane potential, 115, 132
membrane voltage
    see membrane potential, 132
model
    compartmental, 15–22
    intracellular calcium, 42
    minimal, 8
    toy, 8
    two compartment, 32, 56
Morris-Lecar model, 235–248
    excitability, 241
    nullclines, 240
    phase plane, 238, 239

Nernst equilibrium potential, 115
    approximate values, 121
    calculating, 121
neurotransmitter, 65

ODE, *see* ordinary differential equation
Ohm's law, 137
ordinary differential equation
    autonomous, 45, 54
    classification, 54
    first-order, 54
    homogenous, 54
    linear, 54, 55
    nonlinear, 45
    scalar, 55

ordinary differential equation (cont.)
    second-order, 54
    single-parameter family, 98
    superposition property, 55
    two-dimensional, 255
oscillations, 299
    fast/slow analysis, 244
    frequency, 90
    period, 90

parameter
    scale, 83
passive membrane
    equivalent circuit, 143
    exponential time constant, 140
    multiple currents, 142
    steady state voltage, 140
permeation block, 171
phase diagram, 42–51
    isomerization, 63
    passive membrane, 139
phase line, 99
phase plane
    direction field, 255
    nullclines, 256
phase plane analysis, 252–265
    oscillations, 244
    synaptic coupling, 357
physical dimensions, 15
    fundamental, 26
potassium current
    inward rectifying, 171
potential energy, 126

rate constant
    order, 61
reaction, *see* chemical reaction
receptor
    ionotropic, 86
    metabotropic, 86
    postsynaptic, 65

receptor model, 65
    sequential binding, 69, 70
reversal potential, 138

saddle
    stable and unstable manifold, 321
separation of variables, 24, 56
squid giant axon, 216
stability
    asymptotic, 51
    Lyaponov, 45, 51
stability analysis, *see* steady state, classification
steady state, 43
    classification, 47, 49
    multiple, *see* bistability
    semistable, 49
stoichiometric coefficient, 60
straight line solutions, 286
summation, 31
superposition, *see* ordinary differential equation,
        superposition property
synapses
    chemical, 356
    electrical, 353
    escape versus release, 360
    plasticity, 364
    slow, 364
synchrony, 355, 362

T current, 338–346
trace-determinant plane, 277

uncompensated charge, 124

voltage-clamp recording, 199–210, 218
    delayed rectifying potassium current, 202
    persistent currents, 203
    tail current, 211
voltage-gated currents, 171
voltage-gated potassium current, 177
voltage-gated sodium current, 220